7TH EDITION

MATHEMATICS EXPLAINED FOR PRIMARY TEACHERS

DEREK HAYLOCK

S Sage

1 Oliver's Yard
55 City Road
London EC1Y 1SP

2455 Teller Road
Thousand Oaks
California 91320

Unit No 323-333, Third Floor, F-Block
International Trade Tower
Nehru Place, New Delhi – 110 019

8 Marina View Suite 43-053
Asia Square Tower 1
Singapore 018960

Editor: James Clark
Assistant editor: Esosa Otabor
Assistant editor, digital: Ben Hegarty
Production editor: Nicola Marshall
Copyeditor: Martin Noble
Proofreader: Tom Bedford
Indexer: Author
Marketing manager: Lorna Patkai
Cover design: Wendy Scott
Typeset by: C&M Digitals (P) Ltd, Chennai, India
Printed in the UK by Bell and Bain Ltd, Glasgow

Library of Congress Control Number: 2023939382

British Library Cataloguing in Publication data

A catalogue record for this book is available from the British Library

ISBN 978-1-5296-2630-8
ISBN 978-1-5296-2629-2 (pbk)

TABLE OF CONTENTS

About the Author   vii

About the Contributor   viii

Acknowledgements   ix

Guided Tour   x

New to this Seventh Edition   xii

Introduction   1

SECTION A   MATHEMATICAL UNDERSTANDING   3

1   Primary Teachers' Insecurity about Mathematics   5

2   Mathematics in the Primary Curriculum   17

3   Learning How to Learn Mathematics   30

SECTION B   MATHEMATICAL REASONING AND PROBLEM SOLVING   43

4   Key Processes in Mathematical Reasoning   45

5   Modelling and Problem Solving   62

SECTION C   NUMBERS AND CALCULATIONS   77

6   Numbers, Counting and Place Value   79

7   Addition and Subtraction Structures   105

8   Mental Strategies for Addition and Subtraction   125

9   Written Methods for Addition and Subtraction   141

10   Multiplication and Division Structures   161

11   Mental Strategies for Multiplication and Division   180

12   Written Methods for Multiplication and Division   197

SECTION D   FURTHER NUMBER CONCEPTS AND SKILLS   219

13   Natural Numbers: Some Key Concepts   221

14   Integers: Positive and Negative   242

15   Fractions and Ratios   250

16   Decimal Numbers and Rounding   271
17   Calculations with Decimals   293
18   Proportionality and Percentages   313

SECTION E   ALGEBRA   331

19   Algebraic Reasoning   333
20   Coordinates and Linear Relationships   353

SECTION F   MEASUREMENT   367

21   Concepts and Principles of Measurement   369
22   Perimeter, Area and Volume   393

SECTION G   GEOMETRY   409

23   Angle   411
24   Transformations and Symmetry   423
25   Classifying Shapes   437

SECTION H   STATISTICS AND PROBABILITY   453

26   Handling Data   455
27   Comparing Sets of Data   474
28   Probability   492

Answers to Self-assessment Questions   505
References   519
Index   528

ABOUT THE AUTHOR

Derek Haylock worked for over 30 years in teacher education, both initial and in-service. He was Co-Director of Primary Initial Teacher Training and responsible for the mathematics components of the primary programmes at the University of East Anglia (UEA), Norwich. He has taught mathematics at every level and, in particular, has considerable practical experience of teaching and researching in primary classrooms. His commitment to mathematics being learnt with understanding, relevance and enjoyment has underpinned his work as a writer, consultant and professional speaker. As well as his extensive publications in the field of education, he has written seven books of Christian drama for young people and written and composed a Christmas musical (all published by Church House/National Society).

OTHER BOOKS IN EDUCATION BY DEREK HAYLOCK

Browne, A. and Haylock, D. (eds) (2004) *Professional Issues for Primary Teachers*. London: Sage Publications.

Haylock, D. (1991) *Teaching Mathematics to Low Attainers 8–12*. London: Sage Publications.

Haylock, D. (2001) *Numeracy for Teaching*. London: Sage Publications.

Haylock, D. (2024) *Student Workbook for Mathematics Explained for Primary Teachers*, 7th edn. London: Sage Publications.

Haylock, D. and Cockburn, A. (2017) *Understanding Mathematics for Young Children*, 5th edn. London: Sage Publications.

Haylock, D. and D'Eon, M. (1999) *Helping Low Achievers Succeed at Mathematics*. Toronto: Trifolium Books.

Haylock, D. and McDougall, D. (1999) *Mathematics Every Elementary Teacher Should Know*. Toronto: Trifolium Books.

Haylock, D. with Thangata, F. (2007) *Key Concepts in Teaching Primary Mathematics*. London: Sage Publications.

Haylock, D. and Warburton, P. (2013) *Mathematics Explained for Healthcare Practitioners*. London: Sage Publications.

ABOUT THE CONTRIBUTOR

Martin Smith has worked at the University of East Anglia since January 2014 and is currently Course Co-Director on the Primary PGCE course with a shared responsibility for the mathematics and physical education components. Prior to this role he taught across Key Stage 2 in and around Norwich, finding ways to break down barriers to the enjoyment of maths for both children and their adults. Martin has long since enjoyed board games, puzzles and problems and if there is a mathematical element to these, all the better. Outside of education, Martin has recently rediscovered his love of dinghy sailing and manages to combine this with watching and playing football, and spending time with his family.

ACKNOWLEDGEMENTS

My thanks and genuine appreciation are due to the many trainee teachers and primary school teachers with whom I have been privileged to work on initial training and in-service courses in teaching mathematics: for their willingness to get to grips with understanding mathematics, for their patience with me as I have tried to find the best ways of explaining mathematical ideas to them, for their honesty in sharing their own insecurities and uncertainties about the subject – and for thereby providing me with the material on which this book is based. I also acknowledge my indebtedness to James Clark and the team at Sage Publications for their unflagging encouragement and professionalism, and to Ralph Manning who has contributed the website activities for inclusion in lesson plans that are referenced in each of Chapters 6 to 28.

I acknowledge especially the significant contribution to this seventh edition of Martin Smith, Lecturer in Mathematics Education and Co-Director of the Primary PGCE in the School of Education and Lifelong Learning at the University of East Anglia, who has provided advice on recent changes in emphasis in classroom practice and research in primary mathematics. His insights and suggestions have been an invaluable contribution to the seventh edition of this book.

I dedicate this new edition to the memory of my dear wife, Christina, who died in 2022. Without her constant support and encouragement over so many years I could never have achieved what I have as an author in education.

GUIDED TOUR

This book contains a range of features and additional resources to support your learning and understanding. Here is an overview of what you will find inside the book and as part of the online resources https://study.sagepub.com/haylock7e.

FEATURES IN THE BOOK

- **Learning and teaching points** highlight specific points from each chapter for use in teaching.
- **Research focus** sections explore contemporary and seminal academic research on relevant chapter topics.
- **Self-assessment questions** at the end of each chapter give you a chance to test your learning immediately. Check your work against the answers in the back of the book.
- **Links to the Student Workbook** direct you to further self-test maths questions in the Student Workbook (available separately).
- **Glossary of key terms** in each chapter explains important mathematical vocabulary in straightforward language.
- **Suggestions for further reading** highlights more specialist reading material related to topics in each section of the book.

ONLINE RESOURCES

The following online resources (available at: https://study.sagepub.com/haylock7e) are all sign-posted at relevant points in chapters so you can test your own mathematical knowledge as you work through the book.

- **Lesson plans and worksheets** (created by Ralph Manning) give you lesson ideas and ways of approaching concepts for your classroom teaching.
- **'Problem solved!' animated videos** show you how to work through different maths methods to close gaps in your understanding.

- **Curriculum links** highlight where chapter content meets national curriculum requirements in England, Scotland and Wales.
- **Interactive self-assessment questions** offer a further range of multiple-choice testing with feedback on your answers.
- **Knowledge checks** offer a deeper dive on specific mathematical topics and tasks.
- **Section introduction videos** give an overview of each section of the book.

Access the online resources here: https://study.sagepub.com/haylock7e.

NEW TO THIS SEVENTH EDITION

- The Research Focus features and other references have been updated where appropriate to discuss a greater amount of contemporary research.
- Further Reading suggestions at the end of sections have been updated to include more recent relevant literature.
- Expanded content on the development of counting among young children has been brought into Chapter 6.
- Further areas of additional content across the book include: deeper discussion of arrays and technical terms now in greater use, such as subtrahend and minuend in reference to addition and subtraction.
- A new section on the external angles of triangles.
- Increased references to classroom resources and pedagogical tools that support mastery teaching in mathematics, including tens-frames, Dienes blocks and bar-modelling.
- Additional and updated references to curriculum differences across the UK, with reference to curricula documentation from England, Scotland, Wales and Northern Ireland.
- The use of 1p, 10p and £1 coins to demonstrate place value has generally been replaced by place-value counters, given the decreasing use of 1p, 2p, 5p coins and coins in general, in daily life.
- A number of the Learning and Teaching Points have been updated and a number of new ones added.
- End-of-chapter links to questions in the accompanying Student Workbook have been updated to reflect the increased number of questions in the workbook.
- Student Workbook for Mathematics Explained for Primary Teachers.

 Available separately, the *Student Workbook* for *Mathematics Explained for Primary Teachers*, 7th edition (ISBN: 978-1-5296-2631-5) is the perfect companion to your study. It contains 900 questions with detailed solutions and explanatory notes.

- Test your knowledge of key concepts and principles.
- Apply mathematical skills and knowledge in real-life situations and to mathematical problems and investigations.
- Consider how to respond to children's errors and misunderstandings and how to evaluate different teaching approaches.
- Be inspired to develop classroom ideas that promote understanding and mastery.

INTRODUCTION

This seventh edition has been revised to ensure that the content is in line with the mathematics programmes of study for Key Stages 1 and 2 (children aged 5 to 11 years) in England (DfE, 2013) and consistent with the current emphasis on teaching for mastery, promoted by the National Council for Excellence in Teaching Mathematics. As well as ensuring that I address all the statutory requirements for mathematics in primary schools in England that have to teach the National Curriculum, I am aware that many readers of this book will teach mathematics to this age range of children in other countries with a variety of statutory curriculums and also that there are many schools in England that are not required to teach the English National Curriculum. It has been encouraging to hear that teachers, teacher trainees and mathematics educators in many other countries have found the previous editions to be helpful and relevant to their work. I am confident that this will continue to be the case. In particular, I have ensured that the primary national curriculums of the other countries in the United Kingdom are covered comprehensively.

I have continued in my commitment to focus on what has always been the key message of 'Mathematics Explained': the need for priority to be given in initial teacher training and professional development to primary school teachers developing secure and comprehensive subject knowledge in mathematics, characterized by understanding and awareness of the pedagogical implications. Even well-qualified graduates feel insecure and uncertain about much of the mathematics they have to teach, as is demonstrated in Chapter 1 of the book. I know from experience just how much they appreciate a systematic explanation of even the most elementary mathematical concepts and procedures of the primary curriculum. In my long career in teacher education, I have often reflected on what qualities make a good teacher. I have a little list. Top of the list is the following conviction: the best teachers have a secure personal understanding of the structure and principles of what they are teaching. This book is written to help primary teachers, present and future, to achieve this in mathematics. It sets out to explain the subject to primary school teachers, so that they in turn will have the confidence to provide appropriate, systematic and careful explanation of mathematical ideas and procedures to their pupils, with an emphasis on the development of understanding, rather than mere learning by rote. This is always done from the perspective of how children learn, understand and develop mastery of this subject. Implications for learning and teaching are embedded in the text and highlighted as 'Learning and teaching points' distributed throughout each chapter.

Section A (Chapters 1–3) of this book is about mathematical understanding. Chapter 1, drawing on my research with trainee teachers, provides evidence for the need to develop understanding of mathematics and to lower anxiety about this subject in those who are to teach in primary schools. Chapter 2 considers the distinctive contribution that mathematics makes to the primary curriculum; and Chapter 3 – which I consider to be the key chapter in this book – is about pupils learning to learn mathematics with understanding.

Section B explains the key features of mathematical reasoning and problem solving and seeks to give these two major themes in the mathematics curriculum the prominence and priority they warrant.

Sections C–H then focus on the content and principles of various sections of primary school mathematics, covering all you need to understand about: numbers, including the number system and various number properties; calculations, mental and written, including those with fractions, decimals and percentages; algebra; the principles and concepts of measurement and units of measurement; geometry; and statistics and probability.

It is important that those who teach mathematics to children know and understand more mathematics than the children may have to learn. This will help the teacher to feel confident and to teach with authority in this area of the curriculum. It will also help them to be more aware of the significance of what the children are learning. So, in places in the book there will be material that may go beyond the immediate requirements of the primary school curriculum. In particular, some sections – such as some calculations with decimals and percentages and much of the material in Chapter 27 – are specifically provided for the personal development of the primary teacher, to help them to handle, with confidence, some of the encounters with mathematics they will have in their professional role.

Finally: a comment about calculators. Although calculators have effectively disappeared from most mathematics lessons in primary schools that are constrained by the National Curriculum in England, they are still available in primary schools elsewhere in the UK and in countries overseas. For their benefit, and particularly for the benefit of the readers of this book, I have included here and there a number of ways in which simple four-function calculators can be used effectively to provide insights and to promote understanding in learning mathematics. Readers should find that their engagement with these examples enhances their own understanding of the mathematical ideas being explained.

Derek Haylock, Norwich

SECTION A

MATHEMATICAL UNDERSTANDING

1   Primary Teachers' Insecurity about Mathematics   5
2   Mathematics in the Primary Curriculum   17
3   Learning How to Learn Mathematics   30

WATCH THE SECTION OPENER VIDEO AT: HTTPS://STUDY.SAGEPUB.COM/HAYLOCK7E

PRIMARY TEACHERS' INSECURITY ABOUT MATHEMATICS

IN THIS CHAPTER, THERE ARE EXPLANATIONS OF

- the importance of primary school teachers really understanding the mathematics they teach and being able to explain it clearly to the children they teach;
- the relationship of mathematics anxiety to avoidance of mathematical demand, rote learning and low levels of creativity in problem solving;
- attitudes of adults in general toward mathematics;
- mathematics anxiety in primary school teachers;
- the insecurity about mathematics of many primary trainee teachers.

READ THIS CHAPTER'S CURRICULUM LINKS AT: HTTPS://STUDY.SAGEPUB.COM/HAYLOCK7E

UNDERSTANDING AND EXPLAINING

Being a successful learner in mathematics involves constructing *understanding* through exploration, problem solving, discussion and practical experience – and also through interaction with a teacher who has a clear grasp of the underlying structure of the mathematics being learnt. For children to enjoy learning mathematics, it is essential that they should understand it; and that they should make sense of what they are doing in the subject, and not just learn to reproduce learnt procedures and recipes that are low in meaningfulness and purposefulness.

One of the ways for children to learn and understand much of the mathematics in the primary school curriculum is for a teacher who understands it to explain it to them. Those who teach mathematics in primary schools should ensure that the approach they take to organizing children's activities allows sufficient opportunities for them to provide teaching that includes engaging with children in question and answer, discussion and explanation – all aimed at promoting understanding and confidence in mathematics. Of course, there is more to learning mathematics than just a teacher explaining something and then following this up with exercises. The key processes of mathematical reasoning, applying mathematics and problem solving must always be at the heart of learning the subject – and these figure prominently in this book, particularly in Section B (Chapters 4 and 5).

But children do need 'explanation' to help them to understand mathematics, to make sense of it, to give them confidence and to help them have positive attitudes to the subject. There is now in England a greater awareness that primary teachers must organize their lessons and the children's activities in ways that give opportunities for them to provide careful, systematic and appropriate explanation of mathematical concepts, procedures and principles to groups of children. That many primary teachers have in the past neglected this aspect of teaching may possibly have been associated with a prevailing primary ethos which perhaps over-emphasized active learning and children discovering things for themselves. But it seems to me to be often a consequence of the teacher's own insecurity and anxiety about mathematics, which are characteristics of too many primary school teachers. This book is written to equip teachers with the knowledge, understanding and confidence they require to be able to explain mathematical ideas to the children they teach. English primary schools, following the example of some higher-performing countries in terms of mathematics achievement, are now putting an emphasis on teaching for what is termed 'mastery' in mathematics (see Chapter 3). Mastery approaches in teaching mathematics aim for a deep understanding of mathematical procedures, concepts and principles for all children. It is self-evident that a prerequisite for teaching for mastery is that primary school teachers themselves have this deep and secure understanding of the mathematics in the primary school curriculum. Helping teachers and teacher trainees to feel confident in their own understanding and mastery of this subject is the principal purpose of this book.

ATTITUDES TO MATHEMATICS IN ADULTS

There are widespread confusions amongst the adult population in Britain about many of the basic mathematical processes of everyday life. This lack of confidence in basic mathematics appears to be related to the anxiety about mathematics and feelings of inadequacy in this subject that are common amongst the adult population. These phenomena have been demonstrated by surveys of adults' attitudes to mathematics (for example, Coben et al., 2003). Findings indicate that many adults, in relation to mathematical tasks, admit to feelings of anxiety, helplessness, fear, dislike and even guilt. The feeling of guilt is particularly marked amongst those with high academic qualifications, who feel that they ought to be more confident in their understanding of this subject. There is a perception that there are proper ways of doing mathematics and that the subject is characterized by questions to which your answers are either right or wrong. Feelings of failure, frustration and anxiety are identified by many adults as having their roots in unsympathetic attitudes of teachers and the expectations of parents. A project at King's College, London, looking at the attitudes of adults attending numeracy classes, found that the majority of such adults viewed themselves as failures and carried various types of emotional baggage from their schooldays. They spoke of their poor experience of schooling and of feeling that they had been written off by their mathematics teachers, usually at an early stage. Their return to the mathematics classroom as adults was accompanied by feelings of anxiety, even fear (Swain, 2004). Significantly, in a survey of over 500 adults in the UK, Lim (2002) identified three widely claimed myths about mathematics: it is a difficult subject; it is only for clever people (see also Chestnut et al., 2018); and it is a male domain.

MATHEMATICS ANXIETY IN TEACHERS AND LEARNERS

Research over many years into primary school teachers' attitudes to mathematics reveals that many primary teachers experience the same kinds of feelings of panic and anxiety when faced with unfamiliar mathematical tasks (Briggs, 1993), that they are muddled in their thinking about many of the basic mathematical concepts which underpin the material they teach to children, and that they are all too aware of their personal inadequacies in mathematics. The widespread view that mathematics is a difficult subject, and therefore only for clever people, increases these feelings of inadequacy – and the common perception that mathematics is a male domain exacerbates the problem within a subset of the teaching profession that continues to be largely populated by women. The importance of tackling these attitudes to the subject was underlined by the findings of Burnett and Wichman (1997) that primary teachers' (and parents') own anxieties about mathematics can often be passed on to the children they teach. Witt and Mansergh (2008) identify the need to break this anxiety spiral in the course of initial teacher training.

It is important not to generate mathematics anxiety in the children we teach, because high levels of anxiety affect a person's ability to perform to their potential. Boaler (2022) argues that an unhelpful level of anxiety towards mathematics often starts early in the primary years of education and is found even amongst children who otherwise do well at school. She reports the well-established finding that a high level of mathematics anxiety leads to difficulties in learning the subject and avoidance of mathematical demand, and identifies in particular the way in which timed tests in mathematics contribute significantly to anxiety. The research of Ashcraft and Kirk (2001) and Ashcraft and Moore (2009), for example, confirms that raising anxiety about mathematics produces a drop in performance in the subject, particularly in terms of the individual's access to their 'working memory' and a tendency to avoid any kind of mathematical demand. Puteh (1998) has provided a helpful diagrammatic summary of research in this area (see Figure 1.1).

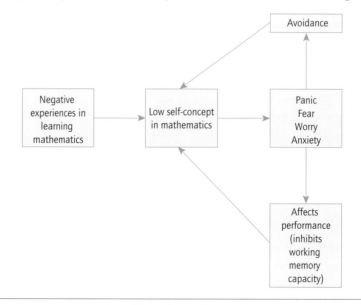

Figure 1.1   Some factors associated with mathematics anxiety
Source: Puteh (1998)

Newstead (1998) found that a teaching method that promoted understanding rather than just the memorization and rehearsal of procedures and recipes was associated significantly with lower levels of mathematics anxiety in primary school children. Khoule et al. (2017) report that the procedural method of teaching mathematics, when applied alone, is 'easy to forget or hard to remember' and is often associated with 'pain and frustration' for learners. Ford et al. (2005), also report that high anxiety towards mathematics has a negative effect on learners' performance. In particular, anxiety leads to reliance on learning mathematics by rote, rather than aiming and expecting to learn with understanding; this is a vicious circle, because reliance on rote memorization of rules rather than understanding them increases anxiety when faced with anything unfamiliar (Figure 1.2). Carey et al. (2016) provide a further analysis of the ways in which mathematics anxiety and mathematics performance are related.

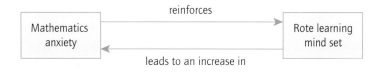

Figure 1.2   Mathematics anxiety and rote learning

In my own research I have found that anxiety about mathematics – which is reinforced when children are not being helped to learn mathematics in a meaningful way – is associated with rigid and inflexible thinking in unfamiliar mathematical tasks and leads to insecurity and caution when faced with a non-routine mathematical challenge, and therefore to low levels of creativity in problem solving (Figure 1.3). Creativity in mathematics is explained further in Chapter 4.

Figure 1.3   Mathematics anxiety inhibits creativity in problem solving

TRAINEE TEACHERS' ANXIETIES

The background for this book is mainly my experience of working with graduates enrolled on a one-year primary initial teacher-training programme. The trainee teachers I have worked with have been highly motivated, good honours graduates, with the subjects of their degree studies ranging across the curriculum. Over a number of years of working with such trainees, it became clear to me that many of them start their course with a high degree of anxiety about having to teach mathematics.

An invitation was given for any trainees who felt particularly worried about mathematics to join a group who would meet for an hour a week throughout their course, to discuss their anxieties and to identify which aspects of the National Curriculum for mathematics appropriate to the age range they would be teaching gave them most concern. A surprisingly large number of trainees turned up for these sessions. Discussions with them revealed both those aspects of mathematics anxiety that they still carry around with them, derived clearly from their own experiences of learning mathematics at school, and the anxiety related to specific areas of mathematics they will have to teach where they have doubts about their own understanding.

Below, I recount some of the statements made by the primary trainee teachers in my group about their attitudes towards and experiences of learning mathematics. In reading these comments, it is important to remember that these are students who have come through the system with relative success in mathematics: all had GCSE grade C, or the equivalent. Yet this is clearly not how they feel about themselves in relation to this subject. The trainees' comments on their feelings about mathematics can be categorized under five headings: (1) feelings of anxiety and fear; (2) expectations; (3) teaching and learning styles; (4) the image of mathematics; and (5) language. These categories reflect closely the findings of other studies of the responses towards mathematics of adults in general and primary teachers in particular.

Feelings of Anxiety and Fear

When these trainee teachers talked freely about their memories of mathematics at school, their comments were sprinkled liberally with such words as 'frightened', 'terror' and 'horrific', and several recalled having nightmares. These memories were very vivid and still lingered in their attitudes to the subject as academically successful adults:

Maths struck terror in my heart: a real fear that has stayed with me from over 20 years ago.

I had nightmares about maths: I really did, I'm not joking. Numbers and figures would go flashing through my head. Times tables, for example. I especially had nightmares about maths tests.

It worried me a great deal. Maths lessons were horrific.

Others recalled feelings of stupidity or frustration at being faced with mathematical tasks:

> I remember that I would always feel stupid. I felt sure that everyone else understood.

> Things used to get hazy and frustrating when I was stuck on a question.

Those of us who teach mathematics must pause and wonder what it is that we do to children that produces successful, intelligent adults who continue to feel like this about the subject.

Expectations

It seems as though the sources of anxiety for some trainee teachers were the expectations of others:

> It was made worse because Dad's best subject was maths.

> My teacher gave me the impression that she thought I was bad at maths. So that's how I was labelled in my mind. When I got my GCE result she said, 'I never thought you'd get an A!' So I thought it must just be a fluke. I still thought I was no good at maths.

But the most common experience cited by these trainees was the teacher's expectation that they *should* be able to deal successfully with all the mathematical tasks they were given. They recalled clearly the negative effect on them of the teacher's response to their failure to understand:

> There were few maths teachers who could grasp the idea of people not being mathematical.

> Teachers expect you to be good at maths if you're good at other things. They look at your other subjects and just can't understand why you can't do maths. They say to you, 'You should be able to do this …'.

Teaching and Learning Styles

The trainee teachers spoke with considerable vigour about their memories of the way mathematics was taught to them, recognizing now, from their adult perspective, that part of the problem was a significant limitation in the teaching style to which they were subjected:

> Surely not everyone can be bad at maths. Is it just that it's really badly taught?

> I remember one teacher who was good because she actually tried to explain things to me.

My primary school teachers tried to help me to make sense of maths, but in secondary school maths teachers just taught us formal written methods without any understanding.

It was clear that most of the trainees in this group felt that they had been encouraged to learn by rote, to learn rules and recipes without understanding, particularly in secondary school. This rote-learning style (see glossary in Chapter 3) was sometimes reinforced by apparent success:

> I was quite good at maths at school but I'm frightened of going back to teach it because I think I've probably forgotten most of what we learnt. I have a feeling that all I learnt was just memorized by rote and now it's all gone.
>
> I could rote learn things, but not understand them.
>
> I got through the exams by simply learning the rules. I would just look for clues in the question and find the appropriate process.

The limitations of this rote-learning syndrome were sometimes apparent to the trainees:

> I found you could do simple problems using the recipes, but then they'd throw in a question that was more complex. Then when the recipe I'd learnt didn't work I became angry.
>
> We would be given a real-life situation but I would find it difficult to separate the maths concepts out of it.

But it seems that some teachers positively discouraged a more appropriate learning style:

> I was made to feel like I was a nuisance for trying to understand.
>
> Lots of questions were going round in my head but I was too scared to ask them.
>
> I always tried to avoid asking questions in maths lessons because you were made to feel so stupid if you got it wrong. There must be ways of convincing a child it doesn't matter if they get a question wrong.

The following remark by one trainee highlights how the role of trainee teacher serves to focus the feeling of anxiety and inadequacy arising from the rote-learning strategy adopted in the past:

> I have a real fear of teaching young children how to do things in maths as I just learnt rules and recipes. I have this dread of having to explain why we do something.

Image of Mathematics

For some trainees, mathematics had an image of being a difficult subject, so much so that it was acceptable to admit that you are not any good at it:

> Maths has an image of being hard. You pick this idea up from friends, parents and even teachers.

> My Mum would tell me not to worry, saying, 'It's alright, we're all hopeless at maths!' It was as if it was socially acceptable to be bad at maths.

> Among my friends and family it was OK to be bad at maths, but it's not acceptable in society or employment.

For some, the problem seemed to lie with the feeling that mathematics was different from other subjects in school because the tasks given in mathematics are seen as essentially convergent and uncreative:

> Maths is not to do with the creativity of the individual, so you feel more restricted. All the time you think you've just got to get the right answer. And there is only one right answer.

> There's more scope for failure with maths. It's very obvious when you've failed, because things are either right or wrong, so you feel a fool, or look a fool in front of the others.

Language

A major problem for all the trainee teachers was that mathematical language seemed to be too technical, too specific to the subject and not reinforced through their language use in everyday life:

> I find the language of maths difficult, but the handling of numbers is fine.

> Most of the words you use in maths you never use in everyday conversation.

> Some words seem to have different meanings in maths, so you get confused.

> I was always worried about saying the wrong things in maths lessons, because maths language seems to be so precise. I worry now that I'll say things wrong to children in school and get them confused. You know, like, 'Which is the bigger half?'

When we discussed the actual content of the National Curriculum programmes of study for mathematics, it became clear that many of the trainees' anxieties were related to language. Often, they would not recognize mathematical ideas that they actually understood quite well, because they appeared in the National Curriculum in formal mathematical language, which they had either never known or forgotten through neglect. This seemed to be partly because most of this technical mathematical language is not used in normal everyday adult conversation, even amongst intelligent graduates:

> I can't remember what prime numbers are. Why are they called prime numbers anyway?
>
> Is a product when you multiply two numbers together?
>
> What's the difference between mass and weight?
>
> What is congruence? An integer? Discrete data? A measure of spread? A quadrant? An inverse? Reflective symmetry? A translation? A transformation?

Even as a 'mathematician', I must confess that it is rare for this kind of technical language to come into my everyday conversation, apart from when I am actually 'doing mathematics'. When this technical language was explained to the students, typical reactions would be:

> Oh, is that what they mean? Why don't they say so, then?
>
> Why do they have to dress it up in such complicated language?

It is clear then that unfamiliar terminology in mathematics inevitably increases anxiety for trainees. This is particularly significant given the current emphasis in teacher-training courses in Britain on 'teaching for mastery' in line with methods used in mathematics teaching in Singapore. This has brought with it new terminology, with which many trainees may be unfamiliar: for example, addend, subtrahend, bar-modelling, manipulatives.

MATHEMATICS EXPLAINED

Recognizing that amongst primary trainee teachers and, indeed, amongst many primary school teachers in general, there is this background of anxiety and confusion, it has always been clear to me that a major task for initial and in-service training is the promotion of positive attitudes towards teaching mathematics in this age range. The evidence from my conversations with trainee teachers suggests that to achieve this we need to shift perceptions of teaching mathematics away from the notion of teaching recipes and more towards the development of understanding. And we need to give time to explaining mathematical ideas, to the ironing out of confusions over the content and, particularly, to the language of mathematics. Some trainees' comments later in the year highlighted the significance to them of having mathematics explained. The emphasis on explaining and understanding paid off in shifts of attitudes towards the subject:

> It's the first time anyone has actually explained things in maths to me. I feel a lot happier about going into the classroom now.

> The course seems to have reawakened an interest in mathematics for me and exploded the myth that maths was something I had to learn by rote for exams, rather than understand.

> I was really fearful about having to teach maths. That fear has now declined. I feel more confident and more informed about teaching maths now.

These kinds of reactions prompted me to write this book. By focusing specifically on explaining the language and content of the mathematics that we teach in the primary age range, this book will help trainee teachers – and primary school teachers in general – to develop this kind of confidence in approaching their teaching of this key subject in the curriculum to children who are at such an important stage in their educational development.

RESEARCH FOCUS: SUBJECT KNOWLEDGE FOR TEACHING

In the context of increasing government concern about the subject knowledge in mathematics of trainee teachers, a group of mathematics educators at the London Institute of Education audited trainee primary trainees' performance in a number of basic mathematical topics. Those topics in which they had the lowest facility were: making algebraic generalizations; Pythagoras's theorem; calculation of area; mathematical reasoning; scale factors; and percentage increase. Significantly, trainees with poor subject knowledge in mathematics were found to perform poorly in their teaching of mathematics in the classroom when assessed at the end of their training (Rowland et al., 2000). Further research in this area (Goulding et al., 2002), which included an audit of subject knowledge of primary teacher trainees and its relationship with classroom teaching, identified weaknesses in mathematical understanding, particularly in the syntactic elements of mathematics, and a link between insecure subject knowledge and poor planning and teaching. Hourigan and O'Donoghue (2013) report similar findings in relation to elementary teachers in Ireland, identifying particular difficulties in knowledge of rational numbers (fractions), conceptual understanding and problem solving in mathematics. Rowland and Ruthven (2011: 1) argue that 'the quality of primary and secondary school mathematics teaching depends crucially on the subject-related knowledge that teachers are able to bring to bear on their work' – a conclusion based on their bringing together of the reflections and implications of a range of mathematics education researchers (see also the Research Focus for Chapter 27).

2

MATHEMATICS IN THE PRIMARY CURRICULUM

IN THIS CHAPTER, THERE ARE EXPLANATIONS OF

- the different reasons for teaching mathematics in the primary school;
- the contribution of mathematics to everyday life and society;
- the contribution of mathematics to other areas of the curriculum;
- the contribution of mathematics to the learner's intellectual development;
- the importance of mathematics in promoting enjoyment of learning;
- how mathematics is important as a distinctive form of knowledge;
- what mathematics is taught in the primary curriculum;
- how mathematics is not just knowledge and skills but also application, reasoning and problem solving;
- the principles of mastery in teaching and learning mathematics.

READ THIS CHAPTER'S CURRICULUM LINKS AT: HTTPS://STUDY.SAGEPUB.COM/HAYLOCK7E

WHY TEACH MATHEMATICS IN PRIMARY SCHOOL?

What is distinctive about mathematics in the primary curriculum? Why is it always considered such a key subject? What are the most important things we are trying to achieve when we teach mathematics to children? To answer questions such as these, we need to identify our aims in teaching mathematics to primary school children. Teachers are normally very good at specifying their short-term objectives – the particular knowledge or skills they want pupils to acquire in a lesson. But reflective teachers will also recognize the value of having a framework for longer-term planning, to ensure that the children receive an appropriate breadth of experience year by year as they progress through their primary education. I find it helpful to identify at least five different

> ## LEARNING AND TEACHING POINT
>
> In shaping, monitoring and evaluating their medium-term planning, teachers should ensure that sufficient prominence is given to each of the five reasons for teaching mathematics:
>
> 1 its importance in everyday life and society;
> 2 its importance in other curriculum areas;
> 3 its importance in relation to the learner's intellectual development;
> 4 its importance in developing the child's enjoyment of learning; and
> 5 its distinctive place in human knowledge and culture.

kinds of **aims of teaching mathematics** in primary school. They relate to the contribution of mathematics to: (1) everyday life and society; (2) other areas of the curriculum; (3) the child's intellectual development; (4) the child's enjoyment of learning; and (5) the body of human knowledge. These are not completely discrete strands, nor are they the only way of structuring our thinking about why we teach this subject.

HOW DOES MATHEMATICS CONTRIBUTE TO EVERYDAY LIFE AND SOCIETY?

This strand relates to what are often referred to as **utilitarian** aims. According to the National Curriculum for England, mathematics is 'essential to everyday life … and necessary for financial literacy and most forms of employment' (DfE, 2013: 3). Similarly, the Scottish Curriculum for Excellence states that 'mathematics is important in our everyday life, allowing us to make sense of the world around us and to manage our lives' (Education Scotland, 2010b: 1). The Welsh Government (2020) Curriculum for Wales likewise notes that 'the development of mathematics has always gone hand in hand with the development of civilisation itself'. Many everyday transactions and real-life problems, and most forms of employment, require confidence and competence in a range of basic mathematical skills and knowledge – such as measurement, manipulating shapes, organizing space, handling money, recording and interpreting numerical and graphical data, using computers and other technology.

Facility with the numerical demands in everyday life is often referred to as 'numeracy'. Jefferey (2011: 6) defines numeracy as 'the knowledge, skills and understanding necessary to move around in the world of numbers with confidence and competence'. My own definition of being 'numerate', also emphasizing *confidence* in numerical situations, is as follows:

> A numerate person can be defined as one who can deal confidently with the numerical situations they encounter in their normal, everyday life. Confidence with numbers, including a good grasp of the relationships between them and how to operate on them in a range of practical situations and contexts, is a key characteristic of numeracy. An innumerate person will lack this confidence and may therefore be inclined to accept unquestioningly the opinions and judgements of others when numbers are involved in a transaction. (Haylock with Thangata, 2007: 133)

The Welsh Government (2020) Curriculum for Wales states that 'numeracy – the application of mathematics to solve problems in real-world contexts – plays a critical part in our everyday lives, and in the economic health of the nation'. Most professional people need confidence in a wide range of such mathematical skills in their everyday working life. For example, a major concern in the field of healthcare is that too many practitioners have worryingly low levels of mathematical skills; if nurses do not understand the relationships between units of measurement, how to read scales, decimal numbers, rounding errors, proportionality and percentages, and rates of flow, for example, then patients' well-being and safety are put at risk (Haylock and Warburton, 2013). Teachers also, regardless of which subjects and which age range they teach, need basic mathematical skills, for example in handling school finances and budgets, in organizing their timetables, in planning the spatial arrangement of the classroom, in processing assessment data, in interpreting inspection reports and in using computer technology in their teaching across the curriculum.

LEARNING AND TEACHING POINT

Learning experiences for children that reflect the contribution of mathematics to everyday life and society could include, for example: (a) realistic and relevant financial and budgeting problems; (b) meeting people from various forms of employment and exploring how they use mathematics in their work; and (c) helping teachers with some of the administrative tasks they have to do that draw on mathematical skills.

The relationship of mathematical processes to real-life contexts is demonstrated in the process of *modelling* which is introduced in Chapter 5 and which forms the basis of the discussion of addition, subtraction, multiplication and division structures in Chapters 7 and 10.

HOW DOES MATHEMATICS CONTRIBUTE TO OTHER AREAS OF THE CURRICULUM?

This strand relates to the **application** of mathematics. We teach mathematics because it has applications in a range of contexts, including other areas of the curriculum. For example, the National Curriculum for England states that mathematics is 'critical to science, technology and engineering' (DfE, 2013). Much of mathematics as we know it today has in fact developed in response to practical challenges in science and technology, in the social sciences and in economics. So, as well as being a subject in its own right, with its own patterns, principles and procedures, mathematics is a subject that can be applied to, and mathematical skills can support, learning across the curriculum. Statutory curriculum guidance for Wales (Welsh Government, 2014), for example, stresses the development of numerical reasoning and the use of number, measuring and data skills across the curriculum – expecting this to be embedded in all subjects taught in schools for ages 5 to 14. In Northern Ireland, 'Using Mathematics' is identified as one of the three cross-curricular skills at the heart of the curriculum (CCEA, Northern Ireland, 2019a).

In contrast to the teachers of primary mathematics in Singapore and Shanghai, for example, in the UK a primary school teacher is often responsible for teaching nearly all the curriculum. This makes their job very demanding, but it does mean that they are well placed to take advantage of the opportunities that arise, for example, in the context of science and technology, in the arts, in history, geography and society, to apply mathematical skills and concepts purposefully in meaningful contexts – and to make explicit to the children what mathematics is being applied. This is a two-way process: these various curriculum areas can also provide meaningful and purposeful contexts for introducing and exemplifying mathematical concepts, skills and principles. In these ways, teachers can enhance children's understanding of mathematics through making links to other areas of learning and wider issues of interest and importance.

HOW DOES MATHEMATICS CONTRIBUTE TO THE CHILD'S INTELLECTUAL DEVELOPMENT?

This strand includes what are sometimes referred to as *thinking skills*, but I am including here a broader range of aspects of the learner's intellectual development. The National Curriculum for England, for example, asserts that 'a high-quality mathematics education … provides a foundation for understanding the world' and 'the ability to reason mathematically' (DFE, 2013). The Northern Ireland Curriculum for Key Stage 2 Mathematics requires pupils to develop through this subject such general intellectual skills as the ability to 'plan and organize their work, learning to work systematically … present information and results clearly … ask and respond to open-ended questions and explain their thinking' (CCEA, Northern Ireland, 2019b).

We teach mathematics because it provides opportunities for developing important intellectual skills in problem solving, deductive and inductive reasoning, systematic organization of data, creative thinking and communication. Mathematics is important for primary school children because it introduces them to some key thinking strategies for solving problems and gives them opportunities to use logical reasoning, to suggest solutions and to try different approaches to problems. These are distinctive characteristics of a person who thinks in a mathematical way.

Sometimes, to solve a mathematical problem we have to reason logically and systematically, using what is called *deductive reasoning* (see glossary,

> **LEARNING AND TEACHING POINT**
>
> Learning experiences for children in mathematics should include a focus on the child's intellectual development, by providing opportunities to foster:
> (a) problem-solving strategies;
> (b) deductive reasoning, which includes reasoning logically and systematically; (c) creative thinking, which is characterized by divergent and imaginative thinking; (d) inductive reasoning that leads to the articulation of patterns and generalizations; and (e) communication of mathematical ideas orally and in writing, using both formal and informal language, and in diagrams and symbols.

Chapter 4). At other times, an insight that leads to a solution may require thinking creatively, divergently and imaginatively. So, not only does mathematics develop logical, deductive reasoning but – perhaps surprisingly – engagement with this subject can also foster creativity. So, mathematics is an important context for developing effective problem-solving strategies that potentially have significance in all areas of human activity. But also in learning mathematics, children have many opportunities to look for and identify patterns. This involves what is called *inductive reasoning* (see glossary, Chapter 4), leading to the articulation of generalizations, statements of what is always the case. The process of using a number of specific instances to formulate a general rule or principle, which can then be applied in other instances, is at the heart of mathematical thinking. All these ideas about the distinctive ways of reasoning in mathematics are explained in more detail in Chapters 4 and 5.

Then finally, in this section, in terms of intellectual development we should note that in learning mathematics children are developing a powerful way of *communicating*. Mathematics is effectively a language, containing technical terminology, distinctive patterns of spoken and written language, a range of diagrammatic devices for organizing and presenting data and a distinctive way of using symbols to represent and manipulate concepts. Children use this language to articulate their observations and to explain and later to justify or prove their conclusions in mathematics. The development and correct use of mathematical language is a key theme throughout this book.

HOW DOES MATHEMATICS CONTRIBUTE TO THE CHILD'S ENJOYMENT OF LEARNING?

This strand relates to what is sometimes referred to as the aesthetic aim in teaching mathematics. According to the National Curriculum for England, one of our aims in teaching mathematics is to develop in the learner 'an appreciation of the beauty and power of mathematics, and a sense of enjoyment and curiosity about the subject' (DfE, 2013: 3). So, we teach mathematics because it has an inherent beauty that can provide the learner with delight and enjoyment. I suspect that there may be some readers whose personal experience of learning mathematics may not resonate with this statement. But there really is potential for genuine enjoyment and pleasure for children in primary school in exploring and learning mathematics. The Scottish Curriculum for Excellence assures us that 'because mathematics is rich and stimulating, it engages and fascinates learners of all ages, interests and abilities' (Education Scotland, 2010b). It can be emotionally satisfying for children to be able to make coherent sense of the numbers, patterns and shapes they encounter in the world around them, for example through the processes of classification (see glossary, Chapter 3) and conceptualization. Children can take delight in using mathematics to solve a problem. Indeed, they will often be seen to smile with pleasure when they get an insight that leads to a solution; when they spot a pattern, discover something for themselves or make connections; and when they find a mathematical rule that always works – or even identify an exception that challenges a rule. The extensive patterns that underlie mathematics can be fascinating, and recognizing and exploiting these can be genuinely satisfying. Mathematics can be appreciated as a creative experience, in which flexibility and imaginative thinking can lead to interesting outcomes or fresh

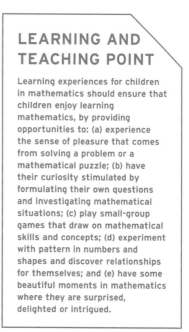

LEARNING AND TEACHING POINT

Learning experiences for children in mathematics should ensure that children enjoy learning mathematics, by providing opportunities to: (a) experience the sense of pleasure that comes from solving a problem or a mathematical puzzle; (b) have their curiosity stimulated by formulating their own questions and investigating mathematical situations; (c) play small-group games that draw on mathematical skills and concepts; (d) experiment with pattern in numbers and shapes and discover relationships for themselves; and (e) have some beautiful moments in mathematics where they are surprised, delighted or intrigued.

avenues to explore for the curious mind. Throughout this book, I aim to increase the reader's own sense of delight in and enjoyment of mathematics, with the hope that this will be communicated to those they teach.

WHY IS MATHEMATICS IMPORTANT AS A DISTINCTIVE FORM OF KNOWLEDGE?

This strand is what the more pretentious of us would call the **epistemological** aim. Epistemology is the theory of knowledge. The argument here is that we teach mathematics because of its cultural significance as a significant and distinctive form of human knowledge, with its own concepts and principles and its own ways of making assertions, formulating arguments and justifying conclusions. This kind of purpose in teaching mathematics is based on the notion that an educated person has the right to be initiated into all the various forms

LEARNING AND TEACHING POINT

Teachers can encourage children to think and reason mathematically by adding the phrase 'convince me' to a question in a class discussion. For example, 'If you count from 51 to 151, will you say the same number of odd numbers as even numbers? Convince me.'

of human knowledge and to appreciate their distinctive ways of reasoning and arguing. For example, an explanation of a historical event, a theory in science, a religious doctrine and a mathematical generalization are four very different kinds of statements, supported by different kinds of evidence and arguments.

In mathematics, as we have indicated above, some of the characteristic ways of reasoning would be to look for patterns, to make and test conjectures, to investigate a hypothesis, to formulate a generalization and then to justify the generalization by means of a deductive argument or proof. The most distinctive quality of mathematical knowledge is the notion of a mathematical statement being incontestably true because it can be deduced by logical argument either from the axioms (self-evident truths) of mathematics or from previously proven truths. Askew et al. (2015: 6) rightly emphasize that learning mathematics involves 'knowing why' as well as 'knowing that' and 'knowing how'. Of course, children in primary school will not normally be able to justify their mathematical conclusions by means of a formal proof, but they can experience many of the other distinctive kinds of mathematical processes and, even at this age, begin to demonstrate and explain why various things are always true. (See glossary, Chapter 4, for definitions of conjecture, hypothesis, generalization, proof and axiom.)

Mathematics is a significant part of our cultural heritage. The National Curriculum for England states that 'mathematics is a creative and highly inter-connected discipline that has been developed over centuries, providing the solution to some of history's most intriguing problems' (DfE, 2013: 3). Not to know anything about mathematics would be as much a cultural shortcoming as being totally ignorant of our musical, artistic and literary heritage. Historically, the study of mathematics has been at the heart of most major civilizations. Certainly, much of what we might regard

> **LEARNING AND TEACHING POINT**
>
> Include as one of your aims for teaching mathematics in the primary school: to promote awareness of some of the contributions of various cultures to the body of mathematical knowledge. This can be a fascinating component of history-based cross-curricular projects, such as the study of ancient civilizations, and can contribute to the 'decolonizing' of the curriculum in the United Kingdom.

> **LEARNING AND TEACHING POINT**
>
> 'Mathematics … is essential to everyday life, critical to science, technology and engineering, and necessary for financial literacy and most forms of employment. A high-quality mathematics education therefore provides a foundation for understanding the world, the ability to reason mathematically, an appreciation of the beauty and power of mathematics, and a sense of enjoyment and curiosity about the subject' (DfE, 2013: 3). Good teachers in primary school will ensure that their planning and teaching of mathematics resonate with this statement of the purpose of studying this area of the curriculum.

as European mathematics was well known in ancient Chinese civilizations. Our number system has its roots in ancient Egypt, Mesopotamia and Hindu cultures. Classical civilization was dominated by great mathematicians such as Pythagoras, Zeno, Euclid and Archimedes. To appreciate mathematics as a subject should also include knowing something of how mathematics as a subject has developed over time and how different cultures have contributed to this body of knowledge. The Williams Review underlined the significance of this aspect of mathematics in the curriculum, suggesting that 'opportunities for children to engage with the cultural and historical story of both science and mathematics could have potential for building their interest and positive attitudes to mathematics' (Williams, 2008: 62).

WHAT MATHEMATICS DO WE TEACH IN PRIMARY SCHOOL?

The mathematics curriculum in primary school – and therefore this book – contains a huge amount of knowledge to be learnt, and a great number of skills to be mastered and concepts and principles to be understood. In learning about numbers, children will develop knowledge and skills in counting, place value and our number system, different kinds of numbers, the structures of the four basic number operations, mental strategies and written methods for calculations, rounding, various properties of numbers, fractions and ratios, calculations with decimals, proportionality and percentages. These are covered extensively in Sections C and D of this book, Chapters 6–18. Even in primary school, children will be introduced to algebraic thinking, through expressing generalizations in words and simple formulas, solving simple equations and using coordinates and linear relationships. Algebra is the focus of Section E of this book, Chapters 19–20.

Children develop many of their numerical skills in the context of various kinds of measurement and practical problems that involve measurements. So, in primary school they learn how to measure and how to estimate length, mass (weight), liquid volume and capacity, time and angle – all of these using both non-standard and standard units. They learn about different metric units and how they are related. They are introduced to the area and perimeter of some simple two-dimensional shapes, and the volume of solid shapes. This mathematics is the subject of Section F of this book, Chapters 21–22.

Geometry is the part of the mathematics curriculum where children learn about shapes and space. They develop understanding of the concept of angle, in both static and dynamic senses, and learn about different kinds of angles. They learn to identify and classify two-dimensional and three-dimensional shapes, to use appropriate language to describe the properties of various shapes, and to recognize various families of shapes. They learn how to transform shapes in various ways, including reflections and rotations, and learn about different kinds of symmetry. Geometry is the focus of Section G, Chapters 23–25.

Statistics in primary school is about learning how to collect, organize, display and interpret various kinds of data, particularly using data-handling software. Children learn about frequency tables and how to use and interpret different ways of representing data pictorially, including Venn diagrams, Carroll diagrams, pictograms, block graphs, bar charts, line graphs, scatter graphs and pie charts. They learn how to calculate and use an average as a representative value for a set of numerical data. In countries other than England or in English schools not constrained by the National Curriculum, they might experience the fascinating concept of the mathematical probability of an event occurring. Statistics and probability are covered in Section H of this book, Chapters 26–28.

Mathematics is a lifeless and purposeless subject if we do not also learn to apply all this knowledge and all these skills, concepts and principles. Williams (2008: 62) stresses the need 'to strengthen teaching that challenges and enables children to use and apply mathematics more often, and more effectively, than is presently the case in many schools'. Williams argues that it is in experiences of using and applying mathematics that we have the best chance of fostering positive attitudes to mathematics: 'if children's interests are not kindled through using and applying mathematics in interesting and engaging ways, and through learning across the full mathematics curriculum, they are unlikely to develop good attitudes to the subject' (Williams, 2008: 62).

So, teachers have to ensure that children get opportunities to learn not just mathematical content but also how to apply their mathematics. Sometimes this will consist of using the mathematical knowledge, skills, concepts and principles they have learnt to solve problems or pursue enquiries within mathematics itself. At other times, it will involve applying mathematics to solve practical problems in the world around them or to support projects within other areas of the school curriculum. It will always involve reasoning in the

> ## LEARNING AND TEACHING POINT
>
> Applying mathematics is not just something for children to do after they have learnt some mathematical content; rather, it should be integrated into all learning and teaching of the subject. Sometimes an appropriate approach to planning a sequence of mathematics lessons might be: introduce some new concept or skill; practise it; and then apply it to various problems. But, at other times, a real-life problem that draws on a wide range of mathematical ideas can be used as a meaningful context in which to introduce some new mathematical concept or to provide a purposeful stimulus for children to extend their mathematical skills.

particular ways that are characteristic of mathematics, looking for patterns, articulating generalizations, making connections, being systematic on the one hand and thinking creatively on the other. So, in Section B of this book, Chapters 4–5, before we get on to all the detail of the curriculum, we focus on mathematical reasoning and problem solving.

WHAT IS MASTERY IN LEARNING MATHEMATICS?

Responding especially to the impressive achievements in teaching mathematics in Singapore and Shanghai, the National Council for Excellence in Teaching Mathematics (NCETM) in England has taken on the promotion of good practice in the learning and teaching of mathematics under the general heading of 'mastery' (see, for example, Newell, 2023). Much of the guidance on teaching for mastery in mathematics focuses on teaching approaches and classroom methodology, with a particular emphasis on whole-class teaching and a commitment to all the children in the class learning the same mathematical ideas together. Mastery approaches are included in various Teaching and Learning Points throughout this book. But what is the quality of learning that the mastery approach seeks to achieve? The key principle is that 'mastering maths means acquiring a deep, long-term, secure and adaptable understanding of the subject' (NCETM, 2017a).

The following is a summary of principles for mastery in mathematics that has been distilled from the NCETM website (www.ncetm.org.uk), Drury (2014) and Newell (2023):

- the aim should be to enhance mathematical understanding, enjoyment and fluency in mathematical procedures for every child;
- this should include a focus on the development of deep structural understanding of mathematical concepts and principles;
- deep understanding involves making connections in mathematics, because making

> **LEARNING AND TEACHING POINT**
>
> Teaching for mastery will require the teacher to spend more time interacting with the whole class in questions and answers, explanation and demonstration than might have been the norm in the past in many primary schools. By working most of the time with the whole class, rather than with smaller groups differentiated by mathematical ability, the teacher seeks to reinforce the expectation that all pupils are capable of learning the same mathematical content and ensures that all pupils get access to the full mathematics curriculum. Differentiation is achieved by variation in questioning so that the more able children develop deeper understanding and the teacher can assess which children require individual support and intervention (NCETM, 2014).

connections makes the learning more secure, ensures what is learnt is sustained over time and cuts down the time required to assimilate and to master further concepts and techniques;

- children should be encouraged to represent and visualize abstract mathematical concepts and principles using **manipulatives** (objects such as blocks and counters) and pictures, alongside representations in symbols;
- mathematical concepts should be explored in a variety of representations and problem-solving contexts to give pupils a richer and deeper learning experience;
- a mathematical concept or skill has been mastered when the learner can represent it in a variety of ways, has the mathematical language to communicate it to others, and can apply the concept to new problems in unfamiliar situations;
- teachers should make children's learning specific and explicit.

RESEARCH FOCUS: CALCULATORS AND LEARNING MATHEMATICS

In recent years, there has been increased recognition of the important part that technology should play in the mathematics curriculum. One of the most useful and accessible electronic devices is, of course, the simple, hand-held calculator. But what should be the place of calculators in the primary mathematics curriculum? Opinions differ. For example, below are the current positions in the mathematics curricula for primary schools in England and Northern Ireland.

> Calculators should not be used as a substitute for good written and mental arithmetic. They should therefore only be introduced near the end of key stage 2 to support pupils' conceptual understanding and exploration of more complex number problems, if written and mental arithmetic are secure. (DfE, England, 2021)

> Children should use calculators in extended investigations in mathematics and in real-life situations. They should explore how a calculator works, appreciate the operations possible on a calculator and their proper order, check calculator results by making an estimate, by repeating the operations in a different order or by using a different operation and learn to interpret calculator results. (CCEA, Northern Ireland, 2019a)

Many mathematics educators would point to the variety of ways in which calculators can be used, particularly across the primary age range, not to do simple calculations, but to promote understanding of mathematical concepts and to explore patterns and relationships between numbers. This particular argument is supported by most of the research evidence. An Australian research project (Groves, 1993, 1994) explored the results achieved by using a calculator-aware number curriculum with young children from the reception class onwards. When these children reached the age of 8–9 years, they were found to perform better in a number of key mathematical tasks than children two years older than them. These included: estimating the result of a calculation; solving real-life problems; understanding place value, decimals and negative numbers; and interpreting calculator answers involving decimals.

GLOSSARY OF KEY TERMS INTRODUCED IN CHAPTER 2

Aims of teaching mathematics   In describing the importance of mathematics in the primary curriculum, a number of different kinds of aims in teaching mathematics can be identified; these can be classified as utilitarian, application, intellectual development, aesthetic and epistemological.

Utilitarian aim in teaching mathematics   Mathematics is useful in everyday life and necessary in most forms of employment.

Application aim in teaching mathematics   Mathematics has many important applications in other curriculum areas.

Intellectual development aim in teaching mathematics   Mathematics provides opportunities for developing important intellectual skills in problem solving, deductive and inductive reasoning, creative thinking and communication.

Aesthetic aim in teaching mathematics   Mathematical experiences in primary school can provide delight, wonder, beauty and enjoyment.

Epistemological aim in teaching mathematics   Mathematics should be learnt because it is a distinctive and important form of knowledge and part of our cultural heritage.

Mastery in learning mathematics   A deep, long-term and structural understanding of mathematical principles, concepts and procedures.

Manipulatives   A general term for any objects such as blocks and counters that children can handle to represent processes with number.

LEARNING HOW TO LEARN MATHEMATICS

IN THIS CHAPTER, THERE ARE EXPLANATIONS OF

- the fundamental importance of children in primary school learning how to learn mathematics;
- the connections model for understanding number and number operations;
- the processes of recognizing equivalences and identifying transformations;
- the process of classification.

READ THIS CHAPTER'S CURRICULUM LINKS AT: HTTPS://STUDY.SAGEPUB.COM/HAYLOCK7E

WHAT IS MEANT BY 'LEARNING HOW TO LEARN MATHEMATICS'?

When the National Curriculum for primary school in England was being reviewed back in 2008, I was invited to take part in a mathematics advisory group. To start the discussion we were asked to say what we considered to be the most important thing for children to learn in mathematics at primary school. Reflecting on this question on the train down to London, I came up with my answer. The most important thing for children to learn in mathematics in the primary years is how to learn mathematics. So, this chapter is, in my view, the most important chapter in this book.

This conclusion is based on my experience of teaching mathematics to children of all ages and to adults, particularly those training to teach in primary school. The biggest problem that I come across is that in learning this subject individuals can develop a **rote-learning mind set**. In essence, this means that they have stopped trying to make sense of what they are taught or asked to do in mathematics; they just sit there waiting for the teacher to tell them what to do with a particular type of question. They no longer want to understand. They see learning mathematics as a matter of learning a whole collection of routines and recipes for different kinds of questions. Sadly, they may even have learnt that to get the teacher's approval and the marks in mathematics tests you do not actually have to understand what is going on, you just have to remember the right procedures.

If children develop a rote-learning mind set in primary schools, then they have not learnt how to learn mathematics. The beauty of the subject is that it does all make sense. It can be understood. It can be learnt meaningfully. Our biggest challenge in teaching mathematics to primary school children is to ensure that they move on to secondary education with a **meaningful-learning mind set**. This means that they are committed to

> **LEARNING AND TEACHING POINT**
>
> The first of three aims for teaching mathematics in Key Stages 1 and 2 in the National Curriculum for England includes the aspiration that pupils will 'develop conceptual understanding' of mathematics (DfE, 2013: 3). The words 'understand' and 'understanding' appear 36 times in the English primary mathematics curriculum. This curriculum is not therefore intended to be taught to children by rote; teachers should help children to learn that mathematics can be learnt meaningfully, with understanding.

> **LEARNING AND TEACHING POINT**
>
> Teachers will help primary school children to learn how to learn mathematics: (a) if they value and reward understanding more highly than mere repetition of learnt procedures and rules; and (b) if they ask questions that promote understanding rather than mere recall of facts and learnt routines.

learning with understanding. They have learnt how to learn with understanding, they expect to understand and will not be content until they do. They have had teachers who have valued children showing understanding more highly than just the short-term reproduction of learnt procedures.

The best teachers in primary school want children to understand what they learn. But for a child to understand, they have to learn how to learn with understanding – and this requires teachers who understand what learning with understanding in mathematics is like and how it is demonstrated in children's responses to mathematical situations.

HOW IS MATHEMATICS UNDERSTOOD?

We focus first on the idea that to understand something in mathematics involves making (cognitive) connections, one of the key principles for mastery in mathematics. In 2016 the Director of the NCETM stated that 'if understanding in any mathematical area is deep (not superficial) then it will mean the learner has recognized and grasped connections between the concept in question and concepts in other areas of maths' (Stripp, 2016).

When I have some new experience in mathematics, if I just try to learn it as an isolated bit of knowledge or a discrete skill, then this is what is called rote learning. All I can do is try to remember it and to recall it when appropriate. If, however, I can connect it in various ways with other experiences and things I have learnt, then it makes sense. For example, if I am trying to learn the 8-times multiplication table and the teacher helps me to connect it with the 4-times table – which I already know – then I feel I am beginning to make some sense of what otherwise would seem to be a whole collection of arbitrary results. So, 7 eights, well, that is just double 7 fours: double 28, which makes 56. And, of course, when I struggle momentarily to recall 7 fours, well, that is just double 7 twos. And so on. So, in learning the multiplication tables, I am not just trying to recall a hundred (or a hundred and forty-four) different results, but I am constructing a network of connections, which helps me to make sense of all these numbers, to see patterns and to use relationships. As teachers, we want to encourage children all the time to make these and other kinds of connections, so this becomes the default setting for how they learn mathematics.

To understand many mathematical ideas – like number, subtraction, place value, fractions – we have to gradually build up these networks of connections, where each new experience is being connected with our existing understanding and is related in some way to other experiences. This is achieved through practical engagement with mathematical materials, through investigation and exploration, through talking about mathematics with the teacher and other learners, and through the teacher's explanations and asking the right kinds of questions – but, above all, through the learner's own cognitive response, which has been shaped by prior successful learning to look for connections and relationships in order to learn in a meaningful way.

In the rest of this chapter, I shall explain four key processes that are at the heart of understanding in primary school mathematics: (1) a connections model of understanding number and number operations; (2) equivalence; (3) transformation; and (4) classification.

WHAT IS THE CONNECTIONS MODEL OF UNDERSTANDING?

Figure 3.1 illustrates a simple **connections model** that I have found helpful for promoting understanding in number and number operations. This model identifies the four kinds of things that children process and employ when doing number work in primary schools: language, pictures, symbols and practical/real-life experience. This diagram represents many of the most important connections to be established in understanding number and number operations. Making any one of these kinds of connections – connecting language with symbols, connecting pictures with language, connecting real-life experience with symbols and so on – contributes to the learner's understanding.

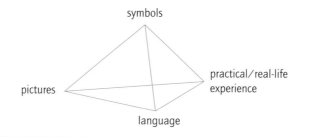

Figure 3.1   A model for understanding number work: making connections

Language in this model includes formal mathematical language: subtract, multiply, divide, equals and so on. It also includes more informal language appropriate to various contexts: taking away, so many lots of so many, sharing, is the same as, makes and so on. In particular, it includes key patterns of language, such as in these examples: 8 is 3 more than 5, and 5 is 3 less than 8; 12 shared equally between 4 is 3 each.

By *pictures*, I have in mind all kinds of charts, graphs, pictograms and sorting diagrams and, especially, the picture of number as provided in number strips and number lines.

Symbols are those we use in mathematics to represent numbers and number operations, equality and inequality: 3, ¾, 0.78, +, −, ×, ÷, =, <, > and so on.

Practical/real-life experiences include any kind of engagement with physical objects (manipulatives), such as counters, coins, blocks, as well as fingers, containers, groups of children, board games or toys. This component also includes any real-life situations, such as shopping, measuring, travelling, cooking or playing in the playground, whether actual or imagined.

So, developing understanding of a concept in number work can be thought of as building up cognitive connections between these four components. Throughout this book, number concepts are explained in this way, with an emphasis on connecting the relevant language, pictures and concrete experiences with the mathematical symbols. For example, in Chapter 6 the concept of place value is explained in terms of: key language, such as 'exchanging one of these for ten of these'; pictures, such as the number line; concrete materials, such as counters and blocks; and then making the connections between all these and the symbols used in our place-value number system. If we as teachers aim to promote understanding in mathematics, rather than just learning by rote, then the key to this is to teach it in a way that encourages children to make connections.

> ## LEARNING AND TEACHING POINT
>
> Mastery through making connections: use question-and-answer sessions with the class specifically to ensure that they are making connections from their experience of doing mathematical tasks. For example, in promoting understanding of subtraction, start with a real-life situation such as: when playing a board game, Tom is on square 13 and wants to land on square 22; what score does he need? Connect it with language, such as what do we add to 13 to make 22? What is 22 subtract 13? How many more than 13 is 22? Connect it with a picture: what do we do on a number line to work this out? Connect it with symbols: what would we enter on a calculator or a calculator app to work this out? (22 − 13 =)? How would we record this subtraction?

WHAT ARE EQUIVALENCES AND TRANSFORMATIONS?

Recognizing similarities and differences are fundamental cognitive processes by means of which we organize and make sense of all of our experiences. They have particular significance in the development of understanding in mathematics. In this context, we refer to forming **equivalences** (by asking the question, what is the same?) and identifying **transformations** (by asking the question, what is different? Or, how has it changed?). So, an equivalence is formed when we identify some mathematical way in which two or more numbers or shapes or sets (or any other kind of mathematical entities) are the same. And a transformation is identified when we specify what is different between two entities and what has to be done to one to change it into the other.

CAN YOU GIVE SOME EXAMPLES OF EQUIVALENCES?

The process of forming equivalences is widespread in the experience of children learning mathematics with understanding. It is part of the process of developing mathematical concepts and is a powerful tool for manipulating mathematical ideas. In the early stages of learning number, for example, children learn to recognize that there is something the same about, say, a set of five beakers and a set of five children. These two collections are different from each other (beakers are different from children), but there is something significantly the same about them, which can be recognized by matching one beaker to each child. Both sets are described by the adjective 'five', which indicates a property they share. Forming equivalences like this contributes to the child's understanding of the number five and numbers in general. Understanding number is explored further in Chapter 6.

This example illustrates how many abstract mathematical concepts are formed by identifying equivalences, recognizing things that are the same or properties that are shared. Geometry provides many such examples (see Chapter 25). In learning the concept of 'square', for example, children may sort a set of two-dimensional shapes into various subsets. When they put all the squares together in a subset, because they are 'all the same shape', they are recognizing an equivalence. The shapes may not all be the same in every respect – they

> **LEARNING AND TEACHING POINT**
>
> To promote the formation of equivalences and the recognition of transformations, frequently ask children the questions: In what ways are these the same? How are they different? How could this change into that? For example, look at the numbers in a set (e.g. 3, 6, 9, 12, 15, 18, 21, 24, 27, 30) and ask, what is the same about them? (They are all multiples of 3.) Select two numbers from the set (for example, 15 and 30) and ask, how are these two different from each other? (15 is smaller than 30, 30 is larger than 15 and so on.) How can one number be changed into the other number? (Double the 15, halve the 30 and so on.) Follow the same approach with sets of shapes.

may differ in size or colour, for example – but they are all the same in some sense; they share the properties that make them squares; there is an equivalence. Any one of them could be used if we wanted to show someone what a square is like, or to do some kind of investigation with squares.

This kind of thinking is powerful and is fundamental to learning and doing mathematics. It enables the learner to hold in their mind one conceptual idea (such as 'five' or 'square') which is an abstraction of their experiences of many specific examples of the concept, all of which are in some sense the same, all of which are equivalent in this respect. In doing this, the learner combines a number of individual experiences of specific exemplars, which have been recognized as being the same in some sense, into one abstraction.

AND EXAMPLES OF TRANSFORMATIONS?

Making sense of the relationship between two mathematical entities (numbers, shapes, sets and so on) often comes down to recognizing that they are the same but different. When we form equivalences by seeing something that is the same about some mathematical objects, we have to ignore temporarily the ways in which they are different. When we take into account what is different, we focus on the complementary process of transformation. We identify a transformation when we specify what we have to do to one thing to change it into another, different thing. Sometimes we focus on the equivalence and sometimes on the transformation. For example, children have to learn that $^2/_3$ and $^4/_6$ are equivalent fractions. They are not identical – two slices of a pizza divided into three equal slices is not in every respect the same as four slices of a pizza divided into six equal slices. But there is something very significantly the same about these two fractions: we do get the same amount of pizza! In making this observation, we focus on the equivalence. But when we observe that you change $^2/_3$ into $^4/_6$ by multiplying both top number and bottom number by 2, then we focus on the transformation. We explore equivalent fractions further in Chapter 15.

In Figure 3.2, the two shapes can be considered equivalent because we can see a number of ways in which they are the same: for example, they are both rectangles with a diagonal drawn, and they are the same length and height. However, one is a transformation of the other, because they are mirror images. So we may identify some differences: for example, if you go from A to B to C to D and back to A, in one shape you go in a clockwise direction and in the other in an anticlockwise direction.

So understanding mirror images and reflections in geometry comes down to recognizing how shapes change when you reflect them (the transformation) and in what ways they stay the same (equivalence). Equivalence and transformation are the major themes of Chapter 24 where we see that a key aspect of mathematical reasoning involves looking at sets of shapes and asking: What is the same about all these shapes? What do they have in common? Or – when we transform a shape in some way – what has changed and what has stayed the same?

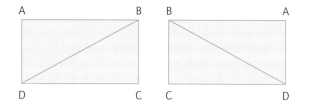

Figure 3.2 What is the same? What is different?

This interaction of equivalence and transformation is also central to understanding the principle of conservation in number and measurement. In the context of measurement, conservation is explained in Chapter 21. In terms of **conservation of number**, the principle is illustrated in Figure 3.3. Young children match two rows of counters, blue and grey, say, and recognize that they are the same number. There is an equivalence established by the one-to-one matching. If one of the rows (B) is then spread out, they have to learn to recognize that, even though the row of counters has been transformed, there is still the same number of counters. This transformation preserves the equivalence. This is a crucial aspect of young children's understanding of number. Children who are in the early stages of understanding number may not yet see number as something that is conserved when the set of objects is transformed – and may perceive that B is now a greater number than A.

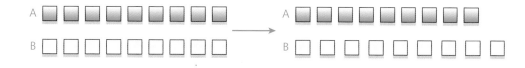

Figure 3.3 Understanding conservation of number

WHAT IS CLASSIFICATION IN MATHEMATICS?

The identification of equivalences is at the heart of the key process of **classification**. Children have to learn to classify numbers and shapes according to a range of criteria and to assign them to various sets. For example, they classify numbers as odd or even, as one-digit or two-digit, as less than 100, as multiples of 3, as factors of 30, as positive or negative and so on. They classify two-dimensional shapes as triangles or quadrilaterals, as squares, as oblongs, as regular or irregular, as symmetric and

LEARNING AND TEACHING POINT

Sorting and naming are often components of young children's play. Teachers should recognize the importance of such experiences in laying the foundation for genuine mathematical thinking through classification.

so on; and three-dimensional shapes as cubes or cuboids, as prisms, as pyramids, as spheres and so on (see Chapter 25).

Figure 3.4 illustrates the process of classification. In each case here, a rule has been used to sort (a) a set of shapes and (b) a set of numbers into exemplars and non-exemplars. Children can be challenged to articulate the rule and to use it to sort some more shapes or numbers. In these cases, the sets of exemplars are: (a) triangles with two sides equal; and (b) single-digit numbers. Sometimes when we do this the set of exemplars is important enough for us to give it a name. For instance, in example (a) we call the triangles that satisfy the rule of having two equal sides 'isosceles triangles' (see Chapter 25). In this way, classification in mathematics enables us to develop and understand new concepts – and then to use these as building blocks for forming higher-order concepts. For example, the concept of 'isosceles triangle' uses as building blocks earlier concepts such as triangle and lines of equal lengths. Classification is also a key process in the early stages of handling data and pictorial representation, as explained in Chapter 26.

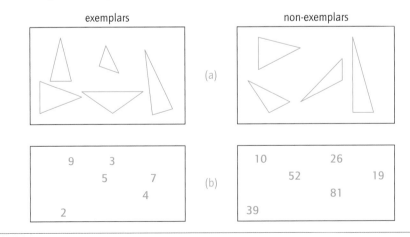

Figure 3.4 Classification using exemplars and non-exemplars

RESEARCH FOCUS: PIAGET'S SCHEMAS

One of the most famous and influential early researchers into how children learn and understand was the Swiss psychologist and philosopher, Jean Piaget (1896–1980). His historic work provides a number of intriguing insights into the learning of key mathematical concepts (see, for example, Piaget, 1952), even though other researchers, such as Donaldson (1978), subsequently challenged some of his conclusions. What I have talked about as 'networks of connections', Piaget called 'schemas' (Piaget, 1953). He investigated how schemas were developed through the learner relating new experiences to their existing cognitive structures. In many cases, the new experience can be adopted into the existing schema by a process that Piaget called 'assimilation': this is when the new experience can be related to the existing schema just as it is. But sometimes, in order to take on board some new experience, the existing schema has to be modified, reorganized. This process Piaget called 'accommodation'. For example, a child may be developing a schema for multiplication, using multiplication by numbers up to 10. When they encounter multiplication by numbers greater than 10, this new experience can usually be assimilated fairly smoothly into the existing multiplication schema. Part of this schema might be the notion that multiplying makes things bigger. When the learner encounters multiplication by a fraction (such as $20 \times \frac{1}{4} = 5$), this does not fit with this aspect of the existing schema. How can you multiply 20 by something and get an answer smaller than 20? To understand this will require a significant accommodation: a reorganization of the existing schema. If this cannot be achieved, the new experience cannot be understood; it can only be learnt by rote. A key role for teachers of mathematics is helping the learner to accommodate new experiences when they provide significant challenges to the learner's existing understanding.

LEARNING RESOURCES

Access activities for your **lesson plans** at: https://study.sagepub.com/haylock7e

Before trying the self-assessment questions below, you should complete the **self-assessment questions** for this chapter at: https://study.sagepub.com/haylock7e

3.1: In developing understanding of addition, what connections might young children make between the symbols, 5 + 3 = 8, and (a) formal mathematical language; (b) practical experience with fingers and informal mathematical language; and (c) the picture of the counting numbers from 1 to 10 shown in Figure 3.5?

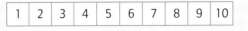

Figure 3.5 A number strip

3.2: Refer to Figure 3.2 and the associated commentary. Suggest two further ways in which the two shapes are the same as each other and two further ways in which they are different.

GLOSSARY OF KEY TERMS INTRODUCED IN CHAPTER 3

Rote-learning mind set A tendency in a learner to learn new material as isolated pieces of knowledge or skills, without making cognitive connections with existing networks of connections; a preference for relying on memorization and recall, rather than seeking to understand.

Meaningful-learning mind set A commitment in the learner to making sense of new material, to understanding it, by making cognitive connections with existing understanding; a preference for understanding rather than just learning by rote.

Connections model A model for understanding number and number operations, expressed in terms of the learner making cognitive connections between language, symbols, pictures and practical/real-life experiences.

Equivalence The mathematical term for any relationship in which one mathematical entity (number, shape, set and so on) is in some sense the same as another; in identifying an equivalence we focus on what is the same, regardless of how the entities are different.

Transformation The mathematical term for any process which changes a mathematical entity (number, shape, set and so on) into another; in identifying a transformation we focus on what is different, and what has changed, even though some things may still be the same.

Conservation of number The principle that a number remains the same under certain transformations; for example, the number of items in a set does not change when the items are rearranged or spread out.

Classification A key process in understanding mathematics, in which some numbers or some shapes (exemplars) are recognized as sharing a specified property or satisfying some criterion, which distinguishes them from other numbers or shapes (non-exemplars); for example, positive whole numbers may be classified as even or odd.

SUGGESTIONS FOR FURTHER READING FOR SECTION A

1   Highly recommended and relevant to the themes of this section is Jo Boaler's inspirational book *The Elephant in the Classroom: Helping Children Learn and Love Mathematics* (Boaler, 2020).

2   A helpful summary of research, evidence and arguments that support teaching for mastery in mathematics is available on the NCETM website (NCETM, 2017b).

3   For a useful study of the theory of mastery teaching with helpful practical examples see Newell (2023).

4   Drury (2014) provides a useful summary of teaching for mastery, on behalf of the Mathematics Mastery partnership.

5   Part 1 of Askew (2015) is entitled 'Thinking through Mathematics' and considers this subject in terms of learning, curriculum and teaching. It is recommended for anyone who wants to engage seriously with issues of mathematics education in primary school.

6   To find out 'how maths illuminates our lives', get hold of Daniel Tammet's accessible and fascinating book *Thinking in Numbers* (Tammet, 2013).

7   Chapter 1 of Haylock and Cockburn (2017) is on understanding mathematics. In this chapter, we outline and illustrate in more detail the connections model for understanding, as well as the ideas of transformation and equivalence, with a particular focus on younger children learning mathematics.

8   Rowland et al. (2009) is a book that will help primary teachers to understand how they can develop their own mathematical subject knowledge in ways that will make their own teaching more effective. Look particularly at Chapter 5 on making connections in teaching.

9   Chapter 1 of Hansen (2020) provides interesting insights into how children in primary school learn mathematics, particularly through making mistakes.

SECTION B

MATHEMATICAL REASONING AND PROBLEM SOLVING

4   Key Processes in Mathematical Reasoning   45
5   Modelling and Problem Solving   62

WATCH THE SECTION OPENER VIDEO AT: HTTPS://STUDY.SAGEPUB.COM/HAYLOCK7E

KEY PROCESSES IN MATHEMATICAL REASONING

IN THIS CHAPTER, THERE ARE EXPLANATIONS OF

- generalization;
- conjecturing and checking;
- the language of generalization;
- counter-examples and special cases;
- hypothesis and inductive reasoning;
- communicating, stem sentences, explaining, convincing, proving and deductive reasoning;
- thinking creatively in mathematics.

READ THIS CHAPTER'S CURRICULUM LINKS AT: HTTPS://STUDY.SAGEPUB.COM/HAYLOCK7E

IS DEVELOPING MATHEMATICAL REASONING IMPORTANT?

In Chapter 2, in considering the epistemological reasons for teaching mathematics, I argued that to understand mathematics we need to have a good grasp not just of the terms, concepts and principles that are used in the subject but also of the distinctive ways in which we reason and make and justify our assertions. These key processes of mathematical reasoning are the focus of this chapter. The reader should note that in order to illustrate these processes I have to use mathematical concepts and results that are explained in later chapters in this book. So do not worry too much at this stage if you struggle a bit with some of the details of the examples used; try to focus on the key processes involved and note the chapters you will need to give special attention to later. You should also return to this chapter when you have studied the rest of the book.

The National Curriculum for England includes as one of its three aims that all pupils should learn to 'reason mathematically by following a line of enquiry, conjecturing relationships and generalizations, and developing an argument, justification or proof using mathematical language' (DfE, 2013: 3). Similarly, curriculum guidance for Scotland states that 'learning mathematics develops logical reasoning, analysis, problem-solving skills, creativity and the ability to think in abstract ways' (Education Scotland, 2010b). So, this aspect of mathematics is at the core of the subject; it is not an optional add-on. The teacher's role will therefore include a focus on the development of these high-level skills. Of course, to be able to reason mathematically the learner also needs fluency in accessing basic knowledge and skills.

We shall see in this chapter that the key processes to be developed in mathematical reasoning include those associated with recognizing patterns and relationships, making conjectures, formulating hypotheses, articulating and using generalizations, explaining and convincing, and thinking creatively when faced with unfamiliar mathematical challenges.

WHAT IS GENERALIZATION IN MATHEMATICS?

Generalization is one of these significant ways of reasoning in mathematics. The Northern Ireland Statutory Requirements for primary mathematics, for example, include that 'pupils should be enabled to interpret, generalize and use simple relationships expressed in numerical, spatial and practical situations' (CCEA, Northern Ireland, 2019b).

To make a generalization is to make an observation about something that is always true or always the case for all the members of a set of numbers, or a set of shapes, or even a set of people. To say that the diagonals of a square always bisect each other at right angles is to make a generalization that is true for all squares. To recognize that every other number when you count is an even number is to make a true generalization about the set of counting numbers. To say that any Member of Parliament is paid more than every primary school teacher in the UK is to make a generalization – which may or may not be true. It is possible to make generalizations that prove to be invalid, of course. In Chapter 19, where I explain the basic ideas of algebra, we will see how the process of making generalized statements is the fundamental process of algebraic thinking. We will see there the central role played by algebraic symbols in representing variables and enabling us to articulate generalizations. To introduce here some of the key components of the process of making generalizations, we will look at the mathematical investigation illustrated in Figure 4.1. In this investigation, we construct a series of square picture frames, using small square tiles, and tabulate systematically the number of tiles needed for different sizes of frame to identify the underlying numerical pattern.

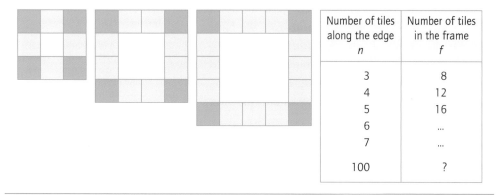

Number of tiles along the edge n	Number of tiles in the frame f
3	8
4	12
5	16
6	...
7	...
100	?

Figure 4.1 How many tiles are needed to make a square picture frame?

Having got to this stage, we might make a **conjecture** that for a frame of side 6 we would need 20 tiles respectively. We would check to see whether our conjecture is true, by actually building the frame. Then we might spot and articulate something that appears to be always true here, a generalization: 'the rule is that to get the next number of tiles you always add 4' – and perhaps build a few more frames to check this.

We may even reason our way to articulating a more sophisticated generalization: 'to find the number of tiles in the frame you multiply the number along the edge by 4 and subtract 4'. This rule would enable us to determine how many tiles are needed for a frame of side 100, for example: multiply 100 by 4 and subtract 4, giving a total of 396 tiles required. We would check this rule against all the examples we have. When we are confident in using algebraic notation (see Chapter 19), we might express this generalization as $f = 4n - 4$ (where n is a whole number greater than 2).

On the way, we might have attempted one or two generalizations that proved to be false. For example, we might have looked at the frames and thought intuitively that the number of tiles must be just 4 times the number along the edge. However, any frame that we use to check this generalization provides a **counter-example**, showing that it is false.

Then we might ask ourselves, what about a frame of side 2? Does that fit the rule? Or is it a **special case**? Well, the rule works, but we could not use the resulting construction as a picture frame. And what about a frame of side 1? The rule tells us that we need 0 tiles, which is a very strange result. These cases will lead us to refine our generalization, perhaps to apply only to frames where the number of tiles along the edge is greater than 2.

After all this, our sophisticated generalization that $f = 4n - 4$ is still only a **hypothesis**. We have not actually provided an *explanation* or a *convincing argument* or what mathematicians would call a **proof** that it must be true in every case. (This is your challenge in self-assessment question 4.3 at the end of this chapter.)

So, from this example of a mathematical investigation, we can identify the following as some of the key processes of reasoning mathematically:

> ## LEARNING AND TEACHING POINT
>
> If teachers attach too much importance to responses being right or wrong in a mathematical investigation, it will discourage children from making conjectures and formulating hypotheses that, after checking, might turn out to be wrong. Incorrect responses are often more useful for learning than correct responses.

making conjectures, using the language of generalization, using counter-examples and recognizing special cases, hypothesizing, explaining, convincing and proving. These are explained further in the rest of this chapter.

WHAT IS A CONJECTURE IN MATHEMATICS?

The word *conjecture* is often used in the context of mathematical problem solving and investigating. It refers to an assertion that something might be true, at a stage where there has not yet been produced the evidence necessary to decide whether or not it is true. A conjecture is usually followed, therefore, by some appropriate mathematical process of checking. This experience of conjecturing and checking is fundamental to reasoning mathematically.

For example, a child might make a conjecture that 91 is a prime number (see Chapter 13). To say this is a conjecture means that they do not know for sure at this stage that 91 is a prime number, but they have a hunch that it might be. After a little bit of exploration, dividing 91 in turn by 2, 3, 5 and then 7, they find that 7 divides into 91, 13 times, so 91 is not prime. So the conjecture was wrong, but we have done some useful mathematics by considering it as a possibility and checking it out. Another child might be given a collection of containers and, by examining them, make a conjecture that the flower vase has the greatest capacity (see Chapter 21). To check this conjecture, the child might carefully fill each of the other containers in turn with water and pour this into the empty flower vase. This might lead to the conclusion that the conjecture was correct.

WHAT IS THE LANGUAGE OF GENERALIZATION?

Another child notices that there are exactly three **multiples** of 3 (3, 6 and 9) in the set of whole numbers from 1 to 10 inclusive. (Multiples of 3 are the numbers in the 3-times table: see Chapter 13.) The child then makes the conjecture that there will be another three of these multiples in the set of whole numbers from 11 to 20 inclusive. Checking leads to the confirmation of this conjecture, since 12, 15 and 18 are the only multiples of 3 in this range. Excited by this discovery, the child then goes on to wonder if there are three multiples of 3 in every decade. (By *decade*, we mean 1–10, 11–20, 21–30, 31–40 and so on.) The child's thinking has moved from specific cases to the general. This is now a *generalization*. This is good mathematical thinking, whether or not it turns out to be a valid generalization. (The reader is invited to check the validity of this generalization in self-assessment question 4.2 at the end of this chapter.)

So, a generalization is an assertion that something is true in a number of cases, or even in every case. To make a generalization in words, we will often use one or other of the following bits of language: always; every; each; any; all;

LEARNING AND TEACHING POINT

Be aware of the range of language available for making generalized statements in mathematics (always, every, each, any, all, if … then … and so on), use it in your own talk and ask questions that encourage children to use it in their observations of patterns and relationships in mathematics.

if … then … It is also possible in English to imply 'always' without actually stating it and to do some nifty things with negatives. Here, for example, is a (true) generalization about multiples (see Chapter 13) stated in nine different ways:

- Multiples of 6 are always multiples of 2.
- Every multiple of 6 is a multiple of 2.
- Each multiple of 6 is a multiple of 2.
- Any multiple of 6 is a multiple of 2.
- All multiples of 6 are multiples of 2.
- If a number is a multiple of 6 then it is a multiple of 2.
- A multiple of 6 must be a multiple of 2.
- There is no multiple of 6 that is not a multiple of 2.
- If a number is not a multiple of 2 then it is not a multiple of 6.

Note that the reverse statement of a true generalization is not necessarily true. For example, it is *not* true to say: if a number is a multiple of 2 then it is a multiple of 6. This statement is still a generalization, but it happens to be a false one.

WHAT ARE COUNTER-EXAMPLES AND SPECIAL CASES?

A *counter-example* is a specific case that demonstrates that a generalization is not valid. For example, to show the falsity of the generalization made at the end of the previous paragraph, we could use the number 14 as a counter-example. This would involve pointing out that 14 is a multiple of 2 (so it satisfies the 'if' bit of the statement), but it is not a multiple of 6 (so it fails the 'then' bit).

Here is an example of a generalization in the context of shape: all rectangles have exactly two lines of symmetry. At first sight, this looks like a sound generalization, but then we may recall that squares are rectangles and they have four lines of symmetry. So a square is a counter-example showing the generalization to be false. Sometimes a counter-example will lead us to modify our generalization rather than discarding it altogether. The generalization above, for example,

LEARNING AND TEACHING POINT

Deliberately make generalizations that are invalid and get children to look for and suggest counter-examples. For example: 'In every year there are exactly four months that have five Sundays.'

could be modified to refer to 'all rectangles except squares' or 'all oblong rectangles' (see Chapter 25).

Here's another example: someone might make the generalization that all prime numbers are odd. Someone else then points out that 2 is a counter-example, being a prime number that is even (see Chapter 13). However, this can be recognized as a *special case*. The generalization can therefore be modified, by excluding the special case, changing the set of numbers to which the generalization applies, as follows: all prime numbers greater than 2 are odd.

Often, 0 or 1 will turn out to be special cases that need checking. For example, when investigating fractions we might notice that $1/3$ is less than $1/2$, $1/4$ is less than $1/3$, $1/5$ is less than $1/4$, $1/6$ is less than $1/5$ and so on, and generalize this by observing that every time you increase the bottom number by 1 the fraction gets smaller. Being very sophisticated we might come up with the statement that for any positive whole number n, $1/n$ is less than $1/(n-1)$. However, we would have to exclude 1 from this generalization, because that would give us $1/1$ is less than $1/0$, which is nonsense because $1/0$ is not a real number (division by zero is not possible). In this example, 1 is a special case. (Fractions are explained in Chapter 15.)

In an investigation about factors (see glossary, Chapter 11), a child discovers that all square numbers have an odd number of factors (see Chapter 13). This is a correct generalization, apart from the special case of zero. The teachers ask, 'What about zero? Is that a square number? If so, how many factors does it have?' Well, 0 is technically a square number, since $0^2 = 0$. However, every positive whole number is a factor of zero. (Note that $0 \times 1 = 0$, $0 \times 2 = 0$, $0 \times 3 = 0$ and so on.) So, it is necessary to exclude zero as a special case by making the generalization as follows: all square numbers greater than 0 have an odd number of factors.

WHAT IS A HYPOTHESIS?

The word *hypothesis* is usually used to refer to a generalization that is still a conjecture and which still has to be either proved to be true or shown to be false by means of a counter-example. Often, a hypothesis will emerge by a process of **inductive reasoning**, by looking at a number of specific instances that are seen to have something in common and then speculating that this will always be the case. For example, some children might investigate what happens when you add together odd and even numbers. They might spot that in all the examples they try, an odd number added to an even number gives an odd number as the answer. So they conjecture that this is always the case. This is a hypothesis, obtained by inductive reasoning. It always seems to be the case, and every example we check seems to work. But, however many cases we check, it is still only a hypothesis – until such time as we produce some kind of a *proof* that it must work in every case.

This is an important point, because sometimes a hypothesis may appear to be correct to begin with but then let you down later. For example, someone might assert that all numbers that are 1 less or 1 more than a multiple of 6 are prime numbers. To start with, this looks like a pretty good hypothesis: 5, 7, 11, 13, 17, 19 and 23 are all prime. You might think that if something works for the first seven numbers you try, it will always work, but the next number, 25, lets us down.

Hypotheses also turn up frequently in the context of statistical data (see Section H). For example, the assertion made by one child that a boy is more likely than a girl to walk to school would be a hypothesis that primary children might investigate, testing it by the collection and analysis of data from a sample of boys and girls. The evidence in this case would only lend support to or against the hypothesis, not prove it or disprove it conclusively, of course.

HOW DO PRIMARY SCHOOL CHILDREN COMMUNICATE WITH MATHEMATICS?

Teachers in primary school will also aim for the children in their classes to develop skills in communicating with mathematics. Principles and practice for Mathematics in the Welsh Curriculum, for example, include children 'explaining their thinking and presenting their solutions to others in a variety of ways' (Welsh Government, 2020). The mathematics curriculum for England recognizes the importance of children 'making their thinking clear to themselves as well as others and teachers' (DfE, 2013: 4). A strategy in teaching for mastery in mathematics is to encourage children to use clearly structured **stem sentences**. An example of a stem sentence might be: 'Seven is five less than twelve and twelve is five more than seven.' Teachers encourage children in class discussions, for example, to respond to diagrams such as Figure 7.5 (in Chapter 7) with sentences using this structure, so that a pattern of words gets linked to the visual image of this particular type of diagram.

Children should be given opportunities to explain why an answer is correct, and learn how to present information and results in a clear and organized way; they should have opportunities to draw simple conclusions of their own and explain their reasoning (for example, in response to the teacher's 'convince me!'). They should get opportunities to learn how to communicate with mathematical language, symbols and diagrams. This will involve explaining their insights, describing the outcomes of an investigation, providing convincing reasons for a conclusion they have drawn, or offering evidence to support a point of view. Older children may begin to move on from explanation of what they have found to be true in their mathematical enquiries to explore the idea of *proving* something to be true.

Proof is a peculiarly mathematical way of reasoning. If a generalization is written in the form of a statement using 'if ... then ...' language, a mathematical proof is a series of logical deductions that start from the 'if ...' bit and lead to the 'then ...' bit. Proof therefore involves **deductive reasoning**. For example, the discovery that all multiples of 6 are multiples of 3 might be rephrased as: if a number is a multiple of 6 then it is a multiple of 3. So, the 'proof' might go something like this:

> If my number is a multiple of 6 then it is, say, n sets of 6.
> But each lot of 6 is 2 sets of 3.
> So my number is n sets, each of which is 2 sets of 3.
> This gives $2n$ sets of 3.
> So my number is a multiple of 3.

Primary children would definitely not be expected to produce proofs of their hypotheses like this. At primary level, children's responses to the question 'Can you convince me this is true?' will be expressed informally, but they may still contain the essence of deductive reasoning. They can be encouraged to try to formulate explanations for their mathematical results and, in some cases, begin to formulate a convincing argument as to why a generalization must be true.

For example, consider the generalization made earlier that an odd number added to an even number always gives an odd number as the answer. A child may be able to provide or at least follow an explanation along these lines. The odd number is made by adding some 2s and a 1 and the even number is made just by adding some 2s. If you add them together you get lots of 2s, plus the extra 1 – which must therefore be

LEARNING AND TEACHING POINT

A key component in children's mastery of mathematical ideas is that they develop the mathematical language to communicate it to others. This requires that teachers give time to class discussions and questions about mathematical problems, patterns and procedures. For this to be effective, teachers themselves must be confident about the meanings of mathematical terminology and use the language of mathematics clearly and accurately: 'The quality and variety of language that pupils hear and speak are key factors in developing their mathematical vocabulary and presenting a mathematical justification, argument or proof' (DfE, 2013: 4).

an odd number. This is not a formal mathematical proof, but it is perhaps a convincing argument.

Some hypotheses can be proved to be true by a method called **proof by exhaustion**. This is a method of proof that can be employed for a generalization that relates to only a finite number of cases. It might then be possible to check every single case – in other words, to exhaust all the possibilities. This is a method of proof that can be accessible to primary school children in appropriate examples, but it requires a high level of systematic thinking. For example, Figure 4.2 shows four shapes with perimeters of 12 units drawn using the lines on a square grid. Of these four shapes, the square has the largest area – just count the number of square units inside each shape. (Area and perimeter are explained in Chapter 22.)

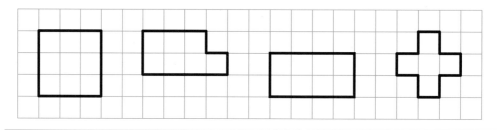

Figure 4.2 *Four shapes with perimeters of 12 units*

Can we prove that whatever shape we draw with a perimeter of 12 units on the grid, it will have an area smaller than that of the square? This could be rephrased as a generalization, using the 'if … then …' format, as follows: if any shape, other than a square, with a perimeter of 12 units is constructed using the lines on a square grid, then the area of the shape will be less than that of the square. Now we can actually *prove* this to be true fairly easily, because the number of different shapes that can be made is finite. So we can exhaust all the possibilities by drawing all the possible shapes and checking that in every case the area is less than that of the square.

> **LEARNING AND TEACHING POINT**
>
> Be prepared to ask children why they think some mathematical generalizations might be true. Use the phrase, 'convince me' to encourage them to work towards offering reasonably convincing explanations. For example, in geometry, ask the class whether it is true that the number of sides in a polygon (see Chapter 25) is always the same as the number of angles; then ask them to try to explain why this must be so.

WHAT ARE AXIOMS?

Some of the generalizations we use in mathematics are called **axioms**. An axiom is a statement that is taken to be true, usually because it is self-evidently true, but which cannot be proved as such. An axiom is one of the building blocks of mathematical reasoning. It is a statement that we have to accept as true, otherwise we would just not be able to get on and do any mathematics. Two examples of axioms that we will discuss in this book are the commutative law of addition and the commutative law of multiplication (see Chapters 7 and 10). These axioms, respectively, tell us, for example, that $3 + 5 = 5 + 3$ and $3 \times 5 = 5 \times 3$, and that statements like these would work whatever two numbers we used. Another example would be the transitive property of the inequality 'greater than', which we will explain in Chapter 21. This axiom allows us to conclude that if number A is greater than some number B, which in turn is greater than some number C, then A must be greater than C. We can explain these laws, we can give examples of what they mean and we can see how they are used – but we never question their truth or feel the need to prove them. These kinds of generalizations are not hypotheses up for discussion and investigation. They are axioms of mathematics.

IS GENERALIZING SOMETHING FOR ALL PRIMARY CHILDREN TO EXPERIENCE?

Our discussion of key processes in mathematical reasoning has led us to some challenging mathematics, so it is timely to remind ourselves that the process of forming generalizations occurs at a range of different levels and at all ages in primary school learning and teaching. At the simplest level, young children are making and using generalizations when they identify a pattern of beads on a string and continue a pattern: blue, red, yellow, yellow, blue, red, yellow, yellow and so on. When young children learn to count beyond 20, they do this by recognizing a pattern for each group of 10 numbers and make a generalization that every time you get to 9 you move on to the next multiple of 10. Once this is established, they generalize the pattern further in order to go beyond 100. When a nine-year-old observes that all the numbers in the 5-times table end in 5 or 0, they are making a generalization.

In teaching for mastery, a teacher-led class discussion about a mathematical pattern of some kind can often allow children of differing levels of attainment in mathematics to make and explain different levels of generalization. For example, looking at a sequence such as 6, 11, 16, 21, 26 …, some ten-year-olds will generalize this by seeing the pattern in the final digits (6, 1, 6, 1, 6 …). Others will formulate a rule for continuing the sequence: such as, you always add 5. Other children will observe that all the numbers in the sequence are the numbers in the 5-times table plus 1.

HOW WOULD I RECOGNIZE CREATIVE THINKING IN MATHEMATICS?

Creative thinking involves being able to break away from routines and stereotype methods, to think flexibly and to generate original ideas and approaches to problems. The opposite of creativity is rigidity and fixation. For example, some ten-year-olds were given the following questions:

a   Find two numbers that have a sum of 10 and a difference of 4.
b   Find two numbers that have a sum of 20 and a difference of 10.
c   Find two numbers that have a sum of 15 and a difference of 3.
d   Find two numbers that have a sum of 19 and a difference of 5.
e   Find two numbers that have a sum of 10 and a difference of 3.

The majority quickly concluded that (e) was impossible. In the previous questions, they had established a procedure that worked. This was essentially to run through all the possible pairs of whole numbers that add to the given sum, until they came to a pair with the required difference. This procedure does not work in (e). Most of the children were unable to break from this mental set, even though they all had sufficient competence with simple fractions and decimals to get the solution (3.5 and 6.5). Some children did, however, get this solution and these were the children who generally showed more inclination to think creatively in mathematics.

Most of the questions we give children in mathematics have one correct answer and therefore require what is called **convergent thinking**. Creativity is usually associated with **divergent thinking**. To give opportunities for flexible and original responses, therefore, we should sometimes give children more open-ended tasks, such as these:

•   Find lots of different ways of calculating 98 × 32.
•   Which numbers could go in the boxes: ($\square$ + $\square$) × $\square$ = 12? Give as many different answers as you can.

- How many different two-dimensional shapes can you make by fitting together six square tiles (not counting rotations and reflections)?
- If I tell you that 33 × 74 = 2442, write down lots of other results you can work out from this without doing any hard calculations.
- Three friends went for a meal. Ali's bill was £12, Ben's was £15 and Cassie's was £19. Make up as many questions as you can that could be answered from this information.
- What's the same about 16 and 36? Write down as many answers as you can think of.
- To answer a mathematics question, Jo arranged 24 cubes as shown in Figure 4.3. What might the question have been?

Figure 4.3   What was the question?

Creativity in mathematics will be shown by: fluency, coming up with many responses; flexibility, using many different ideas; and originality, using ideas that few other children in the group use. For example, the responses of one creative 11-year-old to the question 'What is the same about 16 and 36?' included mathematical ideas such as even, whole numbers, multiples, digits, square numbers, less than, greater than, between, divisibility and factors. Some of the responses were highly original, such as 'they are both greater than 15.9999', 'they are both factors of 144', 'they are both not in the 7-times table' and 'they are both numbers in this question'!

In problem solving, it is usually the teacher who poses the problem. However, the teacher can encourage creative thinking by giving children the

LEARNING AND TEACHING POINT

In your responses to children's ideas in mathematics lessons, show that you value creativity as much as you value accuracy. Reward and encourage children who come up with unusual ideas and imaginative suggestions, or who show willingness to take reasonable risks in their responses to mathematical situations, even if sometimes they get things wrong.

opportunity to ask their own questions or pose their own problems. This is possible in more open mathematical investigations or enquiries. Examples might be: find out as many interesting mathematical things as you can about the windows in the school building; look at this educational supplies catalogue and pose some interesting mathematical problems you might like to enquire into; and think of something you do not like about the school timetable and come up with a way of making it better.

> **LEARNING AND TEACHING POINT**
>
> Remember that primary mathematics does not consist only of the knowledge and skills discussed in Chapters 6-28 of this book, but also involves learning to reason in those ways that are distinctively mathematical: conjecturing and checking; inductive reasoning to formulate hypotheses; generalizing, explaining and convincing; and thinking creatively with mathematics.

RESEARCH FOCUS: CREATIVITY IN MATHEMATICS

Mann (2006) argues that teaching mathematics without providing for creativity denies all students the opportunity to appreciate the beauty of mathematics and, in particular, fails to provide the gifted student with an opportunity to fully develop their talents. Kaufman and Baer (2004) found that students who identified themselves as creative in general also identified themselves as having some specific creativity in all subject areas except mathematics! So what is special about creativity in mathematics? In my own research (Haylock, 1997), I identified mathematical creativity in children in terms of their ability to overcome fixations (mental sets) in problem solving and to produce appropriate but divergent responses in a range of open-ended mathematical tasks. It emerged that children with equal levels of mathematics attainment in conventional terms can show significantly different levels of mathematical creativity, defined in these terms. It was found that the higher the level of attainment, the more possible it is to discriminate between children in terms of the indicators of mathematical creativity. The high-attaining children with high levels of mathematical creativity were distinguished from the high-attaining children with low creativity by a number of characteristics: they had significantly lower levels of anxiety and higher self-concepts; they tended to be broad coders in the way they processed information, seeing similarities rather than differences; and they were more willing to take reasonable risks in mathematical tasks. One implication is that children may be more likely to think more creatively if they learn mathematics in a context in which they are encouraged to take risks and to back their hunches, even if sometimes this results in their getting things wrong.

LEARNING RESOURCES

Access activities for your **lesson plans** at: https://study.sagepub.com/haylock7e

Before trying the self-assessment questions below, you should complete the **self-assessment questions** for this chapter at: https://study.sagepub.com/haylock7e

4.1: For each of these two false generalizations, give a counter-example:

a   Any number less than 10 is greater than 5.
b   If a number is not a multiple of 6, it is not a multiple of 3.

4.2: (a) Show that the child's suggestion that there are exactly three multiples of 3 in every decade is not a correct generalization. (b) Then consider this generalization: there are exactly four multiples of 3 in every third decade (i.e. 21–30, 51–60, 81–90 and so on). True or false?

4.3: In the picture frame investigation (see Figure 4.1 and the accompanying commentary), the hypothesis was formulated that the number of tiles needed is 4 times the number along the edge, subtract 4. Give a convincing explanation as to why this must be true.

4.4: A child makes a chain of equilateral triangles, using matchsticks (see Figure 4.4). The child finds that 3 matchsticks are needed to make one triangle, 5 to make 2 triangles, 7 to make 3 triangles and so on. Formulate a generalization that will enable you to find how many matchsticks are needed to make a chain of 100 triangles. Can you provide a convincing explanation for your generalization? What about zero triangles? Does this fit your generalization or is it a special case?

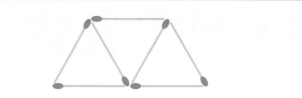

Figure 4.4   A chain of 3 triangles needs 7 matchsticks

4.5: Write down three digits and write them down again, hence making a six-digit number (such as 346,346). Use a calculator to check the hypothesis that all such numbers must be divisible by 7, 11 and 13. Why is this? (Hint, calculate $7 \times 11 \times 13$.)

4.6: In a science laboratory, there are two flasks of water, a small one and a large one. The temperature of the water in the small flask is held steady at 50 °C. When the temperature of the water in the large flask is 60 °C, the temperature difference between the two flasks is 10 °C. When it is 70 °C, the temperature difference is 20 °C. What would be the temperature of the large flask when the difference in the temperatures is (a) 30 °C? (b) 50 °C? (c) 70 °C?

GLOSSARY OF KEY TERMS INTRODUCED IN CHAPTER 4

Generalization   In mathematics, an assertion that something is true for all the members of a set of numbers or shapes or people. A generalization may be true (for example, 'all multiples of 12 are multiples of 3') or false (for example, 'women cannot read maps').

Conjecture   An assertion the truth of which has not yet been established or checked by the individual making it.

Counter-example   A specific instance that shows a generalization to be false.

Special case   A specific instance that does not fit an otherwise true generalization and that may have to be removed from the set to which the generalization is applied.

Hypothesis   A generalization that someone might make, which they have yet to prove to be true in every case.

Proof   A complete and convincing argument to support the truth of an assertion in mathematics, which proceeds logically from the assumptions to the conclusion.

Multiples   For any given (positive whole) number, those numbers that can be divided exactly by the given number. So the multiples of 10 are 10, 20, 30, 40, 50, 60, 70, 80, 90, 100, 110, 120 and so on. Similarly, multiples of 100 are 100, 200, 300 and so on; and multiples of 3 are 3, 6, 9, 12, 15, 18 and so on.

Inductive reasoning   In mathematics, the process of looking at a number of specific instances that are seen to have something in common and then speculating that this will always be the case.

Stem sentence   A clearly structured sentence using accurate mathematical vocabulary that provides learners with a way to communicate their observations with precision and clarity.

Deductive reasoning   Reasoning based on logical deductions.

Proof by exhaustion   A method for proving a generalization by checking every single case to which it applies.

Axiom   In mathematics, a statement that is taken to be true, usually because it is self-evident, but which cannot be proved. For example, $a + b = b + a$ for all numbers a and b.

Convergent thinking   The kind of thinking involved in seeking the one and only correct answer to a mathematical question.

Divergent thinking   The opposite of convergent thinking; thinking characterized by flexibility, generating many different kinds of response in an open-ended task.

Creativity in mathematics   Identified by overcoming fixations and rigidity in thinking; by divergent thinking, fluency, flexibility and originality in the generation of responses to mathematical situations.

MODELLING AND PROBLEM SOLVING

HOW SHOULD CHILDREN DO CALCULATIONS?

This chapter has a focus on two further processes at the heart of the application of mathematics: mathematical modelling and problem solving. First, though, I want to say something about calculations. When we engage in **mathematical modelling** of real-life situations and solving problems, there will often be calculations to do on the way, but we should regard these calculations as mere tools needed to do the real mathematics. So, in starting this chapter with talking about calculation, my intention is not to give the impression that this is what doing mathematics is all about, but to put it firmly in its place, so that we can then focus on more important stuff.

Now, there are essentially three ways in which we can do a calculation, such as an addition, a subtraction, a multiplication or a division. For example, consider the question of finding the cost of 16 items at 25p each. To answer this, we may decide to work out 16×25.

One approach to this would be to use an **algorithm**. The word 'algorithm' (derived from the name of the ninth-century Arabian mathematician, Al-Khowarizmi) refers to a step-by-step process for obtaining the solution to a mathematical problem or, in this case, the result of a calculation. In number work, we use the word to refer to the formal, paper-and-pencil methods that we might use for doing calculations, which, if the procedures are followed correctly, will always lead to the required result. These would include, for example, subtraction by decomposition (see Chapter 9) and long division (see Chapter 12). So, answering the question above using an algorithm might involve, for example, performing the calculation for 16×25 as shown in Figure 5.1, using the method known as long multiplication (see Chapter 12).

$$
\begin{array}{r}
16 \\
\times\ 25 \\
\hline
320 \\
80 \\
\hline
400 \\
\hline
\end{array}
$$

Figure 5.1   Using an algorithm for 16×25

A second approach would be to use one of the many informal methods for doing calculations, which are actually the methods that most numerate adults employ for the calculations they encounter in everyday life. For example, to find the cost of 16 items at 25p each, we might:

- make use of the fact that four 25-pences make £1, so 16 of them must be £4; or
- work out ten 25s (250), four 25s (100) and two 25s (50) and add these up, to get 400; or
- use repeated doubling and reason – 'two 25s is 50, so four 25s is 100, so eight 25s is 200, so sixteen 25s is 400'.

This kind of approach, in which we make ad hoc use of the particular numbers and relationships in the question, I like to call an **adhocorithm** – my own invented word, not yet in the dictionaries! These informal, ad hoc approaches to calculations should be recognized as being equally as valid as the formal, algorithmic approaches. They have the advantage that they are based on our own personal level of confidence with numbers and number operations. They are based on and encourage understanding of the relationships between numbers – because, unlike algorithms, they are not applied mechanically and cannot rely on rote learning. In Chapters 8, 9, 11 and 12, I discuss various algorithms and adhocorithms for each of the four operations.

Then, a third way of doing this calculation is just to use a *calculator*, entering 16, ×, 25, = and reading off the result (400). In this particular example, it would be a little disappointing if, say, Year 5 children needed a calculator to work out 16 multiplied by 25. But, sometimes, having a calculator to do the donkeywork gives the learner the opportunity to focus on the underlying mathematical structure of a situation and to see the relationships within it. Many people are not convinced that there is any mathematics involved in using a calculator; sometimes it is suggested that they should be banned from the classroom. The argument is that all you have to do is to press the buttons and the machine does all the thinking for you. This is a common misconception about calculators. Even in the simple example above, we have to decide what calculation to put into

> ## LEARNING AND TEACHING POINT
>
> Different countries in the UK have different policies regarding the use of calculators in primary mathematics. The English curriculum has no place for them at all, which is a pity because teachers can exploit calculators in the primary classroom in ways that promote understanding of numbers and number relationships, patterns, place value and so on. Also, they might be made available in real-life problems where calculations are particularly difficult and the focus is on the process of problem solving. However, calculators should not be so freely available that children do not develop confidence and fluency in handling simple calculations mentally or by appropriate written methods.

> ## LEARNING AND TEACHING POINT
>
> In your teaching, recognize the validity of the three ways of doing calculations: algorithms, 'adhocorithms' and calculators; do not inadvertently give children the impression that using a formal, written method is somehow the superior or proper way of doing a calculation – or that doing calculations is the most important part of mathematics.

the calculator and interpret the result displayed (400) as meaning £4. There is also here a misconception about mathematics: that the most important things to learn in mathematics are calculations. In fact, using a calculator to do the arithmetic needed in solving a practical problem involves us in a fundamental mathematical process called *mathematical modelling*.

WHAT IS MATHEMATICAL MODELLING?

We are not talking here about making models out of card or other materials. Mathematical modelling is the process whereby we use the abstractions of mathematics to solve problems in the real world. The Scottish curriculum guidelines, for example, emphasize that 'using mathematics enables us to model real-life situations and make connections and informed predictions' (Education Scotland, 2010b: 1). For instance, how would you work out how many boxes you need to hold 150 calculators if each box holds just 18 calculators? You might use one of the calculators to work out 150 divided by 18. This would give you the result 8.3333333. That's a bit more than 8 boxes. So you would actually need 9 boxes. If you only had 8 boxes, there would be some calculators which could not be fitted in, although the calculator answer does not tell you directly how many. The four steps involved in the reasoning here essentially provide an example of the process called mathematical modelling. This process is summarized in Figure 5.2.

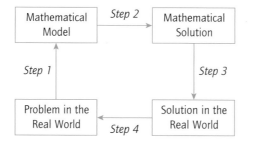

Figure 5.2   The process of mathematical modelling

In step 1 of this process, a problem in the real world is translated into a problem expressed in mathematical symbols. (Strictly, the term 'modelling' is usually used by mathematicians for problems expressed in algebraic symbols, but since the four steps involved are essentially the same, we can adapt the idea and language of modelling and apply them to word problems being translated into number statements.) So, in this example, we shall say that the real-world problem of buying boxes to hold calculators is *modelled* by the mathematical expression 150 ÷ 18. Then, in step 2, the mathematical symbols are manipulated in some way – this could be by means of a mental or written calculation or, as on this occasion, by pressing keys on a calculator – in order to obtain a mathematical solution, 8.3333333. Step 3 is to interpret this mathematical solution back in the real world – for example, by saying that this means '8 boxes and a bit of a box'. The final step is to check the result against the constraints of the original situation. In this case, by considering the real situation and recognizing that 8 boxes would leave some calculators not in a box, the appropriate conclusion is that you actually need 9 boxes.

So, in this process there are basically four steps:

1   Set up the mathematical model.
2   Obtain the mathematical solution.
3   Interpret the mathematical solution back in the real world.
4   Check the solution with the reality of the original situation.

There is potentially a fifth step: if the solution does not make sense when checked against the reality of the original problem, then you may have to go round the cycle again, checking each stage of the process to determine what has gone wrong.

If you have used a calculator in this process, it is important to note that the calculator does only the second of the four steps. You will have done all the others. Your contribution is significant mathematics. In a technological age, in which most calculations are done by machines, it surely cannot be disputed that knowing which

> **LEARNING AND TEACHING POINT**
>
> Provide children with plenty of opportunities to work through and experience the process of mathematical modelling, making the four steps in the process explicit in your class discussion: (1) deciding what calculation has to be done to answer a question in a practical context; (2) doing the calculation by an appropriate method; (3) interpreting the answer back in the original context; and (4) checking the solution against the reality of the original question.

> **LEARNING AND TEACHING POINT**
>
> A particular problem in interpretation arises when a calculator result in a money problem gives only one digit after the decimal point. For example, here is a problem in the real world: how much for 24 marker pens at £1.15 each? We might model this with the mathematical expression, 24 × 1.15. A calculator gives the result 27.6. Primary children have to be taught how to interpret this as £27.60 (not 27 pounds, 6 pence), to draw the conclusion that the total cost of the 24 marker pens is £27.60.

calculation to do is more important than being able to do the calculation. As will be seen in Chapters 7 and 10, recognizing which operations correspond to various real-world situations (step 1 above) is not always straightforward. These chapters explore the range of categories or structures of problems that children should learn to model with each of the operations of addition, subtraction, multiplication and division.

> **LEARNING AND TEACHING POINT**
>
> Recognize that all the steps in the modelling process are important and that step 2 (doing the calculation) is no more important than the others. Allow children to use a calculator when the calculations associated with a real-life problem are too difficult, so that they can still engage in the process and learn to choose the right operation, to interpret the result and to check it against the constraints of the real situation.

WHAT IS PROBLEM SOLVING ALL ABOUT?

With the current concerns in the UK to match high-achieving countries in international comparisons this statement in the Singapore National Curriculum is significant:

> The primary aim of the mathematics curriculum is to enable pupils to develop their ability in mathematical problem solving. Mathematical problem solving includes using and applying mathematics in practical tasks, in real life problems and within mathematics itself … a problem covers a wide range of situations from routine mathematical problems to problems in unfamiliar contexts and open-ended investigations that make use of the relevant mathematics and thinking processes.
>
> (Ministry of Education, Singapore, quoted in Ruddock and Sainsbury, 2008)

One of the three central aims for mathematics in the English National Curriculum for primary school is that children should be able to 'solve problems by applying their mathematics to a variety of routine and non-routine problems with increasing sophistication, including breaking down problems into a series of simpler steps and persevering in seeking solutions' (DfE, 2013: 3). 'Routine' problems here would be those where step 2 of the modelling process is more straightforward and probably involving just one operation. 'Non-routine' problems would be less familiar in structure and context, involving more than one operation. The English National Curriculum therefore specifies that the skills, concepts and principles of mathematics that children master should be used and applied to solve problems, both familiar and unfamiliar in their structure. It is the nature of the subject that applying what we learn in solving problems must always be a central component of mathematical reasoning. The Scottish curriculum guidelines make this point strongly: 'Mathematics is at its most powerful when the knowledge and understanding that have been developed are used to solve problems; problem solving will be at the heart of all our learning and teaching' (Education Scotland, 2010b: 2).

A **problem**, as opposed to something that is merely an exercise for practising a mathematical skill, is a situation in which we have some givens and we have a goal, but the route from the givens to the goal is not immediately apparent. This means, of course, that what is a problem for one person may not be a problem for another. If you tackled some mathematical questions with me, you might think that I am a good problem solver. However, it might just be that I have seen them all before so that for me they are not actually problems. For a task to be a problem, there must be for the person concerned a (cognitive) gap between the givens and the goals, without an immediately obvious way for the gap to be bridged. I call these the three Gs of problem solving: the given, the goal and the gap, as shown in Figure 5.3.

> **LEARNING AND TEACHING POINT**
>
> Problems in mathematics, as defined in this section, should be used: (a) to give children opportunities to apply and therefore to reinforce the knowledge and skills they have already learnt; (b) to develop general problem-solving strategies; and (c) sometimes to introduce a mathematical topic by providing the motivation for the learning of some new skills.

Figure 5.3 The three Gs of problem solving

A typical curriculum target that refers to problem solving is this expectation for Year 2 children in the National Curriculum for England: 'solve simple problems in a practical context involving addition and subtraction of money of the same unit, including giving change' (DfE, 2013: 14). For children of this age, using the toy shop in a role-play situation with plastic coins, buying something for a few pence and working out the change from 50p might be properly described as a *problem* – because they would probably not know immediately how to obtain the solution. For older children, finding change when shopping might be just a routine problem, which with increased familiarity is really just an *exercise* – because they already know what you have to do to work it out. On the other hand, for the older children questions that require two or more calculations could well be non-routine problems, because they have to work out the appropriate sequence of calculations that gets them from the givens to the goal. We shall see later in this chapter, however, that there are much more intriguing and interesting examples of problems that could be used to develop children's problem-solving strategies.

There are many strategies that can be used to help people solve problems, but the most important of these are the most obvious: make sure you understand what you are given; ensure you understand what the goal is. Clarify the givens and clarify the goal. The arrow in Figure 5.3 goes both ways; this is an indication of another problem-solving strategy, which is that, as well as working from the givens to the goal, you can work backwards from the goal to the givens. You can also identify sub-goals, which involves recognizing intermediate steps between the givens and the goal. For example, if the problem is to find out how much it would cost to redecorate the classroom, an intermediate goal might be to find the total area of the walls and ceiling.

WHAT KIND OF PROBLEMS SHOULD PRIMARY SCHOOL CHILDREN TACKLE IN MATHEMATICS?

Problem-solving strategies are used and developed not just in realistic problems set in real-life and cross-curricular contexts, but also through what we might regard as problems that are essentially within mathematics itself. So we aim for children to develop their own strategies for solving problems and to use these strategies both in working within mathematics and in applying mathematics to practical contexts. In practice, it makes little sense to categorize problems as either 'within mathematics' or in 'practical contexts'. There is really a continuum of contexts for applying mathematics.

At one end are problems that are purely mathematical, just about numbers and shapes, in which the outcome is of no immediate practical significance. Problems 1 and 2 are examples of purely mathematical problems, relating to Figure 5.4:

Problem 1: Place three numbers in the boxes so that the top two boxes total 20, the bottom two total 43 and the top and bottom boxes total 37.

Problem 2: Complete the drawing so that the arrowed line is a line of symmetry.

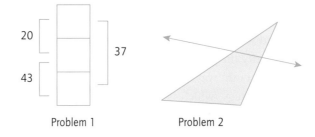

Problem 1   Problem 2

Figure 5.4 Two purely mathematical problems

I am fairly confident that most readers will have all the mathematical knowledge and skills that are required to solve these two problems. Problem 1 requires no more than addition and subtraction. Problem 2 requires nothing more than understanding about reflections and using a ruler. But I am equally confident that, for most readers, it will not be immediately apparent what has to be done to achieve the goal in each case. This is what makes these tasks problems, rather than merely exercises. I shall leave these two problems for the reader to complete (in self-assessment question 5.3) and to reflect on the strategies they use, particularly in terms of clarifying the givens and the goal.

At the other end of the continuum would be problems that arise in genuine, real-life situations that really do need to be solved. Some children will be more motivated by such problems, particularly if they have immediate purpose and practical relevance to them. Problems 3, 4 and 5 are examples of such problems that I have used with children:

Problem 3: How much orange squash should we buy to be able to provide three drinks for each player in the inter-school football tournament?

Problem 4: How should we rearrange the classroom chairs and tables for practical work in groups of four?

Problem 5: Plan our class trip to Norwich Castle Museum and ensure that it all goes smoothly.

Problems such as these require the application of a wide range of numerical and spatial skills. Solving Problem 3, for example, requires understanding the concept of concentration, designing a survey and data-handling skills, measurement of liquid volume and capacity, calculations involving litres and millilitres, multiplication in the context of money, and much more. Problem 4 requires considerable practical measurement, knowledge of units of length and use of decimal notation, making a simple scale drawing, division by 4 and dealing with remainders, as well as a degree of spatial imagination. Problem 5 involves calculations with money, drawing up a budget, calculator skills, timetabling, estimating, average speed (of the coach), as well as a range of communication skills. All three of these problems involve children in clarifying what they were given, what the goal is and what they need to know, and in identifying a number of sub-goals.

Genuine problems like these are clearly very engaging for children and help them to see mathematics as useful and relevant and not just something you learn at school. But we have to recognize that, in practice, many of the problems that we pose for children, although set in real-life and practical contexts, will inevitably be rather artificial in nature. These are problems that come somewhere in the middle of the continuum, such as Problem 6:

> **LEARNING AND TEACHING POINT**
>
> To develop the key processes involved in problem solving, children should have opportunities to apply their mathematics in a range of tasks, including: (a) activities within their everyday experience in the classroom, such as planning their timetable for the day, or grouping children for various activities; (b) identifying and proposing solutions to genuine problems, such as where in the playground staff should park their cars; (c) tackling artificial but realistic problems, such as estimating the cost for a family of four to go on a two-week holiday on the Norfolk Broads; (d) applying mathematics in practical tasks, such as making a box to hold some mathematics equipment; (e) solving mathematical problems, such as finding two-digit numbers that have an odd number of factors; and (f) pursuing open-ended mathematical investigations, such as 'find out as much as you can about the relationships between different paper sizes' (A5, A4, A3 and so on).

Problem 6: Find out as many interesting things as you can about the way the page numbers are arranged on the sheets of a newspaper.

Problem 6 is an example of what might be called an **investigation**. This is where the children are given a more open-ended situation and then have the opportunity to determine their own goals or have a degree of choice in exploring a number of different aspects of the mathematics involved. Many children get very excited about and engage readily with open-ended investigations such as Problem 6. Problem 7 is another example of a more open-ended investigation in a real-life context:

Problem 7: Here is a collection of packets from ten items bought in a supermarket. Look at the numerical information given on the labels and try to work out what it all means.

Children in primary school can also be particularly intrigued by problems where they engage in role-play (like the toy shop) or put themselves in an imaginary quasi-realistic situation, as in Problems 8 and 9:

Problem 8: You are a zookeeper with £1000 budget to buy some snakes, costing £40 each, and some baby alligators, costing £100 each. If you spend all your money, how many snakes and baby alligators might you buy?

Problem 9: You are a teacher planning to take up to 80 children on a camp. What's the best number of children to take if you plan to put them in groups of 3, 4, 5 and 6 for various activities and you do not want any children to be left out of a group for any activity?

These last two problems are left for the reader to solve, as self-assessment question 5.4 below.

RESEARCH FOCUS: MATHEMATICAL MODELLING

The assumption in this chapter has been that there are real benefits for children across the primary school age range in providing them with explicit experiences of the process of mathematical modelling. This has been demonstrated in research by English (English, L., 2004, 2013) and English and Watters (2005). English (2004) observed upper primary school children working collaboratively on authentic problems that could be modelled by mathematics. She identified a number of significant aspects of learning taking place, both mathematical and social. These included: interpreting and reinterpreting given information; making appropriate decisions; justifying reasoning; posing hypotheses; and presenting arguments and counter-arguments. English and Watters (2005: 59) outline similar findings with younger learners and conclude from the evidence of research in this field that 'the primary school is the educational environment where all children should begin a meaningful development of mathematical modelling'. In a later study, English (English, L., 2013) found that, given tasks that involved modelling real-world problems with complex data, children aged 8–9 years used successfully such mathematical processes as ranking and aggregating data, calculating and ranking means, and weighted scores, which were beyond what would be expected in their regular curriculum.

⊂⊃ LEARNING RESOURCES

Access activities for your lesson plans at: https://study.sagepub.com/haylock7e

Before trying the self-assessment questions below, you should complete the **self-assessment questions** for this chapter at: https://study.sagepub.com/haylock7e

5.1: Identify the steps in the process of mathematical modelling that are involved in using a calculator to find the total cost of three books priced at £14.95, £25.90 and £19.95.

5.2: Make up simple two-step problems in the context of money that involve: (a) a division followed by a subtraction; (b) a subtraction followed by a multiplication.

5.3: Solve Problems 1 and 2 given earlier in this chapter (see Figure 5.4).

5.4: Solve Problems 8 and 9 given earlier in this chapter.

FURTHER PRACTICE

Access the website material for Knowledge Check 1: Using a four-function calculator for money calculations at: https://study.sagepub.com/haylock7e

GLOSSARY OF KEY TERMS INTRODUCED IN CHAPTER 5

Mathematical modelling   The process of moving from a problem in the real world, to a mathematical model of the problem, then obtaining the mathematical solution, interpreting it back in the real world, and finally checking the result against the constraints of the original problem.

Algorithm   In number work, a standard, written procedure for doing a calculation, which, if followed correctly, step by step, will always lead to the required result; examples of algorithms are subtraction by decomposition, long multiplication and long division.

Adhocorithm   My term for any informal, non-standard way of doing a calculation, where the method used is dependent on the particular numbers in the problem and the relationships between them.

Problem   In mathematics, a situation consisting of some givens and a goal, with a cognitive gap between them; this constitutes a problem for an individual, as opposed to just an exercise, if the way to fill the gap between the givens and the goal is not immediately obvious.

Investigation   An open-ended problem-solving situation involving mathematical concepts where the learner has the opportunity to determine their own goal(s) or have a degree of choice in exploring a number of different aspects of the mathematics involved.

SUGGESTIONS FOR FURTHER READING FOR SECTION B

1   The entries on 'Creativity in mathematics' and 'Generalization', in Haylock with Thangata (2007), offer further insights into some of the key processes introduced in Chapter 4. The entries on 'Modelling process', 'Problem solving' and 'Using and applying mathematics' offer further insights into some of the key processes discussed in Chapter 5.

2   Pound (2006) shows how young children can be enabled to enjoy thinking mathematically. The book outlines a curriculum for promoting mathematical thinking in the early years and provides guidance on observing, planning and supporting mathematical thinking.

3   Read Chapter 10 of Anghileri (2008) for an interesting analysis of problem solving in primary mathematics, illustrated by some illuminative examples.

4   Fairclough's chapter entitled 'Developing problem-solving skills in mathematics', in Koshy and Murray (2011), provides some lively examples of problems used in the primary classroom to illustrate her analysis of different types of mathematical problems and how to teach primary children to be problem solvers.

5   Chapter 11 of Haylock and Cockburn (2017) is on developing mathematical reasoning and problem solving. In this chapter, we illustrate how mathematical modelling is one of the distinctive ways of thinking mathematically, which can be fostered even in young children.

6   A useful summary of research into problem solving in children's mathematical development and the implications for how we teach mathematics is provided in Chapter 6 of Nickson (2004).

7   There are plenty of good ideas for incorporating problem solving and practical experiences for young children learning mathematics in Skinner and Stevens (2012).

8   Nunes et al. (2015) provide evidence from their research into teaching mathematical reasoning with children aged 9–10 years that primary school teachers can effectively teach problem solving in mathematics. Read just those sections dealing with the problem-solving teaching programme at this stage: save the rest until you have read Chapter 28 on probability.

SECTION C

NUMBERS AND CALCULATIONS

6	Numbers, Counting and Place Value	79
7	Addition and Subtraction Structures	105
8	Mental Strategies for Addition and Subtraction	125
9	Written Methods for Addition and Subtraction	141
10	Multiplication and Division Structures	161
11	Mental Strategies for Multiplication and Division	180
12	Written Methods for Multiplication and Division	197

WATCH THE SECTION OPENER VIDEO AT: HTTPS://STUDY.SAGEPUB.COM/HAYLOCK7E

NUMBERS, COUNTING AND PLACE VALUE

IN THIS CHAPTER, THERE ARE EXPLANATIONS OF

- the difference between numerals and numbers;
- the cardinal and ordinal aspects of number;
- learning to count;
- natural numbers and integers;
- rational, irrational and real numbers;
- the Hindu-Arabic system of numeration and the principles of place value;
- some contrasts with numeration systems from other cultures, including Roman numerals;
- digits and powers of ten;
- two ways of demonstrating place value with manipulatives;
- how the number line supports understanding of place value;
- the role of zero as a place holder;
- comparing and ordering positive integers;
- use of the inequality signs (>, <) for recording 'greater than', 'less than', 'lies between';
- rounding a positive integer to the nearest 10 or 100.

READ THIS CHAPTER'S CURRICULUM LINKS AT: HTTPS://STUDY.SAGEPUB.COM/HAYLOCK7E

WHAT IS THE DIFFERENCE BETWEEN A 'NUMERAL' AND A 'NUMBER'?

A **numeral** is the symbol, or collection of symbols, that we use to represent a number. The number is the concept represented by the numeral, and therefore consists of a whole network of connections between symbols, pictures, language and real-life situations. The same number (for example, the one we call 'three hundred and sixty-six') can be represented by different numerals – such as 366 in our Hindu-Arabic place-value system and CCCLXVI using Roman numerals (see Figure 6.5 later in this chapter and the accompanying commentary). Because the Hindu-Arabic system of numeration is now more or less universal, the distinction between the numeral and the number is easily lost.

WHAT ARE THE CARDINAL AND ORDINAL ASPECTS OF NUMBER?

A numeral, such as 3, together with the associated word 'three', has a wide range of situations and contexts to which it can be connected. The two most significant for young children are the cardinal and ordinal aspects of number.

The learner's first experience of number is likely to be as an adjective describing a small set of objects: two sisters, three sweets, five fingers, three blocks and so on. This idea of a number being a description of a set of things is called the **cardinal aspect of number**. By the process of one-to-one matching between sets containing the same number, as shown in Figure 6.1, the learner is able to recognize that there is something the same about the sets; in other words, they identify an equivalence. The property that is shared by all sets of three things, for example, is then abstracted to form the concept of 'three' as a cardinal number, existing in its own right, independent of any specific context.

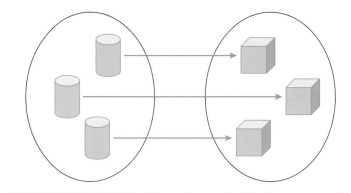

Figure 6.1 One-to-one matching

But this is not by any means the only aspect of number that the young learner encounters. Numbers are much more than just a way of describing sets of things. Young children also encounter numbers used as labels to put things in order. For example, they turn to page 3 in a book. They play games on the number strip in the playground and find themselves standing on the space labelled 3. They learn that they are 3 years old and that next birthday they are going to be 4. One of the tricycles in the playground is labelled 3 and this has to be parked in the space labelled 3, which is once again between 2 and 4. The numerals and words being used here do not represent cardinal numbers, because they are not referring to sets of three things. In these examples, 'three' is one thing, which is labelled three because of the position in which it lies in some ordering process. This is called the **ordinal aspect of number**. Numbers in this sense tell you what order things come in: which thing is first, which is second, which is third and so on. Two important experiences of the ordinal aspect of number are when we represent numbers as locations on a number strip (see Figure 3.5 in Chapter 3) or as points on a number line, as shown in Figure 6.2. We shall make considerable use of these images of number as we explore understanding of number operations in subsequent chapters.

$$0 \quad 1 \quad 2 \quad 3 \quad 4 \quad 5 \quad 6 \quad 7 \quad 8 \quad 9 \quad 10 \quad 11$$

Figure 6.2   Numbers as points on a line

There is a further way in which numerals are used, sometimes called the *nominal* aspect. This is where the numeral is used as a label or a name, without any ordering implied. The usual example to give here would be a number 7 bus. Calling it number 7 is not much different from calling it the East Acton bus. It just identifies the bus and distinguishes it from buses on other routes. When we see a number 7 bus, we do not expect it to be followed by a number 8 and then a number 9 – in fact, we may well expect it to be followed by two more number 7s, as is the habit of buses. Having said that, I should make clear that when various bus services are listed in numerical order in a timetable their numbers are then being used in an ordinal way.

> **LEARNING AND TEACHING POINT**
>
> The fact that two sets of three objects – such as three cups and three spoons – share the property of 'threeness' can be experienced by the process of one-to-one matching; for example, the three spoons can be allocated one to each of the three cups. At snack time a child may count ten children and count out ten pieces of banana and then match one piece of banana to each child. This is a key practical process for the young child that makes explicit what is the same about three cups and three spoons, or about ten pieces of banana and ten children.

WHAT BACKGROUND EXPERIENCE DO CHILDREN DRAW ON IN LEARNING TO COUNT?

In this and subsequent sections on counting I draw substantially on the influential work of Gelman and Gallistel (1986). We should recognize first that learning to count requires a considerable background of experience, much of which many children bring to the Foundation Stage classroom from their life at home. Three kinds of prerequisite experience for counting might be identified.

Top of the list would be *categorization*, the whole idea of sorting objects into sets. In order to engage with counting some objects you clearly need to have the notion of the objects being separated off from those you are not counting. This is an early experience of forming equivalences: identifying a set of objects on the basis of them being in some sense the same; the same colour, the same size, the same family and so on. Young children's play provides many examples of sorting and categorizing. They will sort blocks, toy animals, model cars, dolls, and so on, into different sets and categories. They categorize when they make distinctions between those in their family and those who are not; between children and adults; between boys and girls. In fact, categorization lies at the heart of language development and is the basic process by which we make sense of our experiences. This process is the first stage of learning to count: to be able to identify and separate off the members of a set in order that you can then count just these objects and no others.

Second, we should not underestimate the importance of the child having a rich experience of using language such as 'one more' and 'another one', within the context of family life at home. A basic idea in counting is that numbers increase as you count; so six is more than five, for example. So, when a child learns at such a young age to ask for 'another one', they are getting essential prerequisite experience for learning to count.

Third, there is some evidence to suggest that before they engage in actual counting children are

LEARNING AND TEACHING POINT

Parents may be unaware of the significant experience being provided by the repetition of phrases such as 'one more, then no more' and 'another one'. Teachers, in their own conversations with children, should not be.

able to distinguish between small numbers, such as one, two and three. Three-year-olds will look at picture books and be able to recognize which picture has one cow, two cows or three cows. So, they are beginning to learn that numbers are used to describe sets of objects and to distinguish between sets of various sizes. This is also experienced when small numbers are used in conversation. Children, before they can count, know that they have two feet, two hands and two eyes, but only one nose. They have stories about three little pigs and three bears. All this experience is invaluable preparation for counting.

Perceiving at a glance the number of items in a set without actually counting them is called **subitizing.** The arrangements of dots on dice and dominoes and arranging a set of items in twos, threes or fours are important visual images for the development of young children's ability to subitize. Many of the three-year-olds arriving in a nursery class will already be able to recognize three items at a glance – and four items if they are arranged in a square. The image provided by a tens-frame (see Figure 8.7 in Chapter 8) helps them to visualize six as five and one more, seven as five and two more and so on.

HOW DO YOUNG CHILDREN DEVELOP AN UNDERSTANDING OF COUNTING?

From a mathematical perspective we may see numbers as describing sets or labelling points, but most significantly for young children numbers are words you use to count with. When children start to learn to count they learn first a pattern of noises, memorized by repetition in all sorts of situations both in and out of school: 'One, two, three, four ...' and so on. This set of sounds is probably just as meaningless as many traditional nursery rhymes, demonstrating the young child's amazing capacity for sequential learning. But what they learn from this repetition is the underlying principle of counting that the order of the numbers is fixed. It is invariant: when you are counting, three always comes after two and so on. Gelman and Gallistel call this *the stable order principle.*

Then the child has to learn to co-ordinate the utterance of these noises with the physical movements of a finger and the eye along a line of objects, matching one noise to one object. So the child has to learn that one number name must be matched to one object, until all the objects have been used up. Gelman and Gallistel call this *the one–one principle.*

> **LEARNING AND TEACHING POINT**
>
> Many activities in the early years classroom provide opportunities for young children to participate in the preparation of resources, and often this involves counting; for example, to prepare for playing shops the children may count out five coins (or tokens of some kind) for each child to use to buy items from the play shop. Similarly, children will often count to compare, motivated by their strong sense of what is fair; for example, counting to check that there are three children in each group, or to check that each child has three and no more than three pieces of fruit.

This is a significant skill to learn. Teachers working with children learning to count will know that the two-syllable word 'seven' sometimes poses a difficult problem of co-ordination. Also, children sometimes over-count. A way to see if they have a good understanding of the principle is to arrange the objects to be counted in various ways, such as in a straight line, in a circle and in a heap. This will show whether or not the child is aware that they have to keep track of what has been counted and what remains to be counted.

It is in the process of counting a set of objects that the ordinal and cardinal aspects of number come together. An underlying principle of counting is that the last number that you get to when you are counting a set is the number in the set. Gelman and Gallistel call this *the cardinal principle*. This is where the ordinal and cardinal aspects come together. As each number is spoken it is being used essentially in an ordinal sense, to label the objects and to order them – number one, number two, number three and so on. But then the child has somehow to discover that the ordinal number of the last object is the cardinal number of the set. What a stunning and powerful discovery this is!

A major challenge in learning to do mathematics is to learn to abstract the mathematical concepts from the context in which they are embedded. For example, when a child adds 3 to 4, what actual objects are used makes absolutely no difference to the mathematical process: three sweets and four sweets, three boys and four boys, three counters and four counters, these are all represented by the same abstraction, $3 + 4 = 7$. This is a significant way in which mathematics is so different from so many of the other things children learn at school. In a poem, a story, a picture, a song, a game and so on, the context is the essence of the activity. But in mathematics it is often the case (but by no means always) that the context is of little signifi-cance. This is what Gelman and Gallistel call *the abstraction principle*. It makes no difference what you are counting, whether it be children, animals, counters or fingers, the process is exactly the same. You use the same number names, in the same order, with the same one-to-one matching procedure.

> ## LEARNING AND TEACHING POINT
>
> Teachers of young children can structure children's learning and assess their progress in counting by using this framework:
>
> - subitizing
> - recognizing that the order of numbers is fixed
> - when counting a set of items attaching one number to each item
> - when counting, recognizing that the ordinal number of the last item is the cardinal number of the set
> - realizing that the process of counting is the same whatever objects are being counted
> - awareness that the order in which objects are counted and their arrangement are irrelevant
> - knowing that the next number in counting is 'one more'
> - learning that beyond 20 there is a pattern in counting and it goes on for ever.

Gelman and Gallistel also identify what they call the *order-irrelevance principle*. Children have to learn, for example, that whether they count a row of objects from left to right, or from right to left, makes no difference. In fact, you can count the objects in any order, provided you match them all one-to-one with a number name in the prescribed sequence. This is again a very sophisticated piece of learning and something that teachers should target specifically in the counting experiences of young children and in their assessment of their understanding. An associated principle is that the arrangement of the objects in a set is irrelevant when you are counting them. If a line of seven objects is spread out, there are still seven objects. If they are arranged in a cluster, or a circle, or some other pattern, there are still seven objects. This and the previous principle together constitute what we have identified in Chapter 3 as the principle of conservation of number (see Figure 3.3). We can understand this in terms of the child learning that whatever transformation is applied to the arrangement of the objects in a set, there is an equivalence that is preserved: there is still the same number.

Clearly, at some stage, children have to learn to match the names of numbers to the numerals that represent the numbers. In the early stages of counting they just recite the number names: 'one, two, three, four, five …' as they point to objects. At a later stage they will learn to recite these names as they point to the symbols: 1, 2, 3, 4, 5 and so on, doing this often, until eventually the connection between each name and the corresponding numeral is clearly established.

There is a basic principle of counting that I should make explicit here, so that teachers reading this book are aware of it and ensure that the principle is targeted in their interactions with children. The principle is simply that the next number after any given number is always 'one more'. So seven is one more than six, because it is the next number after six.

Once children go on to count beyond twenty they will learn two important principles: first, that there is a pattern in counting and, second, that counting goes on for ever. In counting, it really is possible always to continue on to one more.

WHAT ARE NATURAL NUMBERS AND INTEGERS?

How many numbers are there between 10 and 20? This is a question I like to ask trainee primary teachers when we start to think about understanding number. The correct answer perhaps should be 'an infinite number'. The most common response is nine: namely, the numbers 11, 12, 13, 14, 15, 16, 17, 18 and 19. Some trainees answer a different question and give the answer ten, which is the difference between 10 and 20. Others give the answer eleven, choosing to include the 10 and the 20, in an unorthodox use of the word 'between'. All of these answers assume that when I say 'number' I mean the numbers we use for counting: {1, 2, 3, 4, 5, 6 ...}, going on forever. These are what mathematicians choose to call the set of **natural numbers**. As we have seen above, natural numbers can have both cardinal and ordinal interpretations. But there are other kinds of numbers that children will encounter in primary school and which will feature in later chapters in this book. Here I introduce them briefly.

How many numbers are there that are less than 10? That's another interesting question I like to discuss with trainee teachers! Some say nine, just counting the natural numbers from 1 to 9. Most include 0 (zero) and give the answer ten. But others have the insight to include negative numbers in their understanding of 'numbers' and give responses such as 'there is an infinite number' or 'they go on forever'. In this way, we can extend our understanding of what constitutes a number to what mathematicians call the set of **integers**: {..., –5, –4, –3, –2, –1, 0, 1, 2, 3, 4, 5 ...} now going on forever in both directions. Integers build on the ordinal aspect of number, by extending the number line in the other direction, as shown in Figure 6.3, labelling the points to the left of zero as negative numbers. Note that for convenience I have chosen to write the negative numbers as –1, –2, –3 and so on; they can also be written as ⁻1, ⁻2, ⁻3, using a superscript for the negative sign.

The mathematical word 'integer' is related to words such as 'integral' (forming a whole) and 'integrity' (wholeness). So the set of integers is simply the set of all whole numbers. But this includes both **positive integers** (whole numbers greater than zero) and **negative integers** (whole numbers less than zero), and zero itself. The integer –4 (or ⁻4) is properly named 'negative four', rather than 'minus four' as is the habit of weather forecasters; **minus** is an alternative word for subtract. Likewise, the integer +4 is named

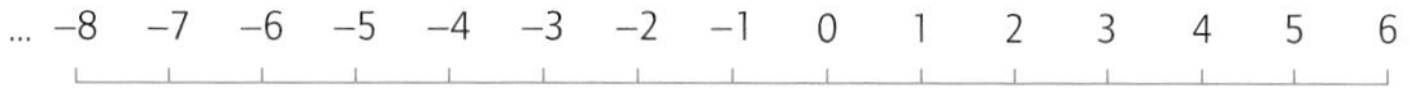

... −8 −7 −6 −5 −4 −3 −2 −1   0   1   2   3   4   5   6 ...

Figure 6.3 Extending the number line

'positive four', not 'plus four'; **plus** is an alternative word for add. Of course, the integer +4 is another way of referring to the natural number 4, so we would not normally write +4, or say 'positive four', but would simply write 4 and say 'four' – unless in the context it were particularly helpful to signal the distinction between the negative and the positive integers. So, we note that the set of integers includes the set of natural numbers: to be precise, natural numbers are positive integers. Integers are explained in greater detail in Chapter 14.

WHAT ARE RATIONAL AND REAL NUMBERS?

When you read earlier in this chapter the question that asked how many numbers there are between 10 and 20, you may have been bursting to say, 'It's an infinite number!' Yes, of course, there is no limit to how many numbers there are between 10 and 20. There's $14\frac{1}{2}$ for a start; and 16.07 and 19.9999999; and endless other numbers using fractions and decimals. So 'number' can also include numbers like these, as well as all the integers. When we extend our concept of what is a number to include fractions and decimal numbers (which, as we shall see, are a particular kind of fraction), we get the set of **rational numbers**.

The term 'rational' derives from the idea that a fraction represents a *ratio*. The technical definition of a rational number is any number that is the ratio of two integers. We shall see in Chapter 10 that the ratio of two numbers is a way of comparing them by dividing one by the other. So, for example, we could say that the ratio of 6 to 2 is 3, because 6 is 3 times larger than 2, and 6 ÷ 2 = 3. Fractions and ratios are explained in Chapter 15 and decimal numbers are explained in Chapters 16 and 17. I include a few examples here that may help to illustrate the concept of a rational number:

$\frac{3}{8}$ is a rational number, because it is the ratio of 3 to 8 (3 divided by 8).
0.8 is a rational number, because it is the ratio of 8 to 10 (8 divided by 10).
$14\frac{1}{2}$ is a rational number, because it is the ratio of 29 to 2 (29 divided by 2).
16.07 is a rational number, because it is the ratio of 1607 to 100 (1607 divided by 100).
23 is a rational number, because it is the ratio of 23 to 1 (23 divided by 1).
−7 is a rational number, because it is the ratio of −7 to 1 (−7 divided by 1).

All the mathematics involved in these examples is explained later in the book. At this stage, the reader should just get hold of the basic idea that the set of rational numbers includes all fractions, including decimal fractions (which are just tenths, hundredths, thousandths and so on), as well as all the integers themselves. Rational numbers enable us to subdivide the sections of the number line between the integers and to label the points in between, as shown in Figure 6.4; here the interval between 6 and 7 is divided first into ten equal parts (tenths) and second into four equal parts (quarters).

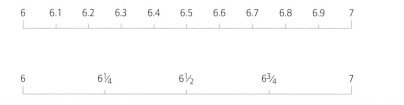

Figure 6.4   Some rational numbers between 6 and 7

Now, the reader may be thinking that the set of rational numbers must include all the numbers there are. But, in fact, there are other **real numbers** that cannot be written down as exact fractions or decimals – and are therefore not rational. Believe it or not, there is a limitless number of points on the number line that cannot be represented by rational numbers. For example, there are loads of square roots like the square root of 50 (see Chapter 13) that do not work out exactly.

I will use the square root of 50 to illustrate this, although some readers should be reassured that the mathematics involved here is all explained later in the book. What I am saying is that there is no fraction or decimal that is exactly equal to the square root of 50 (written as $\sqrt{50}$). This means there is no rational number that when multiplied by itself gives exactly the answer 50. We can get close. In fact, we can get as close as we want. But we cannot get exactly 50. Using a calculator, I could discover that $\sqrt{50}$ is somewhere between 7.07 and 7.08. I find that 7.07 × 7.07 (= 49.9849) is just less than 50 and 7.08 × 7.08 (= 50.1264) is just greater than 50. If we went to further decimal places, we could decide that it lies somewhere between 7.0710678 and 7.0710679. But neither of these rational numbers is the square root of 50. Neither of them when multiplied by itself would give 50 exactly. And however many decimal places we went to – you will just have to believe me about this – we could never get a number that gave us 50 exactly when we squared it. But $\sqrt{50}$ is a real number – in the sense that it represents a real point on a continuous number line, somewhere between 7 and 8. It represents a real length. For example, using Pythagoras's theorem, we could work out that the length of the diagonal of a square of side 5 units is $\sqrt{50}$ units. So this is a real length, a real number, but it is not a rational number. It is called an **irrational number**.

There is no end of irrational numbers, all of them representing real lengths and real points on the number line. Some examples would be: √8, √17.3, ³√50 (the cube root of 50), and that favourite number of mathematicians, π (pi: see Chapter 22). So, what mathematicians call the set of real numbers includes all rational numbers – which include integers, which in turn include natural numbers – and all irrational numbers. We think of it as the set of all numbers that can be represented by real lengths or by points on a continuous number line.

I can imagine that some readers are now wondering if there are numbers other than real numbers. If your appetite for number theory is really that insatiable, you will have to look elsewhere to find out how mathematicians use the idea of an *imaginary number* (like the square root of –1) to construct things called *complex numbers*.

WHAT IS MEANT BY 'PLACE VALUE'?

The system of numeration we use today is derived from an ancient Hindu system. It was picked up and developed by Arab traders in the ninth and tenth centuries and quickly spread through Europe. Of course, there have been many other systems developed by various cultures through the centuries, each with their particular features. Comparing some of these with the way we write numbers today enables us to appreciate the power and elegance of the Hindu-Arabic legacy. There is not space here to go into much detail, but the history of different numeration systems is a fascinating topic, with considerable potential for cross-curriculum work in schools, which will repay further study by the reader.

The Egyptian hieroglyphic system, used as long ago as 3000 BC, for example, had separate symbols for ten, a hundred, a thousand, ten thousand, a hundred thousand and a million. The Romans, some 3000 years later, in spite of all their other achievements, were using a numeration system which was still based on the same principle as the Egyptians, but simply had symbols for a few extra numbers, including 5, 50 and 500. Figure 6.5 illustrates how various numerals are written in these

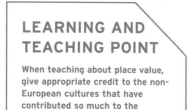

LEARNING AND TEACHING POINT

When teaching about place value, give appropriate credit to the non-European cultures that have contributed so much to the development of numeration.

systems and, in particular, how the numeral 366 would be constructed. Looking at these three different ways of writing 366 demonstrates clearly that the Hindu-Arabic system we use today is far more economic in its use of symbols. The reason for this is that it is based on the highly sophisticated concept of place value.

Egyptian hieroglyphics	Roman numerals	Hindu-Arabic
|	I	1
|||||	V	5
∩	X	10
∩∩∩∩	L	50
9	C	100
99 99 9	D	500
999 ∩∩∩∩∩ ||||||	CCCLXVI	366

Figure 6.5 Some numbers written in different numeration systems

In **Roman numerals**, for example, to represent three hundreds, three Cs are needed, and each of these symbols represents the same quantity, namely, a hundred. Likewise, in the Egyptian system, three 'scrolls' are needed, each representing a hundred. But, in the Hindu-Arabic system we do not use a symbol representing a hundred to construct three hundreds: we use a symbol representing three! Just this one symbol is needed to represent three hundreds, and we know that it represents three hundreds, rather than three tens or three ones, because of the *place* in which it is written. The two sixes in 366, for example, do not stand for the same number: reading from left to right, the first stands for six tens and the second for six ones, because of the places in which they are written.

So, in our Hindu-Arabic place-value system, all numbers can be represented using a finite set of **digits**, namely, 0, 1, 2, 3, 4, 5, 6, 7, 8, 9. Like most numeration systems, no doubt because of the availability of our ten fingers for counting purposes, the system uses ten as a **base**. Larger whole numbers than 9 are constructed using **powers** of the base: ten, a hundred, a thousand and so on. Of course, these powers of ten are not limited and can continue indefinitely with higher powers. This is how some of these powers are named, written as numerals, constructed from tens and expressed as powers of ten in symbols and in words:

> **LEARNING AND TEACHING POINT**
>
> The common use today of IV, IX and XC in Roman numerals instead of IIII, VIIII and LXXXX to represent 4, 9 and 90, respectively, was a later variation introduced to avoid having to write a string of four identical symbols. If you are introducing Roman numerals in the classroom, follow the historical development; so start with the earlier expanded form. For example, 29 will be XXVIIII and 194 will be CLXXXXIIII. When children are confident with this form, show how these can be written in their shorter forms as XXIX and CXCIV. (See self-assessment question 6.4 at the end of the chapter.)

A million	$1,000,000 = 10 \times 10 \times 10 \times 10 \times 10 \times 10 = 10^6$ (ten to the power six)
A hundred thousand	$100,000 = 10 \times 10 \times 10 \times 10 \times 10$ $= 10^5$ (ten to the power five)
Ten thousand	$10,000 = 10 \times 10 \times 10 \times 10$ $= 10^4$ (ten to the power four)
A thousand	$1000 = 10 \times 10 \times 10$ $= 10^3$ (ten to the power three)
A hundred	$100 = 10 \times 10$ $= 10^2$ (ten to the power two)
Ten	10 $= 10^1$ (ten to the power one)

The place in which a digit is written, then, represents that number of one of these powers of ten. So, for example, working from right to left, in the numeral 2345 the 5 represents 5 ones (or units), the 4 represents 4 tens, the 3 represents 3 hundreds and the 2 represents 2 thousands. Perversely, we work from right to left in determining the place values, with increasing powers of ten as we move in this direction. But, since we read from left to right, the numeral is read with the largest place value first: '2 thousands, 3 hundreds, 4 tens and 5 ones'. Certain conventions of language then transform this into the customary form, 'two thousand, three hundred and forty-five'. So, the numeral 2345 is essentially a clever piece of shorthand, condensing a complicated mathematical expression into four symbols, as follows:

$$(2 \times 10^3) + (3 \times 10^2) + (4 \times 10^1) + 5 = 2345.$$

Similarly, the numeral 2,345,678 is made up of 2 millions, 3 hundred-thousands, 4 ten-thousands, 5 thousands, 6 hundreds, 7 tens and 8 ones. Note that the '3 hundred-thousands, 4 ten-thousands, 5 thousands' part of this number is equivalent to 345 thousands. This enables us to read the number more concisely as 'two million, three hundred and forty-five thousand,

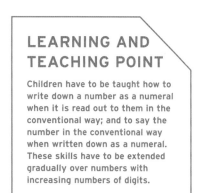

LEARNING AND TEACHING POINT

Children have to be taught how to write down a number as a numeral when it is read out to them in the conventional way; and to say the number in the conventional way when written down as a numeral. These skills have to be extended gradually over numbers with increasing numbers of digits.

six hundred and seventy-eight'. Again, we note that the numeral 2,345,678 is a clever piece of shorthand, condensing a very complicated mathematical expression into just seven symbols:

$$(2 \times 10^6) + (3 \times 10^5) + (4 \times 10^4) + (5 \times 10^3) + (6 \times 10^2) + (7 \times 10^1) + 8 = 2,345,678.$$

Notice that each of the powers of ten is equal to ten times the one below: a hundred equals 10 tens, a thousand equals 10 hundreds and so on. This introduces the principle of exchange. This means that whenever you have accumulated ten in one place, this can be exchanged for one in the next place to the left. This principle of being able to 'exchange one of these for ten of those' as you move left to right along the powers of ten, or to 'exchange ten of these for one of those' as you move right to left, is a very significant feature of the place-value system. It is essential for understanding the way in which we count. For example, when counting in ones the next number after 56, 57, 58, 59 … is 60, because we fill up the units position with ten ones and these are exchanged for an extra ten in the next column.

This principle of exchanging is also fundamental to the ways we do calculations with numbers. It is the principle of 'carrying one' in addition (see Chapter 9). It also means that, when necessary, we can exchange one in any place for ten in the next place on the right, for example when doing sub-traction by decomposition (see Chapter 9). The same principle of exchanging 'one of these for ten of those' extends to decimal numbers, where positions after the decimal point represent tenths, hundredths, thousandths and so on (see Chapter 16).

COMMAS OR SPACES?

I should mention here that the accepted international convention for printing numerals with more than four digits is to group the digits in threes from the right, using *spaces* to separate the groups. So, for example, 6 thousand (with four digits) is written 6000, but six million (with seven digits) would be written as 6 000 000. However, in hand-written work commas (for example: 6,000,000) are usually clearer and preferable. There is potential for confusion here because in some countries commas are used as decimal points. But commas as separators are used widely on the internet: for example, at the time of writing, on one website the population of Wymondham in Norfolk is given as 17,365, and on another the average cost of a house in the UK is given as £323,424.

Significantly, commas, rather than spaces, are used currently for large numbers in the end-of-key-stage mathematics tests in England. For consistency in preparing for the tests and to understand the way they will most often see large numerals written, so children in primary school in the UK will need to use commas when writing them. I have therefore decided in the body of the text in this edition to adopt the comma convention for numerals with more than four digits.

HOW CAN PLACE VALUE BE UNDERSTOOD IN CONCRETE TERMS?

There are two sets of manipulatives that provide particularly effective concrete embodiments of the place-value principle and therefore help us to explain the way our number system works. They are (1) base-ten blocks (**Dienes blocks**) and (2) place-value counters.

Figure 6.6 shows how the basic place-value principle of exchanging one for ten is built into these materials, for ones, tens and hundreds. Note that the ones in the base-ten blocks are sometimes referred to as units, the tens as longs and the hundreds as flats. With these blocks, of course, ten of one kind of block can actually be put together to make one of the next kind. With the counters, it is simply that ten white counters are understood to be *worth* the same as one red counter, and ten reds are worth the same as one orange. This is just the same as the relationships between one penny, ten pence and pound coins; assuming that these are all still in circulation when you read this book, they could be used in the same way as the counters, of course.

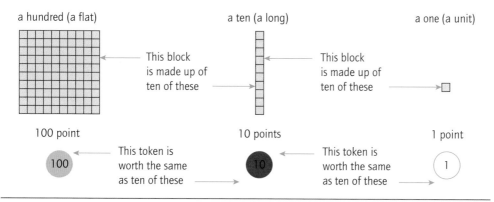

Figure 6.6 Materials for explaining place value

Figure 6.7 shows the number 366 represented with these materials. Notice that with both the blocks and the place-value counters we have 3 hundreds, 6 tens and 6 ones; this collection of blocks is equivalent to 366 units; and the collection of counters is worth the same as 366 white counters. Representing numbers with these materials enables us to build up images that can help to make sense of the way we do calculations such as addition and subtraction by written methods, as will be seen in Chapter 9.

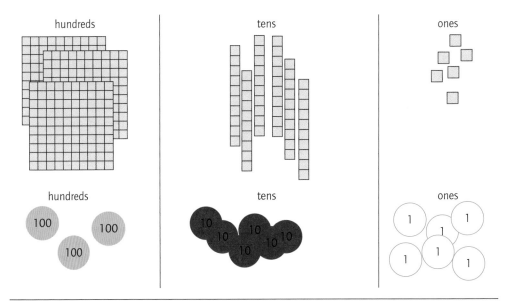

Figure 6.7   The number 366 in base-ten blocks and in place-value counters

HOW DOES THE NUMBER LINE SUPPORT UNDERSTANDING OF PLACE VALUE?

As we have seen already, the **number line** is an important image that is particularly helpful for appreciating where a number is positioned in relation to other numbers. This ordinal aspect of a number is much less overt in the representation of numbers using base-ten materials. Figure 6.8 shows how the number 366 is located on the number line. The number-line image shows clearly: that it comes between 300 and 400; that it comes between 360 and 370; and that it comes between 365 and 367. The significant mental processes involved in locating the position of the number on the number line are: counting in 100s; counting in 10s; and counting in 1s. First, you count from zero in 100s until you get to 300: 100, 200, 300; then from here in 10s until you get to 360: 310, 320, 330, 340, 350, 360; and then in 1s from here until you get to 366: 361, 362, 363, 364, 365, 366. The number-line image is also particularly significant in supporting mental strategies for addition and subtraction calculations, as will be seen in Chapter 8.

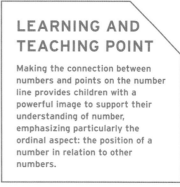

LEARNING AND TEACHING POINT

Making the connection between numbers and points on the number line provides children with a powerful image to support their understanding of number, emphasizing particularly the ordinal aspect: the position of a number in relation to other numbers.

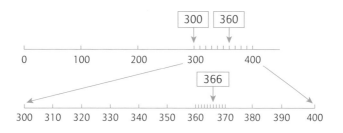

Figure 6.8 The number 366 located on the number line

WHAT IS MEANT BY SAYING THAT ZERO IS A PLACE HOLDER?

The Hindu-Arabic system was not the only one to use a place-value concept. Remarkably, about the same time as the Egyptians, the Babylonians had developed a system that incorporated this principle, although it used sixty as a base as well as ten. But a problem with their system was that you could not easily distinguish between, say, three and three sixties. They did not have a symbol for zero. It is generally thought that the Mayan civilization of South America was the first to develop a numeration system that included both the concept of place value and the consistent use of a symbol for zero.

Figure 6.9 shows 'three hundred and seven' represented in base-ten blocks. Translated into symbols, without the use of a zero, this would easily be

LEARNING AND TEACHING POINT

Incorporate some study of numeration systems into history-focused topics such as Roman, Egyptian and Mayan civilizations, and use this to highlight the advantages and significance of the place-value system we use today.

confused with thirty-seven: 37. The zero is used therefore as a **place holder**; that is, to indicate the position of the tens' place, even though there are no tens there: 307. It is worth noting, therefore, that when we see a numeral such as 300, we should not think to ourselves that the 00 means 'hundred'. It is the *position* of the 3 that indicates that it stands for 'three hundred'; the function of the zeros is to make this position clear whilst indicating that there are no tens and no ones. This may seem a little pedantic, but it is the basis of the confusion that leads some children to write, for example, 30045 for 'three hundred and forty-five'.

LEARNING AND TEACHING POINT

Give particular attention to the function and meaning of zero when writing and explaining numbers to children. The zero in 307 does not say 'hundred'. The 3 says 'three hundred' because of the position it is in. The zero indicates an absence of tens; it says 'no tens'.

LEARNING AND TEACHING POINT

In teaching for mastery of place value, consider the use of arrow cards to construct numerals with three digits from hundreds, tens and ones; and numerals with four digits from thousands, hundreds, tens and ones. Discussion with the class of how these cards make 3- and 4-digit numerals and what gets hidden when the cards are stacked enables the children to make helpful connections between the visual image of the cards, language and symbols.

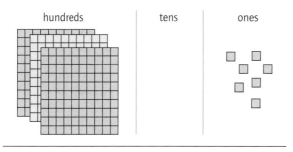

Figure 6.9 Three hundred and seven in base-ten blocks (Dienes blocks)

I have found that arrow cards, as shown in Figure 6.10, provide another strong visual image to help children to understand how a number like 452 is made up of 400 (4 hundreds), 50 (5 tens) and 2 (2 ones). By placing the three cards for 2, 50 and 400 one on top of the other, as shown, so that the arrows at the ends of the cards line up, the numeral 452 is constructed. Clearly, this idea can be extended to four-digit numerals.

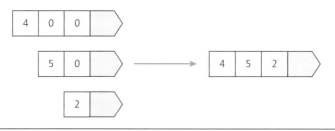

Figure 6.10 Arrow cards showing the values represented by the digits in the numeral 452

WATCH THE PROBLEM SOLVED! VIDEO AT: HTTPS://STUDY.SAGEPUB.COM/HAYLOCK7E

HOW IS UNDERSTANDING OF PLACE VALUE USED IN ORDERING NUMBERS?

Being able to put numbers in order requires a good understanding of the basic principles of place value. For example, given the set of numbers {906, 2345, 97, 967} a child who is developing mastery of place value should be able to rearrange them in order from smallest to largest: {97, 906, 967, 2345}. They will also be able to visualize these numbers as being positioned from left to right on a conventional horizontal number line.

To do this, you need to be able to decide instantly which is the greater or which is the smaller of two numbers. For example, think about how it is that you immediately know that 2345 is greater than 967? I would guess that you just use the fact that any 4-digit positive integer is greater than any 3-digit positive integer. This probably seems very obvious but it is actually a very sophisticated idea, showing just how powerful is the place-value system for numeration. It is always the first digit in a numeral that is most significant in determining the size of the number. So, the first digits in these two numerals tell us that the 4-digit number is into the thousands, whereas the 3-digit number is only into the hundreds.

Similarly, how do you know that 487 is less than 609? Again, the first digits tell us all we need to know: the first number has 4 complete hundreds – which must be smaller than the second number, which has 6 complete hundreds. But what about comparing, say, 3456 and 3701, where they are both 4-digit numbers and the first digits are the same? In this case, they both have 3 complete thousands. You will now find yourself looking at the next most significant digit, which tells you how many complete hundreds there are to go with the 3 complete thousands. The number 3456 has 3 complete thousands and 4 complete hundreds; the 3701 is greater than this, having 3 complete thousands and 7 complete hundreds.

A statement that one number is greater than another (for example, 25 is greater than 16) or less than another (for example, 16 is less than 25) is called an **inequality**. Children can record

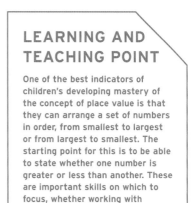

LEARNING AND TEACHING POINT

One of the best indicators of children's developing mastery of the concept of place value is that they can arrange a set of numbers in order, from smallest to largest or from largest to smallest. The starting point for this is to be able to state whether one number is greater or less than another. These are important skills on which to focus, whether working with younger children and numbers up to 20 or with older children and numbers into the millions.

such inequalities using the mathematical symbols > (greater than) and < (less than). For example, 25 > 16 (25 is greater than 16) and 16 < 25 (16 is less than 25). These two symbols are called *inequality signs*. Inequality signs can also be used to record that one number **lies between** two others. For example, '48 lies between 40 and 50' can be recorded as 40 < 48 < 50. This is stating that 40 is less than 48, which is less than 50. The fact that 48 lies between 40 and 50 can also be recorded using greater than signs: 50 > 48 > 40. This is stating that 50 is greater than 48, which is greater than 40.

Extending this idea further, a string of numbers that have been put in order (such as 40, 48, 50, 59, 70) can be connected using inequality signs: from smallest to largest as 40 < 48 < 50 < 59 < 70, or from largest to smallest as 70 > 59 > 50 > 48 > 40. Note that in these orderings we use only 'less than' signs or only 'more than' signs, but never a mixture of the two. The extensive language used to compare quantities in a wide range of measuring contexts is discussed further in Chapter 7. Significant digits are discussed further in Chapter 16. Putting three or more numbers in order also makes use of the principle of transitivity, which is explained in the context of measurement in Chapter 21.

HOW ARE NUMBERS ROUNDED TO THE NEAREST 10 OR THE NEAREST 100?

Rounding is an important skill in handling numbers, particularly, as we shall see in Chapter 16, in handling the results of calculations involving decimals and in many practical contexts. One skill to be learnt is to round a number or quantity to the nearest something. For example, someone might state their annual salary to the nearest thousand pounds, or a parent might measure a child's height to the nearest centimetre (see Chapter 21).

The first step in developing this skill is to be able to round a 2-digit number to the nearest ten. To find what 67 is when rounded to the nearest ten is to find which multiple of 10 is nearest to 67. The word 'nearest' implies that this is a spatial idea, so the image of a number as a point on a number line is important here. Figure 6.11 – with multiples of 10 marked on a number line – shows how 67 lies between 60 and 70, but is nearer to 70. This is exactly what we mean when we say that 67 rounded to the nearest ten is 70. This can extend to numbers with more digits of course: so, for example, 167 rounded to the nearest ten is 170, and 4163 rounded to the nearest ten is 4160. I should just mention that if the number ends in a 5 and therefore lies halfway between two multiples of 10 then there just is not a nearest multiple of 10! Note also that a positive integer less than 5 (that is, 1, 2, 3 or 4) rounded to the nearest ten is 0.

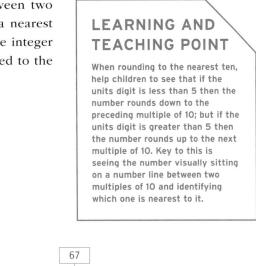

LEARNING AND TEACHING POINT

When rounding to the nearest ten, help children to see that if the units digit is less than 5 then the number rounds down to the preceding multiple of 10; but if the units digit is greater than 5 then the number rounds up to the next multiple of 10. Key to this is seeing the number visually sitting on a number line between two multiples of 10 and identifying which one is nearest to it.

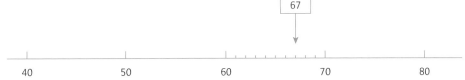

Figure 6.11   The nearest multiple of 10 to 67 is 70

We can then round numbers to the nearest hundred, using the same two steps. First, between which two multiples of 100 does the number lie? Then, which of these is it nearest to? So, for example, 765 lies between 700 and 800, and it is nearer to 800 than 700, so we could say that it is 800 rounded to the nearest hundred. The principle is that if the last two digits are less than 50, you round down to the previous multiple of 100, and if they are greater than 50, you round up to the next multiple of 100. For example, 4126 (ending in 26) is rounded down to 4100; and 4156 (ending in 56) is rounded up to 4200. If the final two digits are actually 50, then there is no nearest hundred.

A tricky example would be to round, say, 8967 to the nearest hundred. The two multiples of 100 that this number lies between are 8900 and 9000. Since it ends in 67, it must be rounded up. So 8967 is 9000 to the nearest hundred.

Clearly, these principles can extend to rounding to the nearest thousand, the nearest ten thousand, the nearest hundred thousand, or the nearest million. For example, at the time of writing, rounded to the nearest thousand there are about 17,000 state-funded primary schools in England; and rounded to the nearest ten thousand, the population of the beautiful county of Norfolk, where I am writing this book, is about 910,000.

RESEARCH FOCUS: PLACE VALUE

Since the essence of our number system is the principle of place value, it seems natural to assume that a thorough grasp of place value is essential for young children before they can successfully move on to calculations with two- or three-digit numbers. Thompson has undertaken a critical appraisal of this traditional view (Thompson, 2000). Considering the place-value principle from a variety of perspectives, Thompson concludes that the principle is too sophisticated for many young children to grasp. He argues that many of the mental calculation strategies used by children for two-digit addition and subtraction are based not on a proper understanding of place value but on what he calls *quantity value*. This is being able to think of, say, 47 as a combination of 40 and 7, rather than 4 tens and 7 units. Similar findings are reported by Price (2001), whose research focused on the development of place-value understanding in Year 3 children. He proposed an *independent-place* construct: that children tend to make single-dimensional associations between a place, a set of number words and a digit, rather than taking account of groups of 10. A later research study by Thompson and Bramald (2002) with 144 children aged 7 to 9 years demonstrated that only 19 of the 91 children who had successful strategies for adding two-digit numbers had a good understanding of the place-value principle. Approaches to teaching calculations with young children that are consistent with these findings would include: delaying the introduction of column-based written calculation methods; emphasis on the position of numbers in relation to other numbers through spatial images such as hundred squares and number lines; practice of counting backwards and forwards in 1s, 10s, 100s; and mental calculation strategies based on the idea of quantity value. Similar observations were made by Harcourt-Heath and Borthwick (2014) who noted the growing success of primary school children in England in working with nonstandard calculation methods, such as the use of the number line for additions and subtractions, the grid method for multiplication and chunking for division. (These are all explained in later chapters in this book.)

┍┑ LEARNING RESOURCES

Access activities for your **lesson plans** at: https://study.sagepub.com/haylock7e

Before trying the self-assessment questions below, you should complete the **self-assessment questions** for this chapter at: https://study.sagepub.com/haylock7e

6.1: What is the next number after 199?

6.2: (a) How many numbers are there between 0 and 20? (b) How many integers are there between 0 and 20?

6.3: Arrange these numbers in order from the smallest to the largest, without converting them to Hindu-Arabic numbers: DCXIII, CCLXVII, CLVIII, DCC, CCC. Then convert them to Hindu-Arabic, repeat the exercise and note any significant differences in the process.

6.4: In the later form of Roman numerals, writing a lower-value symbol in front of another symbol meant that this value had to be subtracted. For example, IV represents 4 (= 5 − 1), XC represents 90 (= 100 − 10) and CD represents 400 (= 500 − 100). What years are represented by (a) MCMLIV? (b) MCDXCII?

6.5: Add one to four thousand and ninety-nine.

6.6: Write these numbers in Hindu-Arabic numerals, and then write them out in full using powers of ten: (a) five hundred and sixteen; (b) three thousand and sixty; and (c) two million, three hundred and five thousand and four.

6.7: Playing various mathematics games a child has won 34 white counters (worth 1 point each), 29 red counters (worth 10 points each) and 3 orange counters (worth 100 points each). Apply the principle of 'exchanging ten of these for one of those' to reduce this collection of counters to the smallest number of counters.

6.8: Insert the correct inequality signs (> or <) in the gaps in the following: (a) 101 ... 98; (b) 998 ... 1001; (c) 48 ... 38 ... 28.

6.9: Connect together all five of these numbers using inequality signs for 'less than': 4002, 3998, 499, 3500 and 500.

FURTHER PRACTICE

FROM THE STUDENT WORKBOOK

Questions 6.01–6.20: Checking understanding (numbers and place value)

Questions 6.21–6.30: Reasoning and problem solving (numbers and place value)

Questions 6.31–6.42: Learning and teaching (numbers and place value)

GLOSSARY OF KEY TERMS INTRODUCED IN CHAPTER 6

Numeral The symbol used to represent a number; for example, the number of children in a class might be represented by the numeral 30.

Cardinal aspect of number The idea of a number as representing a set of things. This idea of number has meaning only in terms of non-negative integers.

Ordinal aspect of number The idea of a number as representing a point on a number line. This idea of number as a label for putting things in order has meaning for negative as well as positive numbers.

Natural numbers The set of numbers that we use for counting, 1, 2, 3, 4, 5 and so on, going on forever.

Subitizing Recognizing the number of items in a set without counting them.

Integer A whole number, positive, negative or zero.

Positive integer An integer greater than zero. The integer +4 is correctly referred to as 'positive four'. Usually, the + sign is understood and the integer is just written as 4 and referred to as 'four'.

Negative integer A number less than zero. The integer –4 is correctly referred to as 'negative four'.

Minus and **Plus** Synonyms for 'subtract' and 'add' respectively. Strictly speaking, it is incorrect to refer to negative integers and positive integers as 'minus numbers' and 'plus numbers', as is often done by weather forecasters.

Rational number A number that can be expressed as the ratio of two integers (whole numbers). All whole numbers and fractions are rational numbers, as are all numbers that can be written as exact decimals.

Real number Any number that can be represented by a length or by a point on a continuous number line. The set of real numbers consists of all rational and all irrational numbers.

Irrational number A number that is not rational; for example, $\sqrt{2}$ is irrational because it cannot be written exactly as one whole number divided by another.

Place value The principle underpinning the Hindu-Arabic system of numeration, in which the position of a digit in a numeral determines its value; for example, '6' can represent six, sixty, six hundred, six tenths, six hundredths and so on, depending on where it is written in the numeral.

Roman numerals A system of numeration that does not use the principle of place value, so the value represented by one of the symbols used (I, V, X, L, C and so on) is not dependent on the position in which it is written; C represents a hundred, for example, wherever it is written.

Digits The individual symbols used to build up numerals in a numeration system; in our Hindu-Arabic system, the digits are 0, 1, 2, 3, 4, 5, 6, 7, 8 and 9.

Base The number whose powers are used for the values of the various places in the place-value system of numeration; in our system, the base is ten, so the places represent powers of ten, namely, units, tens, hundreds, thousands and so on.

Power A way of referring to a number repeatedly multiplied by itself; for example, $10 \times 10 \times 10 \times 10$ is referred to as '10 to the power 4', abbreviated to 10^4.

Dienes blocks The base-ten version of these are wooden or plastic blocks equivalent to ones (units), tens (longs) and hundreds (hundred-blocks), where a hundred-block can be constructed from ten longs, and a long from ten units; an invention of the brilliant Hungarian educationist, Zoltan Dienes (1916–2014).

Exchange The principle at the heart of our place-value system of numeration, in which ten in one place can be exchanged for one in the next place to the left, and vice versa; for example, 10 hundreds can be exchanged for 1 thousand, and 1 thousand can be exchanged for 10 hundreds.

Number line A straight line in which points on the line are used to represent numbers, emphasizing particularly the order of numbers and their positions in relation to each other.

Place holder   The role of zero in the place-value system of numeration; for example, in the numeral 507 the 0 holds the tens place to indicate that there are no tens here. Without the use of zero as a place holder, there would just be a gap between the 5 and the 7.

Inequality   A statement that one number is greater than another (>) or less than another (<). For example, $80 < 87$ (80 is less than 87) and $100 > 87$ (100 is greater than 87).

Lies between   This phrase, when used for comparing numbers or quantities, can be expressed using two 'less than' symbols or two 'greater than' symbols. For example, '87 lies between 80 and 100' could be written $80 < 87 < 100$ or $100 > 87 > 80$.

Rounding (to the nearest 10 or 100)   This is to find which multiple of 10 (or 100) is nearest to the given number and to use this as an approximate value for the number.

ADDITION AND SUBTRACTION STRUCTURES

IN THIS CHAPTER, THERE ARE EXPLANATIONS OF

- two different structures of real-life problems modelled by addition;
- the key language associated with these structures;
- the contexts in which children will meet these structures;
- the commutative law of addition;
- four different structures of real-life problems modelled by subtraction;
- the key language associated with these structures;
- the contexts in which children will meet these structures.

READ THIS CHAPTER'S CURRICULUM LINKS AT: HTTPS://STUDY.SAGEPUB.COM/HAYLOCK7E

WHAT ARE THE DIFFERENT KINDS OF ADDITION SITUATIONS THAT PRIMARY CHILDREN MIGHT ENCOUNTER?

This chapter is not about doing addition and subtraction calculations, but focuses first on understanding the mathematical structures of these operations. Essentially, it is concerned with step 1 of the modelling process introduced in Chapter 5 (see Figure 5.2): setting up the mathematical model corresponding to a given situation. The approach taken for each of addition and subtraction in this chapter is to identify the range of situations that children have to learn to connect with the operation. The following two chapters then consider step 2 of the modelling process – the mental and written methods for doing the actual calculations.

There are two basic categories of real-life problems that are modelled by the mathematical operation we call addition. The problems in each of these categories may vary in terms of their content and context, but essentially they all have the same structure. I have coined the following two terms to refer to the structures in these two categories of problems:

- the aggregation structure;
- the augmentation structure.

Distinguishing between these two addition structures is not always easy, nor is it necessarily helpful to try to do so. But I find it is useful in teaching to have them in mind to ensure that children have opportunities to experience the full range of situations and, most importantly, the associated language that they have to learn to connect with addition.

> ## LEARNING AND TEACHING POINT
>
> The word 'addend' (meaning 'being added') is sometimes used for each of the numbers in an addition calculation. The result of the addition is the sum of the addends. The word 'sum' should not be used for just any calculation; for example, '103 – 87' and '(256 + 14) ÷ 30' are not sums.

WHAT IS THE AGGREGATION STRUCTURE OF ADDITION?

I use the term **aggregation** to refer to a situation in which two (or more) quantities are combined into a single quantity and the operation of addition is used to determine the total. For example, there are 15 marbles in one circle and 17 in another: how many marbles altogether? This idea of 'how many (or how much) altogether' is the central notion in the aggregation structure (see Figure 7.1). In this example, notice that the two sets do not overlap. They are called **discrete sets**. When two sets are combined into one set, they form what is called the **union of sets**. So another way of describing this addition structure is 'the union of two discrete sets'. This notion of addition builds mainly on the cardinal aspect of number, the idea of number as a set of things (see Chapter 6).

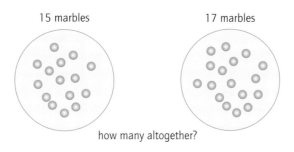

15 marbles   17 marbles

how many altogether?

Figure 7.1   Addition as aggregation

WHAT ARE SOME OF THE CONTEXTS IN WHICH CHILDREN WILL MEET ADDITION IN THE AGGREGATION STRUCTURE?

First and most simply, children will encounter this structure whenever they are putting together two sets of objects into a single set, to find the total number; for example, combining two discrete sets of children (25 boys and 29 girls, how many altogether?) or combining two separate piles of counters (46 red counters, 28 blue counters, how many altogether?). Second, children will encounter the aggregation structure in the context of money. This might be, for example, finding the total cost of two or more purchases, or the total bill for a number of services. The question will be, 'How much altogether?' Similarly, any measurement context will provide examples

LEARNING AND TEACHING POINT

Key language to be developed in the aggregation structure of addition includes: *how many altogether? How much altogether? The total*. 'How many' is used with countable sets, such as 'how many beakers?' The response might be, for example, that the *number* of beakers is 8. 'How much' is used for quantities in various measuring contexts, such as 'how much water?' The response to this question might be, for example, that the *amount* of water is 2 litres.

of this addition structure; for example, when finding the total volume of water in two containers holding 150 millilitres and 75 millilitres respectively (150 + 75).

> **LEARNING AND TEACHING POINT**
>
> Ensure that children experience the aggregation structure in a range of relevant contexts: money (shopping, bills, wages and salaries), length and distance, mass, capacity and liquid volume, and time (see Chapter 21). For example, addition would be the operation required to find the total time taken on a journey if the outward journey takes 65 minutes and the return journey takes 55 minutes (65 + 55).

WHAT IS THE AUGMENTATION STRUCTURE OF ADDITION?

I use the term **augmentation** to refer to a situation where a quantity is increased by some amount and the operation of addition is required in order to find the augmented or increased value. For example, the price of a bicycle costing £164 is increased by £25: what is the new price? This is the addition structure which lies behind the idea of counting on along a number line and which we might use

> **LEARNING AND TEACHING POINT**
>
> The key language to be developed in the augmentation structure of addition includes: *start at and count on, increase by, go up by.*

with young children for experiencing simple additions such as, say, 7 + 5: 'start at 7 and count on 5' (see Figure 7.2). Because it is connected so strongly with the image of moving along a number line, this notion of addition builds particularly on the ordinal aspect of number (see Chapter 6).

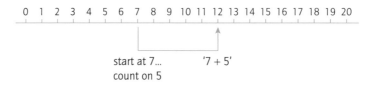

Figure 7.2 Addition as augmentation

WHAT ARE SOME OF THE CONTEXTS IN WHICH CHILDREN WILL MEET ADDITION IN THE AUGMENTATION STRUCTURE?

The most important and relevant context for experiencing the augmentation structure is again that of money, particularly the idea of increases in price or cost, wage or salary. Another context that has relevance for children is temperature, where addition would model an increase in temperature from a given starting temperature. A significant context for use with younger children is their age: 'You are 6 years old now, how old will you be in 4 years' time?' This is a good way for younger children to experience 'start at six and count on 4'. The key language that signals the operation of addition is that of 'increasing' or 'counting on'. This idea may also be encountered occasionally, but not often, in other measurement contexts, such as length (for example, stretching a length of elastic by so much), mass (for example, putting on so many kilograms over Christmas) and time (for example, increasing the length of the lunch break by so many minutes).

> **LEARNING AND TEACHING POINT**
>
> As well as counting on from a point on a number line, ensure that children experience the augmentation structure in a range of relevant contexts. Some key contexts for experiencing addition as augmentation are increasing costs or wages, increasing temperatures and increasing ages.

WHAT IS THE COMMUTATIVE LAW OF ADDITION?

It is clear from Figure 7.1 that the problem there could be represented by either 15 + 17 or by 17 + 15. Which set is on the left and which on the right makes no difference to the total number of marbles. The fact that these two additions come to the same result is an example of what is called the **commutative law of addition**. To help remember this technical term, we could note that *commuters* go both ways on a journey. So the commutative law of addition is simply the principle that an addition can go both ways: for example, 17 + 15 = 15 + 17. The principle is an axiom (see Chapter 4) – a self-evident fact – one of the fundamental building blocks of arithmetic. We can state this commutative law formally by the following generalization, which is true whatever the numbers a and b: $a + b = b + a$.

The significance of this property is twofold. First, it is important to realize that subtraction does *not* have this commutative property. For example, 10 – 5 is not equal to

5 – 10. Second, it is important to make use of commutativity in addition calculations. Particularly when using the idea of counting on, it is nearly always better to start with the bigger number. For example, it would not be sensible to calculate 3 + 59 by starting at 3 and counting on 59. The obvious thing to do is to use the commutative law mentally to change the addition to 59 + 3, then start at 59 and count on 3.

WHAT ARE THE DIFFERENT KINDS OF SUBTRACTION SITUATIONS THAT PRIMARY CHILDREN MIGHT ENCOUNTER?

There is a daunting range of situations in which we have to learn to recognize that the appropriate operation is subtraction. I find it helpful to categorize these into at least the following four categories:

- the partitioning structure;
- the reduction structure;
- the comparison structure;
- the inverse-of-addition structure.

Each of these has its own characteristic language patterns, all of which have to be connected in the learner's mind with subtraction. It is important for teachers to be aware of this range of structures, to ensure that children get the opportunity to learn to apply their number skills to all of them. Being able to connect subtraction with the whole range of these situations and to switch freely from one to the other is also the basis for being successful and efficient at mental and informal strategies for doing subtraction calculations. For example, to find out how much taller a girl of 167 cm is than a boy of 159 cm (which, as we shall see, is the comparison structure), a child may recognize that this requires the subtraction '167 – 159', but then do the actual calculation by interpreting it as 'what must be added to 159 to get 167?' (which, as explained below, is the inverse of addition).

It helps us to connect problems that incorporate these mathematical structures with the operation of subtraction if we ask ourselves the question: what is the calculation I would enter on a calculator (or a calculator app on a phone or computer) in order to solve this problem? In each case, the answer will involve using the subtraction key. It is one of the baffling aspects of mathematics that the same symbol, as we shall see particularly with the example of the subtraction symbol, can have so many different meanings.

> **LEARNING AND TEACHING POINT**
>
> Familiarity with the range of subtraction structures and the associated language patterns will enable children to interpret a subtraction calculation in a number of ways and hence increase their ability to handle these calculations by a range of methods.

WHAT IS THE PARTITIONING STRUCTURE OF SUBTRACTION?

The **partitioning** structure refers to a situation in which a quantity is partitioned off in some way or other and subtraction is required to calculate how many or how much remains. For example, there are 17 marbles in the box, and 5 are removed, so how many are left? (See Figure 7.3.) The calculation corresponding to this situation is '17 – 5'. Partitioning is the structure that teachers (and consequently their children) most frequently con-

> **LEARNING AND TEACHING POINT**
>
> Key language to be developed in the partitioning structure of subtraction includes: *take away … how many left? How many are not? How many do not?*

nect with the subtraction symbol. Because it is linked in the early stages so strongly with the idea of a set of objects, it builds mainly on the cardinal aspect of number.

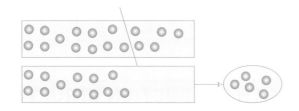

Figure 7.3 Subtraction as partitioning

The word **subtrahend** (meaning 'being subtracted') is sometimes used for the second number in a subtraction calculation, particularly when the calculation is related to partitioning or reduction. The number from which it is subtracted is called the minuend. It cannot be stressed too strongly that subtraction is not just 'take away'. As we shall see, partitioning is only one of a number of subtraction structures. So teachers should not overemphasize the language of 'take away, how many left' at the expense of all the other important language identified below that has to be associated with subtraction.

WHAT ARE SOME OF THE CONTEXTS IN WHICH CHILDREN WILL MEET THE PARTITIONING SUBTRACTION STRUCTURE?

Partitioning occurs in any practical situation where we ask how many are left, or how much is left. For example, this structure is encountered whenever we start with a given number of things in a set and a subset is taken away (removed, destroyed, eaten, blown up, lost or whatever). In each case, the question being asked is, 'how many are left?' It also includes situations where a subset is identified as possessing some particular attribute and the question asked is, 'how many are not?' or 'how many do not?' For instance, there might be 58 children from a year group of 92 going on a field trip. The question, 'how many are not going?' has to be associated with the subtraction, 92 – 58. (See the discussion on complements of a set in Chapter 26.)

The structure also has a number of significant occurrences in the context of money and shopping. For example, we might plan to spend £72 from our savings of £240 and need to work out how much would be left (that is, carry out the subtraction, 240 – 72). Then there are various practical situations in the context of measurement where we encounter the partitioning subtraction structure: for example, when we have a given length of some material, plan to cut off a length of it and wish to calculate how much will be left; or where we have some cooking ingredients measured by mass or volume, plan to use a certain amount in a recipe and wish to calculate how much will be left.

WHAT IS THE REDUCTION STRUCTURE OF SUBTRACTION?

The reduction structure is similar to 'take away' but it is associated with different language. It is simply the reverse process of the augmentation structure of addition. It refers to a situation in which a quantity is reduced by some amount and the operation of subtraction is required to find the reduced value. For example: if the price of a bicycle costing £164 is reduced by £25, what is the new price? The calculation that corresponds to this problem is '164 – 25'. The essential components of this structure are a starting point and a reduction or an amount to go down by. It is this subtraction structure that lies behind the idea of counting back along a number line, as shown in Figure 7.4.

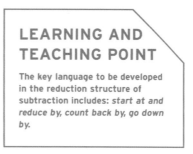

LEARNING AND TEACHING POINT

The key language to be developed in the reduction structure of subtraction includes: *start at and reduce by, count back by, go down by.*

Because of this connection, the idea of subtraction as reduction builds on the ordinal aspect of number.

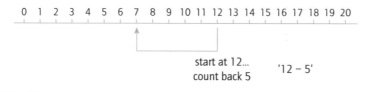

Figure 7.4 Subtraction as reduction

WHAT ARE SOME OF THE CONTEXTS IN WHICH CHILDREN WILL MEET SUBTRACTION IN THE REDUCTION STRUCTURE?

Realistic examples of the reduction structure mainly occur in the context of money. The key idea which signals the operation of subtraction is that of 'reducing', for example reducing prices and costs, or cutting wages and salaries. For example, if a person's council tax of £625 is reduced by £59, the resulting tax is determined by the subtraction, 625 – 59. The structure may also be encountered occasionally in other measurement contexts, such as a reduction in mass or a falling temperature.

LEARNING AND TEACHING POINT

Target specifically children's understanding and correct use of prepositions in mathematical statements. For example, ensure they are confident about the subtle difference between 'reduced by £25' and 'reduced to £25'.

WHAT IS THE COMPARISON STRUCTURE OF SUBTRACTION?

The **comparison** structure refers to a completely different set of situations, namely those where subtraction is required to make a comparison between two quantities, as for example in Figure 7.5. How many more blue cubes are there than red cubes? The calculation corresponding to this situation is '12 – 7'. Subtraction of the smaller number from the greater enables us to determine the *difference*, or to find out *how much greater* or *how much smaller* one quantity is than the other.

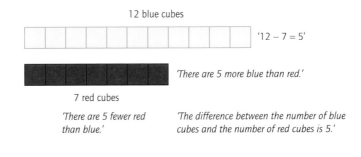

Figure 7.5   Subtraction as comparison

Because making comparisons is such a fundamental process, with so many practical and social applications, the ability to recognize this subtraction structure and the confidence to handle the associated language patterns are particularly important. Comparison can build on both the cardinal aspect of number (comparing the numbers of objects in two sets) and the ordinal aspect (finding the gap between two numbers on a number line).

Many people have difficulty in deciding when to use the word 'fewer' rather than the word 'less'. A simple rule is to use 'fewer' when we use the plural verb 'are' and to use 'less' when we use the singular 'is'. For example, in the winter there *are fewer* sunny days and there *is less* sunshine. Interestingly, there is no corresponding distinction with the word 'more': we would say both that there are more sunny days in the summer and that there is more sunshine. This is an odd convention in the English language, but I have tried to adhere to it pedantically in this book. Language conventions change over time and it is becoming increasingly acceptable to use 'less' instead of 'fewer' (for example, less sunny days) but never the other way round (no one would say 'fewer sunshine'). Note that we use 'less' when comparing abstract numbers, so we would say '4 is less than 7'. Abstract numbers are treated as singular nouns, as, for example, in '7 is my favourite number' and '4 is 2 add 2'. But if Jack has 4 teddies and Jenny has 7 teddies, we should say that Jack has *fewer* teddies than Jenny – although I might tolerate it if you said '*less* teddies'.

WHAT ARE SOME OF THE CONTEXTS IN WHICH CHILDREN WILL MEET SUBTRACTION IN THE COMPARISON STRUCTURE?

Wherever two numbers or quantities occur, we will often find ourselves wanting to compare them. The first step in this is to decide which is the larger and which is the smaller and to articulate this using the appropriate language.

The process of comparison is a central idea in all measurement contexts, as we shall see in Chapter 21. These contexts require an extensive range of language to be used to make comparisons. To illustrate this important point, Figure 7.6 provides examples of some of the language that children might need in various contexts in primary mathematics to compare and order quantities. The statements of comparison are arranged in two columns, so that alongside a comparison with the greater quantity as the subject of the sentence ($A > B$, 'A is greater than B') is the equivalent statement with the lesser quantity as the subject ($B < A$, 'B is less than A'). For example, when children compare two objects by balancing them on some weighing scales, their observations would be both 'the bottle is heavier than the book' and 'the book is lighter than the bottle'. Figure 7.6 illustrates the huge significance of the language and experience of comparison in children's learning.

A > B	B < A
A set of 21 has more items than a set of 19.	A set of 19 has fewer items than a set of 21.
21 is greater than 19.	19 is less than 21.
£3.50 is more than £2.95.	£2.95 is less than £3.50.
1/2 is larger than 1/3.	1/3 is smaller than 1/2.
Australia is bigger than Britain.	Britain is smaller than Australia.
Anna is taller than Ben.	Ben is shorter than Anna.
The pencil is longer than my finger.	My finger is shorter than the pencil.
The ceiling is higher than the light.	The light is lower than the ceiling.
London is further than Ipswich.	Ipswich is nearer than London.
The corridor is wider than the door.	The door is narrower than the corridor.
The teddy is fatter than the rabbit.	The rabbit is thinner than the teddy.
The bottle is heavier than the book.	The book is lighter than the bottle.
The jug holds more than the bottle.	The bottle holds less than the jug.
Oranges cost more than bananas.	Bananas cost less than oranges.
Fruit juice is dearer than milk.	Milk is cheaper than fruit juice.
A maths lesson takes longer than music.	Music takes less time than a maths lesson.
Music is later than literacy.	Literacy is (earlier) sooner than music.
Music happens after literacy.	Literacy happens before music.
Mrs Jones is older than Ben.	Ben is younger than Mrs Jones.
Ben is faster (quicker) than Mrs Jones.	Mrs Jones is slower than Ben.
Inside is hotter (warmer) than outside.	Outside is colder (cooler) than inside.

Figure 7.6   The language of comparison

The next stage of comparison is then to go on to ask: how many more? How many fewer? How much greater? How much less? How much heavier? How much lighter? How much longer? How much shorter? And so on. Answering these questions is where subtraction is involved.

First, the child might compare the numbers of items in two sets (for example, the numbers of marbles in two bags, the numbers of cards in two packs, the numbers of children in two classes, the numbers of counters in two piles, the numbers of pages in two books and so on). To do this, they have to connect the situation and the associated language of 'difference', 'how many more?', 'how many fewer?' with the operation of subtraction.

Then, in the context of money they would encounter this subtraction structure whenever they are comparing the prices of articles or the costs of services. If holiday package *A* costs £716 and package *B* costs £589, then we would ask questions such as: 'What is the difference in price?', 'How much cheaper is *B*?', 'How much dearer is *A*?', 'How much more does *A* cost than *B*?', 'How much less does *B* cost than *A*?' Note the

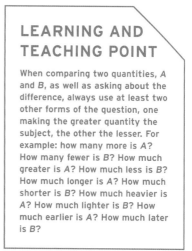

LEARNING AND TEACHING POINT

When comparing two quantities, *A* and *B*, as well as asking about the difference, always use at least two other forms of the question, one making the greater quantity the subject, the other the lesser. For example: how many more is *A*? How many fewer is *B*? How much greater is *A*? How much less is *B*? How much longer is *A*? How much shorter is *B*? How much heavier is *A*? How much lighter is *B*? How much earlier is *A*? How much later is *B*?

range of language patterns used here – and that in each case the question is answered by the same subtraction, 716 – 589. Also, subtraction might be used to compare salaries and wages: for example, how much more does the police officer earn than the teacher? Or, to put it another way, how much less does the teacher earn than the police officer?

If a child has measured the heights of Anna and Ben, there would be a subtraction involved in comparing their heights, to determine how much taller Anna is than Ben and how much shorter Ben is than Anna. The calculation enables us to say, for example, 'Anna is 2.5 cm taller than Ben' or 'Ben is 2.5 cm shorter than Anna'. If the child had measured the masses of the bottle and the book, they could compare them by asking 'how much heavier is the bottle?' or 'how much lighter is the book?' – and, again, subtraction is required to be able to say, for example, 'the bottle is 65 g heavier than the book' or 'the book is 65 g lighter than the bottle'.

WHAT IS THE INVERSE-OF-ADDITION STRUCTURE OF SUBTRACTION?

The **inverse-of-addition** structure refers to situations where we have to determine what must be added to a given quantity in order to reach some target. The phrase 'inverse of addition' underlines the idea that subtraction and addition are **inverse processes**. The concept of inverse turns up in many situations in mathematics, whenever one operation or transformation undoes the effect of another one. For example, moving 5 units to the left on a number line is the inverse of moving 5 units to the right: do one and then the other and you finish up back where you were. When we think of subtraction as the inverse of addition, we mean, for example, that since 28 + 52 comes to 80, then 80 – 52 must be 28. The subtraction of 52 undoes the effect of adding 52. Hence, to solve a problem of the form 'what must be added to x to give y?' we subtract x from y.

An example of an everyday situation with this structure would be: the entrance fee is 80p, but I have only 52p, so how much more do I need? Even though the question is about adding some-

thing to the 52p, the calculation that corresponds to this is a subtraction, 80 – 52. Figure 7.7 shows how this subtraction structure might be interpreted as an action on the number line: starting at 52 we have to determine what must be added to get to 80. This is a particularly important structure to draw on when doing subtraction calculations by mental and informal strategies.

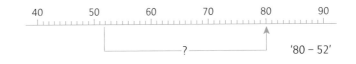

Figure 7.7   Subtraction as inverse of addition

WHAT ARE SOME OF THE CONTEXTS IN WHICH CHILDREN WILL MEET SUBTRACTION IN THE INVERSE-OF-ADDITION STRUCTURE?

This subtraction structure is often the most difficult for primary children to recognize, because the language associated with it – such as 'how much more is needed?' and 'what must be added?'– signals the idea of addition rather than subtraction.

There are many commonplace situations where we encounter this structure: for example, any situation where we have a number of objects or a number of individuals and we require some more in order to reach a target. The most convincing examples for many will be in the context of sport. For example, if I have scored 180 in darts, how many more do I need to reach 501? This corresponds to the subtraction, 501 – 180. If we are chasing a score of 235 in cricket and we have scored 186 so far, how many more runs do we need? The calculation that corresponds to this is 235 – 186.

Other examples of the inverse-of-addition structure occur in the context of measurement, such as: how much further do you have to drive to complete a journey of 345 miles if you have so far driven 196 miles?

Perhaps the most relevant instances of the inverse-of-addition subtraction structure occur in the context of money. For example, if we have saved £485 towards a holiday costing £716 then to calculate how much more we need to save we will need to do the subtraction, 716 – 485. Again, if the reader is uncertain about the assertion that this situation is an example of subtraction, it will help to ask: what is the calculation you would enter on a calculator or calculator app to work this out?

> **LEARNING AND TEACHING POINT**
>
> The language used in problems with the inverse-of-addition structure often signals addition rather than subtraction, so that some children will automatically add the two numbers in the question. Such children will need targeted help to recognize the need for a subtraction.

> **LEARNING AND TEACHING POINT**
>
> Representing a numerical problem using bar modelling, as shown in Figure 7.9, provides a simple and effective image of the problem, which, supported by appropriate questioning by the teacher, will help children to identify the calculations required.

Of course, when calculating a subtraction like 716 – 485 we might very well use a number-line image and the idea of adding on from 485 to get to 716. For example, we could add 15 to get from 485 to 500, another 200 to get to 700, and another 16 to get to 716, as shown in Figure 7.8. This is a powerful mental strategy, as we shall see in the next chapter.

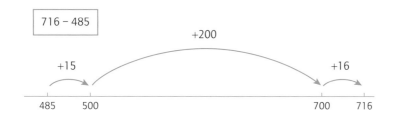

Figure 7.8 What must be added to 485 to get to 716?

Figure 7.9 shows a representation of this problem using **bar-modelling.** This is a simple way of representing a numerical problem that is widely used in Singapore, and now in England, as part of the commitment to teaching for mastery (see NCETM, 2022). The numbers in the problem are represented by rectangular bars arranged to show how they relate to each other. In this example the top bar represents the whole amount required (£716) and the bar underneath the amount that has been saved so far (£485). This image helps children to think in terms of adding appropriate extensions to the lower bar (for example, £15, £200, then £16; or £100, £100, £15, £16) in order to match the target shown in the upper bar.

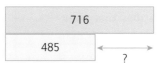

Figure 7.9   Bar model representing what must be added to 485 to get 716

The main point of this chapter has been to stress the importance of children associating subtraction with the full range of structures discussed. This is not just so that they will know when a situation requires a subtraction calculation, but also so that when a subtraction calculation is required they can interpret it in a number of different ways in order to deal with it.

RESEARCH FOCUS: WORD PROBLEMS, ADDITION AND SUBTRACTION

Greer (1997), reviewing research into children's responses to word problems in mathematics, identifies a widespread tendency for children to disregard the reality of the situations described by the text of the problem. His analysis suggests that an explanation is not to be found in some cognitive deficit of the children, but rather in the culture of the classroom where word problems are presented in a stereotyped fashion. Children learn that the solution involves the application of one of the basic arithmetical operations to the numbers mentioned in the text and look for clues as to which operation they should use. One of the first studies to identify this tendency, using word problems involving addition and subtraction, was undertaken in Israel by Nesher and Teubal (1975). They analysed the responses of about 85 children aged 7 years to four word problems: a problem using 'more' requiring addition (87% successful), one using 'less' requiring addition (64% successful), one using 'less' requiring subtraction (81% successful) and one using 'more' requiring subtraction (43% successful). These results suggested that there were particular difficulties for children in recognizing the correct operation in problems where the verbal cues 'more' (suggesting addition) and 'less' (suggesting subtraction) appeared to lead the children to select the wrong operation. The lowest success rate was in the subtraction problem where the word 'more' was used: 'The milkman brought 11 bottles of milk on Sunday; that was 4 more than he brought on Monday; how many did he bring on Monday?' Nesher and Teubal concluded that children tended to look for key words and to respond to them, rather than try to understand and grasp the logical structure of the problem. Their findings suggest, for example, that there is a need for teachers to focus particularly on children's grasp of the logical structure of situations incorporating subtraction as the inverse of addition. Greer (1997) proposes that teachers can address these problems by putting a more specific emphasis on the process of modelling (see Chapter 5) and by taking into account the children's real-world knowledge.

Daroczy and colleagues (2015) undertook an analysis of the complexities of mathematical word problems and reveal the ways in which these can vary in terms of (a) linguistic factors, such as the way the wording of the problem is structured, the linguistic embedding of the operation, semantic complexities, and the presence of verbal cues that relate to various arithmetic operations; and (b) numerical complexity and number properties. Their study makes clear that there is much more to learn here than a straightforward relationship between words and numerical operations. Spencer and Fielding (2015) conducted a research project in a large junior school where children, normally confident with calculation, were experiencing difficulties with the interpretation of word problems. They used the Singapore bar-model approach to provide a clear visual representation of a problem. They report that children valued the model more in areas of mathematics that were difficult and new to them, and where they felt less confident.

LEARNING RESOURCES

Access activities for your **lesson plans** at: https://study.sagepub.com/haylock7e

Before trying the self-assessment questions below, you should complete the **self-assessment questions** for this chapter at: https://study.sagepub.com/haylock7e

7.1: Make up a problem that corresponds to the addition, 5.95 + 6.95, using the aggregation structure in the context of shopping.

7.2: Make up a problem that corresponds to the addition, 1750 + 145, using the augmentation structure in the context of salaries.

7.3: Make up a problem that corresponds to the addition, 15 + 25 + 55 + 20 + 65, using the aggregation structure in the context of time.

7.4: Make up a problem that corresponds to the subtraction, 6.95 – 4.99, using the comparison structure in the context of shopping.

7.5: Make up a problem that corresponds to the subtraction, 250 – 159, using the partitioning structure and the phrase 'how many do not?'

7.6: Make up a problem corresponding to 989 – 650 that uses the inverse-of-addition structure in the context of shopping.

FURTHER PRACTICE

FROM THE STUDENT WORKBOOK

Questions 7.01–7.15: Checking understanding (addition and subtraction structures)

Questions 7.16–7.24: Reasoning and problem solving (addition and subtraction structures)

Questions 7.25–7.44: Learning and teaching (addition and subtraction structures)

GLOSSARY OF KEY TERMS INTRODUCED IN CHAPTER 7

Addend   Each of the numbers in an addition calculation.

Sum   The result of doing an addition; for example, 25 is the sum of the addends 17 and 8. The word 'sum' should not be used as a synonym for 'calculation'.

Aggregation   The process modelled by addition in which two quantities are combined into a single quantity and addition is used to determine the total. The key language is 'how many altogether?'

Discrete sets   Two (or more) sets that do not overlap, having no members in common; for example, the set of boys and the set of girls in a class are discrete sets.

Union of sets   The set formed when two (or more) sets are combined to form a single set. The union of two discrete sets is an example of the aggregation structure of addition.

Augmentation   The process modelled by addition in which a given quantity is increased by a certain amount and addition is used to determine the result of the increase. This structure includes 'start at … count on by …'.

Commutative law of addition   The principle that the order of two numbers in an addition calculation makes no difference to their sum. In symbols, the commutative law of addition states that, whatever the numbers a and b, $a + b = b + a$.

Partitioning (subtraction structure)   The process modelled by subtraction in which a quantity is partitioned off or taken away from a given quantity and subtraction is used to determine how many are left (or how much is left). The key idea is 'take away … how many (much) left?'

Subtrahend   The second number in a subtraction calculation, the number being subtracted. For example, in 76 – 48, the subtrahend is 48.

Minuend   In a subtraction calculation, the number from which the subtrahend is being subtracted. For example, in 76 – 48, the minuend is 76.

Reduction   The process modelled by subtraction in which a given quantity is reduced by some amount and subtraction is used to determine the result of the reduction. This structure includes 'start at … count back by …'.

Comparison   The process modelled by subtraction in which two quantities are compared and subtraction is used to find the difference, or how much greater or less one is than the other.

Inverse of addition   The process modelled by subtraction in which the question asked is 'what must be added?' in order to reach some target.

Inverse processes   Two processes, one of which has the effect of undoing the effect of the other. For example: add 7 and subtract 7; double and halve; turn clockwise through a right angle and turn anticlockwise through a right angle.

Bar-modelling   A visual representation of the numbers involved in a problem, both those known and those to be found, as rectangular bars or parts of bars, arranged in a way that demonstrates how the numbers relate to one another. A bar model, supported by teacher questioning, can help children identify the calculations required for problems involving addition, subtraction, multiplication and division.

MENTAL STRATEGIES FOR ADDITION AND SUBTRACTION

IN THIS CHAPTER, THERE ARE EXPLANATIONS OF

- the associative law for addition;
- counting forwards and backwards in ones, tens, hundreds;
- addition and subtraction on a hundred square;
- using multiples of 10 and 100 as stepping stones;
- addition and subtraction on an empty number line;
- front-end addition and subtraction;
- compensation in addition and subtraction calculations;
- the correct use of the symbol for 'equals';
- using multiples of 5 in additions and subtractions;
- relating additions and subtractions to doubles;
- using 'friendly' numbers;
- mental estimation for addition and subtraction calculations.

READ THIS CHAPTER'S CURRICULUM LINKS AT: HTTPS://STUDY.SAGEPUB.COM/HAYLOCK7E

WHAT IS THE ASSOCIATIVE LAW OF ADDITION?

Like the commutative law of addition ($a + b = b + a$) discussed in the previous chapter, the **associative law of addition** is a fundamental property of addition and an axiom of arithmetic. Written formally, as a generalization, it is the assertion that for any numbers a, b and c: $a + (b + c) = (a + b) + c$.

Using a particular example, this might be: $7 + (13 + 18) = (7 + 13) + 18$. The brackets indicate which addition should be done first. In simple terms, the associative law says that if you have three numbers to add together you get the same answer, whether you start by adding the second and third or start by adding the first and second. In the example above, it's probably easier to start by adding the 7 and 13, but you get the same answer if you start with $13 + 18$. I like to remember the associative law by thinking of it as a picture of three political parties: sometimes the party in the centre associates with the right and sometimes it associates with the left, but it does not make any difference.

This law allows us to write down $7 + 13 + 18$, without using any brackets to indicate which two numbers should be added first. We can choose whichever we prefer. The commutative and associative laws of addition together give us the freedom to add a string of numbers together in any order we like. For example, $7 + (13 + 18)$ could be changed as follows:

$$7 + (13 + 18)$$
$$= 7 + (18 + 13) \quad \text{(using the commutative law of addition)}$$
$$= (7 + 18) + 13 \quad \text{(using the associative law of addition)}$$
$$= (18 + 7) + 13 \quad \text{(using the commutative law of addition)}$$
$$= 18 + (7 + 13) \quad \text{(using the associative law of addition)}$$
$$= 18 + (13 + 7) \quad \text{(using the commutative law of addition)}$$
$$= (18 + 13) + 7 \quad \text{(using the associative law of addition)}$$
$$= (13 + 18) + 7 \quad \text{(using the commutative law of addition).}$$

We shall see below that deciding on the most efficient way of adding up various bits of numbers is an important strategy for informal calculations, so the associative and commutative laws of addition are important – even though most people use them without realizing it or without referring to them explicitly.

An important point to make about associativity is that subtraction does *not* have this property. For example, $25 - (12 - 8)$ is not equal to $(25 - 12) - 8$. This means that we cannot write $25 - 12 - 8$ to mean both of these. The convention is that $25 - 12 - 8$ means $(25 - 12) - 8$, that is, that the subtractions are done in order from left to right, unless brackets are used to indicate otherwise.

HOW IMPORTANT IS MENTAL CALCULATION?

The ability to calculate mentally using a range of strategies is an important component of numeracy. It is a reasonable expectation that most children in primary school should be able to learn to add and subtract using informal, mental strategies with three-digit numbers. This does not mean, of course, that they do not write anything down. They may need to write the question down for a start, so they do not forget it – and it may be helpful to support their mental calculation with a few jottings along the way or with a picture such as a number line.

Many of the problems that children encounter in calculations in primary school are associated with them being introduced too early to formal algorithms, written in a vertical format. (See Chapter 5 for what is meant by 'algorithm' in the context of calculations.) Vertical layouts for additions and subtractions especially lead children to treat the digits in the numbers as though they are individual numbers and then to combine them in all kinds of bizarre and meaningless ways. Mental strategies by their very ad hoc nature lead you to build on what you understand and to use methods that make sense to you.

The decision as to which strategy to employ is guided by the actual numbers in the problem. For example, very few of us would use the same strategy for calculating 201 – 20 and 201 – 197. The first I would do essentially by counting back in tens from 201, and the second by counting on from 197. Both of these are very simple subtractions, of course, when written down horizontally and done by mental methods. But the potential for error when these are written down as vertical calculations and tackled by the conventional algorithm is considerable, as illustrated in Figure 8.1 (see self-assessment question 8.2 at the end of this chapter). Many mathematics educators would support the view that children should be thoroughly confident in additions and subtractions written in horizontal format, using mental and informal

LEARNING AND TEACHING POINT

Make it clear to children and parents (and grandparents) that in mathematics there are many acceptable ways of doing any given calculation and that an effective mental or informal method is as valid as a formal written method.

LEARNING AND TEACHING POINT

When formal written methods have been taught, encourage children still to look for an easier way to do additions and subtractions by mental and informal methods first, before resorting to formal written algorithms.

LEARNING AND TEACHING POINT

To be able to encourage children to be creative in their use of a variety of mental methods to handle calculations, teachers themselves need to be familiar with the key mental and informal strategies described in this chapter and in Chapter 11, and to be confident and enthusiastic in using them.

strategies, before they are introduced to the vertical layout algorithms. Greater sharing of such methods, through open discussion and specific teaching of some of the key mental strategies, will undoubtedly lead to greater confidence with number.

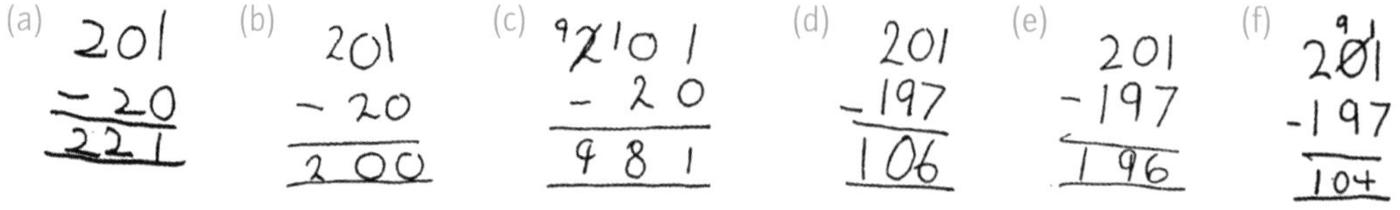

Figure 8.1 Examples of children's errors in vertical layout of subtractions

HOW DOES COUNTING FORWARDS AND BACKWARDS HELP IN MENTAL CALCULATIONS?

The numbers 10, 20, 30, 40, 50, 60, 70, 80, 90, 100, 110, 120 and so on, are the multiples of 10. (Multiples are discussed more generally in Chapter 13.) Just as children learn to add by counting on in ones from any number, they will also learn to add multiples of 10 by counting on in 10s from any number. For example, starting with 7, counting on in 10s, we get 7, 17, 27, 37, 47, 57 and so on. And in the same way as they learn to subtract by counting back in ones from any number, they learn to subtract multiples of 10 by counting back in 10s from any number. So, for example, counting back in 10s from 97, we get: 97, 87, 77, 67, 57 and so on.

In Chapter 7 we saw how addition could be understood as counting on and subtraction as counting back, and that these ideas were strongly linked with movements along a number line. These ideas are central to much mental and informal calculation. Doing additions and subtractions on a hundred square (see Figure 8.2) provides children with a strong image that supports the process of counting on and back in ones and tens. So, 57 + 3, done by counting on in ones, is associated with a movement to the right along a row: 57 … 58, 59, 60. And 57 − 3, done by counting back in ones, is associated with a movement to the left along a row: 57 … 56, 55, 54. Then, 57 + 30, done by counting on in tens, is seen as a movement down a column: 57 … 67, 77, 87. And 57 − 30, done by counting back in tens, is seen as a movement up a column: 57 … 47, 37, 27.

These are important strategies that can be extended to counting on and back in hundreds, for adding or subtracting multiples of 100 (100, 200, 300, 400, 500 and so on), and which we combine with other strategies when we become more proficient mental calculators.

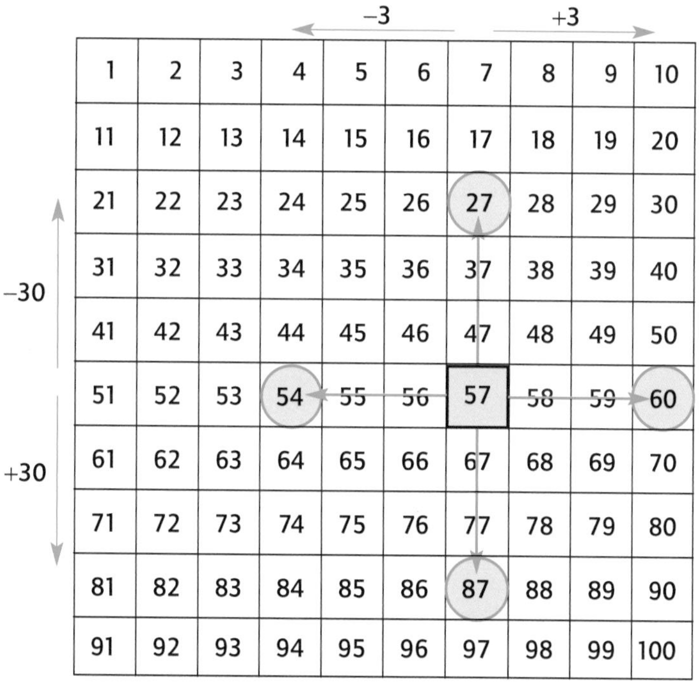

Figure 8.2 Using a hundred square

HOW DO WE USE MULTIPLES OF 10 AND 100 AS STEPPING STONES?

Notice what happens when we add 5 to 57 on a hundred square. We have to break the 5 down into two bits, 3 and 2. The 3 gets us to the next multiple of 10 (60) and then we have 2 more to count on. This process of using a multiple of ten (60) as a **stepping stone** is an important mental strategy for addition and subtraction. Some writers refer to this process as 'bridging'.

Here is how we might use this idea of a stepping stone for calculating, say, 57 + 28. First, we could count on in 10s, to deal with adding the 20: 57 … 67, 77. Then break the 8 up into 3 and 5, to enable us to use 80 as a stepping stone: 77 + 8 = 77 + 3 + 5 = 80 + 5 = 85.

LEARNING AND TEACHING POINT

Confidence in counting backwards and forwards in ones, tens and hundreds is an essential prerequisite for effective mental calculation, so these skills should be taught specifically and reinforced frequently. These particular processes are central in the mastery of the principles of place value.

A number-line diagram is a very useful image for supporting this kind of reasoning. Children can be taught to use an **empty number line**, which is simply a line on which they can put whatever numbers they like, not worrying about the scale, just ensuring that numbers are in the right order relative to each other. Figure 8.3 shows an empty number-line representation of the calculation of 57 + 28, using 80 as a stepping stone.

> ### LEARNING AND TEACHING POINT
>
> Take every opportunity to explain mental strategies with reference to hundred squares and empty number lines, in order to provide children with mental images that will underpin their manipulation of numbers.

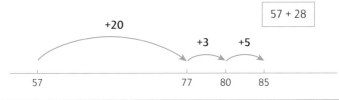

Figure 8.3   Using a multiple of 10 as a stepping stone on an empty number line

 WATCH THE PROBLEM SOLVED! VIDEO AT: HTTPS://STUDY.SAGEPUB.COM/HAYLOCK7E

Figure 8.4 shows how we might use multiples of 100 as stepping stones for calculating 542 − 275, using an empty number line. Using the inverse-of-addition structure, the subtraction can be interpreted as, 'What do you add to 275 to get 542?' This is done in three steps (25 + 200 + 42), with 300 and 500 as convenient stepping stones lying between the 275 and the 542.

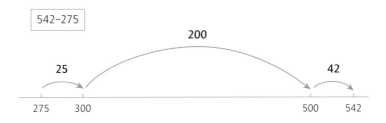

Figure 8.4   Using multiples of 100 as stepping stones in a subtraction

 WATCH THE PROBLEM SOLVED! VIDEO AT: HTTPS://STUDY.SAGEPUB.COM/HAYLOCK7E

WHAT IS FRONT-END ADDITION AND SUBTRACTION?

Most formal written algorithms for addition and subtraction work with the digits from right to left, starting with the units. In mental calculations, it is much more common to work from left to right. This makes more sense, because you deal with the biggest and most significant bits of the numbers first. One strategy is to mentally break the numbers up into hundreds, tens and ones, and then to combine them bit by bit, starting at the front end – that is, starting by adding (or subtracting) the hundreds. So, for example, given $459 + 347$, we would think of the 459 as $(400 + 50 + 9)$ and the 347 as $(300 + 40 + 7)$. This process is sometimes called **partitioning into hundreds, tens and ones**. (Note that this is a different use of the word 'partitioning' from that used in Chapter 7 for a subtraction structure.) We would then use the freedom granted to us by the associative and commutative laws to add these bits in any order we like. The **front-end approach** would deal with the hundreds first $(400 + 300 = 700)$, then the tens $(50 + 40 = 90$, making 790 so far) and then the ones (for example, $790 + 9 = 799$; followed by $799 + 7 = 799 + 1 + 6 = 806$). Notice that I have used 800 as a stepping stone for the last step here.

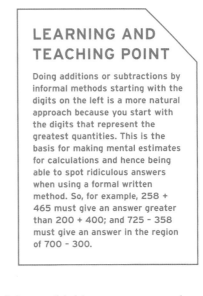

LEARNING AND TEACHING POINT

Doing additions or subtractions by informal methods starting with the digits on the left is a more natural approach because you start with the digits that represent the greatest quantities. This is the basis for making mental estimates for calculations and hence being able to spot ridiculous answers when using a formal written method. So, for example, $258 + 465$ must give an answer greater than $200 + 400$; and $725 - 358$ must give an answer in the region of $700 - 300$.

Writing this out in full, in a way which might explain my thinking to someone else:

$$459 + 347 = (400 + 50 + 9) + (300 + 40 + 7)$$
$$= (400 + 300) + (50 + 40) + (9 + 7)$$
$$= 700 + 90 + 9 + 7$$
$$= 799 + 7 = 799 + 1 + 6 = 800 + 6 = 806.$$

We will quite often use the front-end approach to get us started in a subtraction done mentally. For example, for $645 - 239$, we would immediately deal with the hundreds $(600 - 200 = 400)$, leaving us simply to think about $45 - 39$. This gives us 6, so the answer is $400 + 6 = 406$.

WHAT IS COMPENSATION IN ADDITION AND SUBTRACTION?

You can often convert an addition or subtraction question into an easier question by temporarily adding or subtracting an appropriate small number. For example, many people would evaluate 673 + 99 by adding 1 temporarily to the 99, so the question becomes 673 + 100. This gives 773. Now take off the extra 1, to get the answer 772. This strategy is sometimes called **compensation**. Figure 8.5(a) shows how this way of finding 673 + 99 looks when carried out on an empty number line.

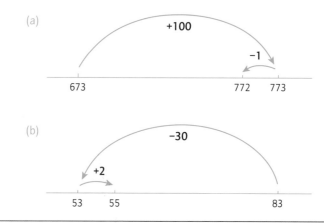

Figure 8.5 Using compensation to calculate (a) 673 + 99 and (b) 83 - 28

The trick in the strategy is always to be on the lookout for an easier calculation than the one you have to do. This will often involve temporarily replacing a number ending in a 9 or an 8 with the next multiple of 10. For example, the subtraction, 83 – 28, can temporarily be changed into the much easier calculation, 83 – 30, which gives 53, with 2 to be added to this to compensate for the fact that we've actually taken away 2 more than required. This is illustrated on an empty number line in Figure 8.5(b).

The strategy can be used to change any subtraction into an easier one. For example, 453 – 178 looks a bit daunting. I would rather do 453 – 180. (But remember: I will have taken away 2 more than I should and this will have to be added on again later.) If I'm still struggling with 453 – 180, I could instead choose to do 453 – 200, which is 253. (Now I have to remember there's an extra 20 to add on.) So the answer is 253 + 20 + 2, which is a relatively easy addition, giving 275. Compensation can be tricky when working entirely mentally, so a quick sketch of a number line like this can help in understanding whether to add or subtract the extra bits at the end.

This approach is particularly effective with precisely those subtractions that cause most problems using the formal decomposition-algorithm (see Chapter 9): those with zeros in the first number. For example, Figure 8.6(a) shows a typical error made by a nine-year-old boy attempting to calculate 101 – 97 set out in vertical format. He was then given the question in horizontal format and encouraged to work it out mentally. This he did successfully by first dealing with 100 – 97 by counting back and then compensating. Figure 8.6(b) is his response to the invitation to 'write down how you did it in a way that shows your thinking to someone else'.

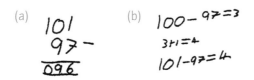

Figure 8.6   (a) A nine-year-old's unsuccessful attempt to calculate 101 – 97; and (b) his successful use of compensation

There are other ways of using the strategy of compensation, all of which amount to changing one or more of the numbers in order to produce an easier calculation. We can use this approach to exploit our confidence in handling multiples of five, to relate additions and subtractions to doubles, or simply to replace one of the numbers with a more 'friendly' number. These strategies are explained below.

> **LEARNING AND TEACHING POINT**
>
> Making numbers up to the next multiple of 10 and to the next multiple of 100 are really important skills in mental addition and subtraction. They should therefore be very specific teaching focuses with primary school children.

HOW SHOULD THE SYMBOL FOR 'EQUALS' BE USED IN RECORDING CALCULATIONS?

As children begin to make more detailed written records of their calculations, they can be encouraged to use the equals sign (=) to link only those things that are actually equal, as has been done in all the examples in this chapter. There is a tendency for children (and some teachers) to abuse the equals sign by employing it rather casually just to link the steps in a calculation, without it having any real meaning. For example, it is not uncommon to see this kind of thing written down: 103 – 87 = 103 – 90 = 13 + 3 = 16.

Only the final equals sign here connects two expressions that are actually equal. Although this kind of recording is acceptable behaviour in the privacy of your own scribbling pad, you should be warned that it is likely to seriously upset pedantic mathematicians if done in public.

HOW DO MULTIPLES OF 5 HELP IN MENTAL ADDITIONS AND SUBTRACTIONS?

Multiples of 5 (5, 10, 15, 20, 25, 30, …) are particularly easy to work with. Young children quickly learn to relate simple additions and subtractions to fives. For example, a five-year-old might think of 6 + 7 as '5 and 1 and 5 and 2', and proceed by first combining the fives. This tendency to relate numbers to multiples of 5 is no doubt related to the children's early experience of counting on their fingers, which leads them to perceive 6, 7, 8 and 9 as '5 and some more'. This is then reinforced by images and manipulatives, such as **ten-frames** (see Figure 8.7), in which seven is presented in one frame as a row of five counters and two more; and nine is seen as a row of five counters and an additional four. So 7 + 9, for example, is easily transformed into 5 + 5 and 2 + 4, which is 10 + 6, or 16.

We can similarly exploit this confidence with multiples of 5 in many additions and subtractions done mentally. For example, 37 + 26 could be related to 35 + 25, which we would probably find much easier. This can be taught as a specific strategy. Here are some examples, written out in full to show what would, of course, be a mental process:

$$37 + 26 \ = \ (35 + 25) + 2 + 1 \ = \ 60 + 2 + 1 \ = \ 63$$
$$77 + 24 \ = \ (75 + 25) + 2 - 1 \ = \ 100 + 2 - 1 \ = \ 101$$
$$174 - 46 \ = \ (175 - 45) - 1 - 1 \ = \ 130 - 1 - 1 \ = \ 128$$

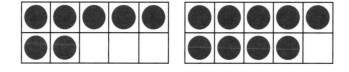

Figure 8.7   Ten-frames, showing 7 as 5 and 2; and 9 as 5 and 4

HOW DO YOU RELATE ADDITIONS AND SUBTRACTIONS TO DOUBLES?

Sometimes in additions and subtractions we can exploit the fact that most people are fairly confident with the processes of doubling and halving. Even quite young primary school children will quickly show confidence with doubles, no doubt because they have two sets of fingers on which to learn to calculate. So it is found that young children will often exploit their facility with doubles to calculate 'near-doubles'. For example, many five-year-olds would think of $6 + 7$ as 'double 6 and 1 more'. This is another example of

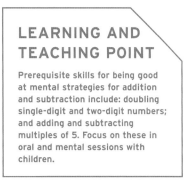

LEARNING AND TEACHING POINT

Prerequisite skills for being good at mental strategies for addition and subtraction include: doubling single-digit and two-digit numbers; and adding and subtracting multiples of 5. Focus on these in oral and mental sessions with children.

compensation, of course, turning the calculation into what most of us would find to be an easier one.

With larger numbers, for example $36 + 37$, using our facility for doubling we could look at this and think 'double 36 and 1 more'. Here are some more examples of how we might use our confidence with doubling and halving to get us started on some additions and subtractions:

> $48 + 46$ could be related to double 46: $46 + 46 = 92$, so $48 + 46 = 92 + 2 = 94$.
> $62 + 59$ could be related to double 60: $60 + 60 = 120$, so $62 + 59 = 120 + 2 - 1 = 121$.
> $54 - 28$ could be related to half 54 (27): $54 - 27 = 27$, so $54 - 28 = 27 - 1 = 26$.
> $54 - 28$ could be related to half 56 (28): $56 - 28 = 28$, so $54 - 28 = 28 - 2 = 26$.

HOW DO YOU USE 'FRIENDLY' NUMBERS?

Most of us would much prefer to deal with $742 - 142$ than $742 - 146$, because the 142 and 742 are much more **friendly** in their relationship to each other than the 146 and the 742. We always have as an option in addition and subtraction to use the compensation approach and temporarily replace one of the numbers in a calculation with one that is 'more friendly'. This is especially useful in subtraction:

LEARNING AND TEACHING POINT

Teach children specifically the strategies outlined in this chapter and give them opportunities to discuss different ways of tackling additions and subtractions by mental methods supported by jottings and empty number-line diagrams. These discussions are particularly important in enabling children to recognize and employ a suitable method at the right time.

To calculate   742 – 146
Change the 146 to 142: 742 – 142 = 600
Now compensate: 742 – 146 = 600 – 4 = 596
Or,
Change the 742 to 746: 746 – 146 = 600
Now compensate: 742 – 146 = 600 – 4 = 596

HOW ARE MENTAL METHODS USED IN ESTIMATIONS?

Confidence in handling mental calculations for addition and subtraction and a facility in rounding numbers to the nearest 10, 100, 1000, or higher power of ten (see Chapter 6) are prerequisites for being able to make reasonable estimates for the answer that should be expected from an addition or a subtraction. In everyday life, we often require no more than an approximate indication of what the result should be for many of the calculations we engage with – particularly since, in practice, difficult calculations will usually be done on a piece of technology where the likeliest error is that we enter the numbers incorrectly. It is also particularly important to have some rough idea of what size of answer should be expected when using written calculation methods (see Chapter 9), where a small slip in applying a procedure might produce a huge error.

For example, a ten-year-old who gets the result 78,215 for 5128 + 2707 should spot immediately that this result is seriously wrong and thus seek to correct it. Mentally rounding the two numbers being added here to the nearest thousand, the calculation is approximately 5000 + 3000, which suggests that the result should be around 8000, not nearly 80,000. A closer estimate for this addition would be obtained by rounding the numbers to the nearest hundred: 5100 + 2700. That's 51 hundreds add 27 hundreds. Mentally adding 51 and 27 (= 78) gives an estimate of 78 hundreds, which is 7800. (The exact result is 7835.)

A child who enters the subtraction 5128 – 2707 on a calculator and gets the result 4841 should spot that an error has been made, if they are in the habit of checking whether the answer is reasonable. We would expect the answer for 5128 – 2707 to be about 5000 – 3000 (= 2000) or, using a better estimate, about 5100 – 2700 (= 2400). (The exact answer here is 2421.)

Estimation techniques can become quite sophisticated. For example, consider the addition of 25,378 and 46,235. Rounded to the nearest ten thousand, the numbers are approximately 30,000 and 50,000, so we can expect an answer around 80,000.

But, since both numbers here were rounded up, we can be sure that this is an over-estimate and that the actual sum of the two numbers will be less than 80,000. Rounded to the nearest thousand, the numbers are 25,000 and 46,000, which have a sum of 71,000. In this case, since both numbers have been rounded down we can be sure that this is an under-estimate and that the actual sum will be greater than the 71,000. (The exact answer is actually 71,613.)

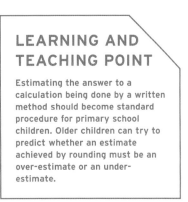

LEARNING AND TEACHING POINT

Estimating the answer to a calculation being done by a written method should become standard procedure for primary school children. Older children can try to predict whether an estimate achieved by rounding must be an over-estimate or an under-estimate.

So, here's a generalization: (a) if two numbers are rounded up to estimate their sum then the result will be an over-estimate; (b) if two numbers are rounded down to estimate their sum then the result will be an under-estimate.

The reader may like to check the corresponding generalizations for subtraction: (a) if the first number in a subtraction is rounded up and the second number rounded down then the result will be an over-estimate; (b) if the first number in a subtraction is rounded down and the second number rounded up then the result will be an under-estimate. These ideas are considered further in relation to calculations with decimal numbers in Chapter 17.

RESEARCH FOCUS: INFORMAL ADDITION AND SUBTRACTION

An interesting finding by Peters et al. (2012) was that children aged 8–10 years in Belgium, given subtraction questions presented in horizontal format, tended to use an informal adding-on approach when the number being subtracted was larger than the difference (for example, 87 – 59), but a direct subtraction method when the opposite was the case (for example, 87 – 29). This shows a degree of flexibility and mathematical good sense, no doubt related to the fact that children in Belgium spend two years developing ways of doing calculations presented in horizontal format before being introduced to vertical layout. In a research project with children aged 6–9 years in Queensland, Australia, Heirdsfield and Cooper (1997, 2004) found that introducing children too early to formal written algorithms for addition and subtraction limits their willingness and ability to develop a range of mental strategies and a good number sense. When presented with a calculation written in horizontal format and invited to find the answer mentally, the majority of such children tended to use a mental strategy based on the formal written method. Interestingly, the children showed a greater range of mental strategies when the calculation was presented as a word problem. Clearly, the range of addition and subtraction structures embedded in real-life situations suggests different ways of doing the calculations, some of which they had not specifically been taught in school. The researchers also found that children who were most successful in mental addition and subtraction had two key skills: very secure knowledge of number facts and a good sense of computational estimation. In their later work (2004) they report that students who were flexible in mental computation employed efficient number facts strategies (derived facts strategies) in the number facts test. Further, some of the number facts strategies were applied to mental computation strategies (e.g., 9 + 7: add 1 to 9, take 1 from 7, so 10 + 6 = 16; cf. 148 + 99 is the same as 147 + 100). On the other hand, the inflexible students did not possess efficient number fact strategies. They resorted to count if the number fact was not known by recall, particularly for the interim calculations in the mental computation tasks.

LEARNING RESOURCES

Access activities for your **lesson plans** at: https://study.sagepub.com/haylock7e

Before trying the self-assessment questions below, you should complete the **self-assessment questions** for this chapter at: https://study.sagepub.com/haylock7e

8.1: What sign should go in the box to make this true: (a) $67 - (20 - 8) = (67 - 20) = \Box 8$? Try this with some other numbers and state a general rule. (b) What sign should go in the box to make this true: $67 - (20 + 8) = (67 - 20) = \Box 8$? Try this with some other numbers and state a general rule.

8.2: Identify the errors made by the children in the examples in Figure 8.1.

8.3: Find the answer to 538 + 294 by the front-end approach, partitioning the numbers into hundreds, tens and ones, then starting with the hundreds and working from left to right.

8.4: Calculate 423 + 98 mentally, using compensation.

8.5: Calculate 297 + 304 mentally, by relating it to a double.

8.6: Calculate 494 + 307 mentally, using 500 as a stepping stone.

8.7: Calculate 26 + 77 mentally, by relating the numbers to multiples of 5.

8.8: Calculate 819 – 523 mentally, by making one of the numbers more friendly.

8.9: Calculate 732 – 389 mentally, by adding on from 389, using 400 and 700 as stepping stones.

8.10: Do these calculations, using any mental strategies that seem appropriate: (a) 974 – 539; (b) 400 – 237; (c) 597 + 209; (d) 7000 – 6; (e) 7000 – 6998.

FURTHER PRACTICE

Access the website material for Knowledge Check 2: Mental calculations, adding lists at: https://study.sagepub.com/haylock7e

FROM THE STUDENT WORKBOOK

Questions 8.01–8.16: Checking understanding (mental strategies for addition and subtraction)

Questions 8.17–8.28: Reasoning and problem solving (mental strategies for addition and subtraction)

Questions 8.29–8.45: Learning and teaching (mental strategies for addition and subtraction)

GLOSSARY OF KEY TERMS INTRODUCED IN CHAPTER 8

Associative law of addition   The principle that if there are three numbers to be added it makes no difference whether you start by adding the first and second, or by adding the second and third. In symbols, this law states that, for any three numbers a, b and c, $(a + b) + c = a + (b + c)$.

Stepping stone   Usually a multiple of 10 or 100 used to break down an addition or subtraction into easier steps. For example, to find what has to be added to 37 to get to 75, the numbers 40 and 70 might be used as stepping stones. This process is also called 'bridging'.

Empty number line   A number line without a scale, used to support mental and informal additions and subtractions; numbers involved in the calculation can be placed anywhere on the line provided they are in the right order relative to each other.

Partitioning into hundreds, tens and ones   Breaking a number up into hundreds, tens and ones as an aid to using it in a calculation. For example, 476 when partitioned is 400 + 70 + 6.

Front-end approach   A method for doing a calculation that focuses first on the digits at the front of the number. For example, to add 543 and 476, a front-end approach would start by adding the 500 and the 400.

Compensation   A strategy that involves replacing a number in a calculation with an easier number close to it and then compensating for this later. For example, to subtract 38 you could subtract 40 instead and compensate by adding on the additional 2 at the end.

Tens-frames   Rectangular frames with ten squares arranged in two fives used to help younger children in additions and subtractions (see Figure 8.7). The squares in a frame can be filled either with counters or by shading.

Near-double   When two numbers involved in an addition are nearly the same, such as 46 + 48; or when one number involved in a subtraction is nearly double or half of the other, such as 87 – 43. Such calculations can be done by treating them as exact doubles and then compensating.

'Friendly' numbers   Two numbers that are related to each other in a way that makes a calculation particularly easy; for example, 457 – 257. Often, a calculation can be made easier by replacing one of the numbers with a more friendly number close to it and then compensating later.

WRITTEN METHODS FOR ADDITION AND SUBTRACTION

IN THIS CHAPTER, THERE ARE EXPLANATIONS OF

- how column addition and subtraction might be introduced;
- 'exchange' and 'carrying' in the formal addition algorithm;
- the decomposition method for doing subtraction calculations;
- the equal additions method for subtraction;
- how the two methods differ and why decomposition is preferred;
- the problem with zeros in the top number in a subtraction calculation;
- the constant difference method for subtraction;
- using addition to check subtraction.

READ THIS CHAPTER'S CURRICULUM LINKS AT: HTTPS://STUDY.SAGEPUB.COM/HAYLOCK7E

HOW MIGHT CHILDREN BE INTRODUCED TO COLUMN ADDITION?

Most of us find that for additions with numbers containing three or more digits we need to use a written method of **column addition**: this is a way of laying out an addition calculation that lines up the hundreds, tens and ones in columns. These build on some of the mental strategies outlined in the previous chapter, particularly the idea of partitioning the numbers into hundreds, tens and ones. Figure 9.1(a) shows how one nine-year-old recorded the calculation, 372 + 247, using this strategy of partitioning and lining up the hundreds, tens and ones in columns. Figure 9.1(b) shows an alternative layout for recording the same thinking, which can then be abbreviated to the version in Figure 9.1(c). These ways of recording are a useful, informal introduction to what we might call 'the formal addition algorithm', which is shown in Figure 9.1(d). The major source of error in using this format is that it encourages children to think of the digits as separate numbers, losing any sense that they represent hundreds, tens or ones. The layouts in Figures 9.1(a), (b) and (c) have the advantage that they do not obscure the meaning of the digits and children should still be consciously aware that they are handling hundreds, tens and ones. Note that these layouts also use the more natural procedure of working from left to right, dealing with the largest bits of the numbers first, as in the 'front-end' approach discussed in Chapter 8. The traditional columnar method shown in Figure 9.1(d) requires that we start by adding the ones, the least significant parts of the numbers.

> ## LEARNING AND TEACHING POINT
>
> Consider introducing children to column addition using some of the methods shown in Figure 9.1 that build on the idea of partitioning the numbers into hundreds, tens and ones, and which do not obscure the meaning of the digits.

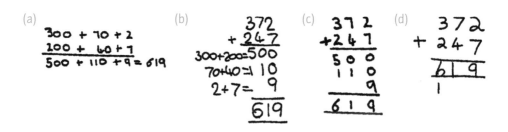

Figure 9.1 Four ways of recording 372 + 247 as a column addition

HOW DO YOU EXPLAIN WHAT'S GOING ON WHEN YOU 'CARRY ONE' IN ADDITION?

The conventional, formal written algorithm (shown in Figure 9.1(d)) is a very condensed and abstract record of a calculation. Teachers might therefore help children to understand what is going on here by making clear links between the written record and the manipulation of some form of base-ten materials that incorporate place-value principles (see Chapter 6). These could be, for example, base-ten Dienes blocks (units, longs and flats) or base-ten counters (white, red and orange) to represent ones, tens and hundreds.

To explain addition, then, I will use orange, red and white place-value counters, which will be referred to as 'hundreds', 'tens' and 'ones' respectively. Some teachers prefer to refer to the 'ones' as 'units'. I have no strong views about this and tend to switch freely between the two words. Clearly, the principle of **exchange**, that ten of 'these' can be exchanged for one of 'those', applies to the ones and the tens, and to the tens and the hundreds. The process could equally well be experienced with base-ten blocks.

So, let's take the example: 356 + 267. This calculation is set out with place-value counters, as shown in Figure 9.2, with 356 interpreted as 3 hundreds, 5 tens and 6 ones, and the 267 interpreted as 2 hundreds, 6 tens and 7 ones. These two addends now have to be combined to find the total. So, where do you start? The standard algorithm usually involves working from right to left, that is, starting with the ones. To some extent, this procedure of working from right to left conflicts with the natural mental strategy of starting with the digits with the greatest value and so working from left to right. When it comes to using the standard addition algorithm, all I can say is that with experience you find that it's easier to be systematic and to avoid getting in a muddle if you work from right to left. But, in fact, it really does not matter as long as you remember and apply correctly the principle that 'ten of these can be exchanged for one of those'.

> **LEARNING AND TEACHING POINT**
>
> To promote mastery of the formal written method of addition, teachers should help their children to understand the process by making strong links between the manipulation of materials, such as base-ten counters and base-ten Dienes blocks, and the written record. If 1p coins are still in circulation then 1p, 10p and £1 coins (or tokens) can also be used in the same way as place-value counters.

> **LEARNING AND TEACHING POINT**
>
> The important language to use when explaining the addition algorithm includes: *hundreds, tens, ones (units), 'ten of these can be exchanged for one of those', carrying one.*

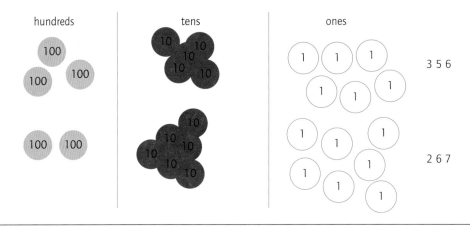

Figure 9.2   The addition 356 + 267 set out with place-value counters

So, in Figure 9.2, we first put together all the ones, making thirteen in all. Ten of these can then be *exchanged* at the 'bank' for a ten. Since this ten is literally 'carried' from the bank and placed in the tens column, I find the language of '**carrying one**' to be helpful and appropriate – provided it is clear that we are carrying 'one of these' ('a ten') and not carrying 'a one'.

The counters at this stage are arranged as shown in Figure 9.3. This also shows the recording so far, in which there is a direct relationship between what is done with the symbols and what has been done with the counters. The 3 written in the ones column corresponds to the three remaining white counters. The 1 written below the line in the tens column corresponds to the one ten that has been carried from the bank in exchange for ten ones.

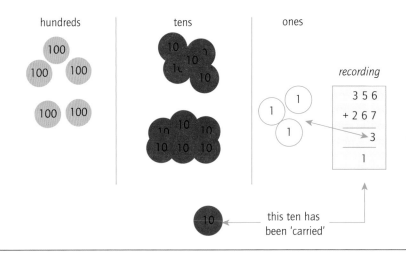

Figure 9.3   Carrying a ten

Next, all the tens (red counters) are combined: that's the 5 tens in the top row, plus the 6 tens in the next row, plus the 1 ten that has been carried. This gives a total of 12 tens. Ten of these are then exchanged for a hundred. So, once again, we are 'carrying one', but this time, of course, it is 'one hundred', an orange counter. Figure 9.4 shows the situation at this stage and, once again, the direct correspondence between the recording in symbols and the manipulation of the place-value counters.

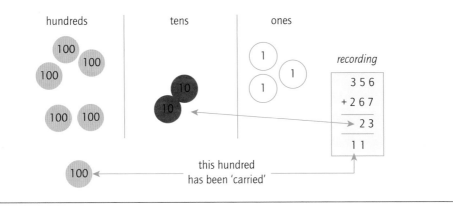

Figure 9.4 Carrying a hundred

The final stage in this calculation is to combine the hundreds: that's the 3 hundreds (orange counters) in the top row, plus the 2 hundreds in the next row, plus the 1 hundred that has been carried, giving a total of 6 hundreds. Figure 9.5 shows the final arrangement of the counters, with the 6 hundreds, 2 tens and 3 ones corresponding to the answer to the addition, namely 623.

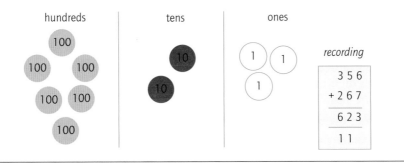

Figure 9.5 The result of adding 356 and 267

When the learner has understood the formal written column method for additions of numbers with up to three digits, it is then possible to build on their understanding and – using the same principles – to extend their mastery of the method to numbers with more digits, as shown in Figure 9.6(a); or to the addition of more than two numbers, as shown in Figure 9.6(b). The key principle is always that whichever column we are working in, ten in one column can be exchanged for one in the next column to the left. When there are more than two numbers being added, as in Figure 9.6(b), note that it is possible to get more than one ten in the total for a column, so we may need to carry two, as in this example, or more.

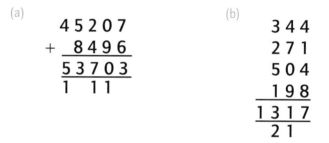

Figure 9.6 Extending the method of column addition

WHAT ABOUT INTRODUCING COLUMN SUBTRACTION?

Subtractions are straightforward when each digit in the first number is greater than the corresponding digit in the second. For example, in calculating 576 – 324 we just take the 3 hundreds away from the 5 hundreds, the 2 tens away from the 7 tens, and the 4 units away from the 6 units, to get the answer 252. The problem comes when one (or more) of the digits in the first number is smaller than the corresponding digit in the second number, for example 448 – 267. Of course, this is not a difficult calculation when tackled by some of the mental strategies outlined in the previous chapter. But, as the numbers get bigger, children will need to develop some kind of standard written procedure for **column subtraction**, lining up the hundreds, tens and ones in columns.

Children can be introduced to the idea of lining up the hundreds, tens and ones in columns for subtraction using the same format as is suggested in Figure 9.1(a) for introducing addition. Figures 9.7(a)–(c) show ways in which some children recorded their own calculations of 448 – 267, all of which start by partitioning the numbers into hundreds, tens and ones. It is instructive to reflect on the understanding shown by these children.

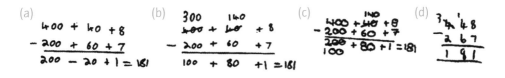

Figure 9.7   Children calculating 448 - 267 using column subtraction

In Figure 9.7(a), the child has worked from left to right, subtracting the hundreds, then the tens and then the ones. The recording of –20 in the tens column does not require a sophisticated understanding of negative numbers. It can be understood simply as meaning that we have taken away 40 of the 60 in the second number, so we still have 20 to be taken away. The final step of calculating 200 – 20 + 1 is done mentally.

In Figure 9.7(b), the child has been encouraged to work from right to left, and has been introduced to the idea of transferring something from the next column when you need to. The thought process is as follows: starting with the ones, 8 – 7 = 1; then on to the tens: 40 take away 60 is a problem; so take 100 from the next column and add this to the 40, making 140; then 140 – 60 = 80; then deal with the hundreds: 300 – 200 = 100. The final calculation, 100 + 80 + 1, is again done mentally.

Figure 9.7(c) shows a novel front-end approach, working from left to right. Having dealt with 400 – 200 = 200, the child then encounters the problem of 40 – 60. To deal with this, 100 is taken from the 200 in the answer in the hundreds column and added on to the 40. This gives 140 – 60 = 80.

The layout of Figure 9.7(b) especially is a helpful introductory procedure prior to the development of the formal algorithm, which is shown in Figure 9.7(d). As with addition, this formal algorithm for subtraction is a highly condensed and abstract form of recording and can become a meaningless routine in which digits are manipulated without any thought as to what they represent. Again, to promote understanding of what is going on here, teachers might discuss with children the corresponding manipulation of some base-ten materials to represent hundreds, tens and ones. This would involve putting out a pile of hundreds, tens and ones to represent the first number, then taking away the second number, exchanging a hundred for ten tens, or a ten for ten ones when necessary. As with addition, children can be helped to connect the manipulation of the materials with the written record, step by step.

Historically, in Britain, there have been essentially two formal written algorithms for subtraction. Nowadays, most primary schools, when they teach children a subtraction algorithm, use the method known as subtraction by **decomposition**, which is the procedure introduced in Figure 9.7(b) and set out in the traditional format in Figure 9.7(d). The other method, taught to people of my age when we were in school, is called **equal additions**. These methods are explained later in this chapter. Although the equal additions method may not be taught in many schools, the reader will find it informative to try to understand it and, as we shall see later, it can be the basis for a novel approach to subtraction calculations. The reason for the widespread adoption of the method of decomposition is that it is much easier to be learnt with *understanding*, in the sense of making connections between the manipulation of concrete materials, the manipulation of the symbols and the corresponding language. Subtraction by equal additions can only really be taught to children by rote, as a procedure to be followed blindly with little real understanding of what is going on. The shift towards decomposition coincided with a greater emphasis in teaching on learning mathematics with understanding.

> **LEARNING AND TEACHING POINT**
>
> When children need a formal algorithm for subtraction, the method of decomposition is preferred because this method can be explained in a way that encourages understanding of the process, which is essential for mastery; it is not learnt just as a recipe without meaning.

SO HOW DOES SUBTRACTION BY DECOMPOSITION WORK?

As with addition calculations, the key to achieving mastery of this method is a sound grasp of place value and the use of some appropriate concrete embodiments of number, such as place-value counters or base-ten Dienes blocks.

I will explain the method of decomposition with base-ten blocks, using the example of 443 – 267. First, the 443 is set out with base-ten blocks, as shown in Figure 9.8: 4 hundreds, 4 tens and 3 units. The task is to take 267 away from this collection of blocks, that is, to remove 2 hundreds, 6 tens and 7 units.

As with addition, the natural place to start might be to take away the biggest blocks first, that is, working from left to right, and this is how most of us would deal with a calculation of this kind if doing it mentally or by informal written methods. But, again, the standard written algorithm actually works from right to left. This is certainly not essential, but it is usually tidier to do it this way.

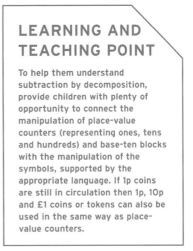

LEARNING AND TEACHING POINT

To help them understand subtraction by decomposition, provide children with plenty of opportunity to connect the manipulation of place-value counters (representing ones, tens and hundreds) and base-ten blocks with the manipulation of the symbols, supported by the appropriate language. If 1p coins are still in circulation then 1p, 10p and £1 coins or tokens can also be used in the same way as place-value counters.

So we start by trying to remove 7 units from the collection of blocks in Figure 9.8. Since there are only 3 units there we cannot do this – yet. So we pick up one of the tens, take it to the box of blocks and exchange it for ten units.

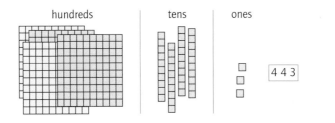

Figure 9.8   The number 443 set out with base-ten blocks

Figure 9.9 shows the situation at this stage and the corresponding recording. Notice how the recording in symbols corresponds precisely to the manipulation of the materials. We have crossed out the 4 tens in the top number and replaced it by 3, because one of these tens has been exchanged for units and we do indeed now have 3 tens in our collection. The little 1 placed beside the 3 units in the top number is to indicate that we now have 13 units. We are now in a position to take away the 7 units as required, leaving 6 units. This is recorded as in Figure 9.10.

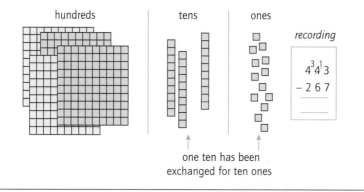

Figure 9.9 Exchanging one ten for ten ones

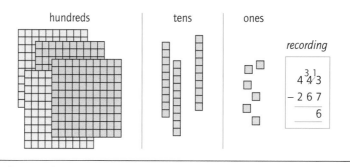

Figure 9.10 After 7 units have been taken away

The next step is to deal with the problem of removing 6 tens when we have only 3 of them. So we take one of the hundreds and exchange it for 10 tens, producing the situation shown in Figure 9.11. The recording indicates that after the exchange process we now have 3 hundreds and 13 tens. We can now complete the subtraction, taking away first the 6 tens and then the 2 hundreds.

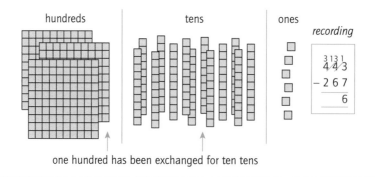

Figure 9.11 Exchanging one hundred for 10 tens

Figure 9.12 shows the final arrangement of the blocks, with the remaining 1 hundred, 7 tens and 6 units corresponding to the result of the subtraction, namely 176.

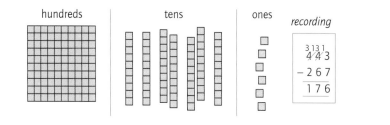

Figure 9.12   The result of subtracting 267 from 443

The method of subtraction by decomposition extends quite easily to subtractions involving numbers with more than 3 digits, using the same principle throughout of exchanging one from the column for ten whenever you need to. Three examples are shown in Figure 9.13. In Figure 9.13(b), where the top number (4004) has 0 hundreds and 0 tens, it is necessary to start by exchanging one of the thousands for 10 hundreds, then one of these for 10 tens, and then one of these for 10 units. Note in Figure 9.13(c), where we are subtracting a three-digit number from a four-digit number (5916 – 824), it might help children to make explicit the absence of any thousands in the number being subtracted by writing in a zero in this place (5916 – 0824). This will also help to ensure they write the digits of this number in the correct columns.

LEARNING AND TEACHING POINT

Encourage children to set out subtraction calculations in vertical format generously, keeping the columns well separated, to give themselves plenty of room for their working.

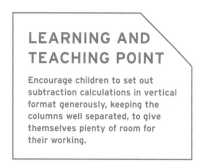

Figure 9.13   Extending subtraction by decomposition to larger numbers

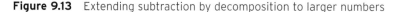

WATCH THE PROBLEM SOLVED! VIDEO AT: HTTPS://STUDY.SAGEPUB.COM/HAYLOCK7E

There are three important points to note about the method of decomposition. First, there is the quite natural idea of exchanging a block in one column for ten in the next column to the right when necessary. Second, there is the strong connection between the manipulation of the materials and the recording in symbols, supported by appropriate language. Third, notice that all the action in the recording takes place in the top line, that is, in the number you are working on, not the number you are subtracting.

LEARNING AND TEACHING POINT

Help children to understand that you only exchange one in the next column for ten when you need to. If the method of decomposition is learnt as a process without understanding, children will start exchanging one in one column for ten in the next, even when it is not required, and in this way get in a terrible muddle.

HOW DOES THE METHOD OF EQUAL ADDITIONS DIFFER FROM DECOMPOSITION?

The method of equal additions differs in all three of these respects. It does not use the principle of exchange, it is not naturally rooted in the manipulation of materials and the method involves working on both numbers simultaneously. Although the method is not often taught these days, it might actually prove to be quite useful for the reader for me to explain it. My explanation will be merely in terms of numbers, without the support of counters or blocks, simply because the method is not easily understood in these terms.

The method is based on the comparison structure of subtraction (see Chapter 7) and uses the principle that the difference between two numbers remains the same if you add the same number to each one. For example, the difference between your height and my height is the same if we stand together on a table or on the floor. Faced with the subtraction, 443 – 267, the person using equal additions would manipulate the symbols as shown in Figure 9.14.

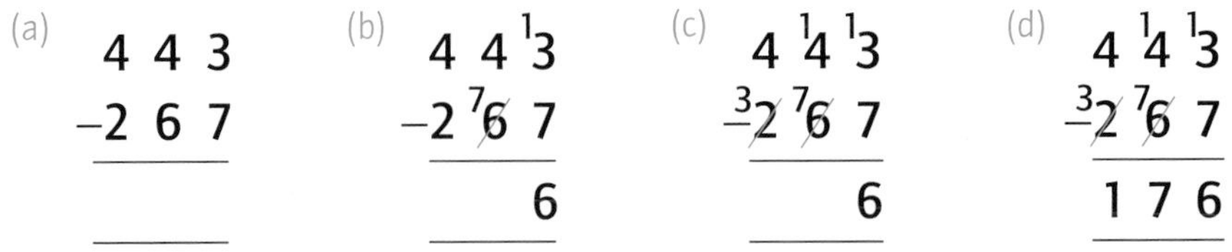

Figure 9.14 The method of equal additions

Unable to deal with '3 take away 7' in Figure 9.14(a), we add 10 to both numbers, as shown in Figure 9.14(b). But we do this in a subtle way. In the top number, this 10 is added to the units digit, increasing the 3 to 13. This is shown by writing a little 1 in front of the 3. In the bottom number, the 10 is added to the tens column, increasing the 6 to 7. I show this by striking through the 6 and writing 7.

In Figure 9.14(b), we can now deal with the units, writing in the 6 in the answer; but then we are faced with the problem of '4 subtract 7' in the tens column. We apply the same principle of adding the same thing to both numbers, but this time we

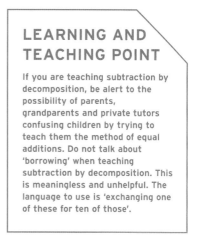

LEARNING AND TEACHING POINT

If you are teaching subtraction by decomposition, be alert to the possibility of parents, grandparents and private tutors confusing children by trying to teach them the method of equal additions. Do not talk about 'borrowing' when teaching subtraction by decomposition. This is meaningless and unhelpful. The language to use is 'exchanging one of these for ten of those'.

actually add 100, as shown in Figure 9.14(c). This appears as 10 tens in the top number, increasing the number of tens from 4 to 14. Simultaneously, in the bottom number we add one to the hundreds digit, increasing it from 2 to 3. We are then able to complete the subtraction, as shown in Figure 9.14(d). The reader may be encouraged to note that the two methods produce the same answer.

WHAT IS 'BORROWING' IN SUBTRACTION?

I really do not know. Back in the last century when I was at school and we did equal additions, we were taught to say, 'Borrow one and pay it back'. I have never understood this phrase, since it is not at all clear from whom we are borrowing or even whether we are paying it back to the same source. I have therefore come to the conclusion that it was merely something to say to yourself, with no meaning, simply as a reminder that there were two things to do. We could equally well have said, 'Pick one up and put one down.' The intriguing thing is, though, that this language of 'borrowing' has actually survived the demise of the method for which it was invented. Some teachers still talk about 'borrowing one' when explaining the method of decomposition to children. I find this unhelpful. We are not 'borrowing one' in decomposition, we are '*exchanging* one of these for ten of those'.

Of course, in practice no one ever thought consciously that what they were doing, for example, was adding ten or a hundred to both numbers: they were simply 'borrowing one and paying it back'. Thus, subtraction by equal additions was always taught by rote, with no real attempt to understand what was going on. This is the real advantage of decomposition: that there is the potential in the method for children to understand it in terms of concrete experiences of place-value counters or blocks, connected to the meaningful language of exchange.

The reader may wonder therefore why everyone did not always teach the method of decomposition. The reason is that, unfortunately, there is a difficulty in using the decomposition method when there is a zero in the top number in any position other than the units column. A modification in the process of decomposition is then required, whereas with equal additions zeros in the top number make no difference to the routine. However, the modification is a natural process, still easily understood if related strongly to concrete materials and the appropriate language of exchange.

WHAT IS THE PROBLEM IN DECOMPOSITION WITH A ZERO IN THE TOP NUMBER?

Figure 9.15 shows the steps involved in tackling 802 – 247 by decomposition. In Figure 9.15(a), the person doing the calculation is faced with the problem of '2 subtract 7'. The decomposition method requires a ten to be exchanged for ten units, but in the 802 the zero indicates that there are no tens. This is the problem! However, it is not difficult to see that the thing to do is to go across to the hundreds column and exchange one of these for 10 tens, as shown in Figure 9.15(b), then to take one of these tens and exchange it for ten units, as shown in Figure 9.15(c). The subtraction can then be completed, as in Figure 9.15(d). Of course, all this can be carried out and understood easily in terms of base-ten blocks or counters, representing hundreds, tens and ones (units).

(a)
```
  8 0 2
– 2 4 7
_____
```

(b)
```
  7
  8 ¹0 2
– 2 4 7
_____
```

(c)
```
  7 9
  8 ¹0 ¹2
– 2 4 7
_____
```

(d)
```
  7 9
  8 ¹0 ¹2
– 2 4 7
_____
  5 5 5
```

Figure 9.15   The problem of a zero in the top number

WATCH THE PROBLEM SOLVED! VIDEO AT: HTTPS://STUDY.SAGEPUB.COM/HAYLOCK7E

However, let me remind you that these subtractions with zeros in the first number, which are potentially problematic when done by the formal decomposition algorithm, are often very straightforward when tackled using mental strategies such as compensation, as explained in Chapter 8. For example, 802 – 247 is a cinch if you start with 802 – 250 and then compensate for the additional 3 taken away. The method of constant differences explained below is an alternative approach that is also particularly effective when there are zeros in the first number.

WHAT IS THE CONSTANT DIFFERENCE METHOD?

Surprisingly, in view of what I have said above, the principle of equal additions is not redundant but is actually an important tool in the collection of strategies that we might use for mental or informal methods of doing subtraction calculations. But it does not have to be just 10 or 100 that you add. For instance, to work out 87 – 48 we could simply add 2 to both numbers and change it to 89 – 50, thus converting it into a much easier calculation. This adhocorithm can almost be developed into an algorithm that some children might find more to their liking than the formal method of decomposition. I call it the **constant difference method**, because as we change the subtraction into easier subtractions, we keep the difference between the numbers constant.

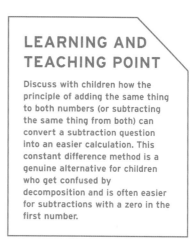

LEARNING AND TEACHING POINT

Discuss with children how the principle of adding the same thing to both numbers (or subtracting the same thing from both) can convert a subtraction question into an easier calculation. This constant difference method is a genuine alternative for children who get confused by decomposition and is often easier for subtractions with a zero in the first number.

So, for example, returning to 802 – 247, we could proceed like this:

The problem is 802 – 247
Add 3 to both numbers: 805 – 250 (that makes it easier)
Add 50 to both numbers: 855 – 300 (that makes it really easy)
So the answer is 555.

This approach can be combined, where appropriate, with subtracting the same thing from both numbers. For example, to calculate 918 – 436:

The problem is 918 – 436 (no problem with the units)
Subtract 6 from both: 912 – 430 (getting easier)
Add 70 to both numbers: 982 – 500 (really easy)
So the answer is 482.

With a bit of practice, this method of convert-ing the second number into a multiple of 10 or a multiple of 100 by adding the same thing to both numbers, or subtracting the same thing from both numbers, can become extremely efficient and just as quick as any other.

HOW IS ADDITION USED TO CHECK A SUBTRACTION CALCULATION?

Because subtraction is the inverse of addition, as was explained in Chapter 7, a subtrac-tion calculation can always be checked by doing an addition. In general, if $a - b = c$, then $c + b = a$. This is illustrated in the number diagram shown in Figure 9.16. Putting this into words, if you add the answer to the second number in a subtraction you should get back to the first number. If you do not, then you have made a mistake in your calculations somewhere. Since the subtraction is likely to be the more difficult calculation, that is where the error probably lies.

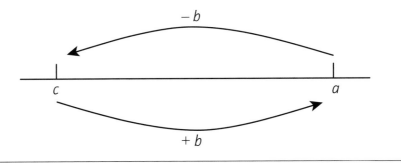

Figure 9.16   If $a - b = c$, then $c + b = a$

For example, if I have calculated $6513 - 3918$ and got the result 2595, then I would check by adding the 2595 to the 3918 and expect to get back to the first number, 6513. In this case, the calculation is shown to be correct: $6513 - 3918 = 2595$ and $3918 + 2595 = 6513$. See Figure 9.17.

Figure 9.17   Using addition to check subtraction

▶ **WATCH THE PROBLEM SOLVED! VIDEO AT: HTTPS://STUDY.SAGEPUB.COM/HAYLOCK7E**

The reader should note how this same principle is used in missing-number questions, such as finding the number in the box in the statement: ☐ – 12 = 13. If the result of subtracting 12 is 13 then the number we started with must be 13 + 12, which equals 25. Similarly, to find the missing number in ☐ + 17 = 25, we would recognize that this is equivalent to 25 – 17 = ☐.

RESEARCH FOCUS: ERRORS AND SUCCESS IN SUBTRACTION ALGORITHMS

What kinds of errors are most common in children's written subtraction calculations? A classic study of the errors made by 2500 children in the USA in the decomposition method for doing subtraction (Brown and Burton, 1978; Burton, 1981) identified the most common 'bugs'. These included: subtracting the smaller digit in a column from the larger, regardless of which one is on the top line; and writing zero in the answer in any column where the digit in the top line is smaller than the one in the second line (for example, '3 take away 5, cannot do it, so put down 0'). All the other most common bugs were in subtractions with a zero in the top number: crossing out the 0, replacing it with 9, but leaving the digit in the next column to the left unchanged; and using incorrect rules for subtracting from zero, for example, 0 – 4 = 4 or 0 – 4 = 0. If zero occurs in the tens column and exchange is needed to deal with the units, then common bugs were: skipping the zero and exchanging one hundred for ten ones; just inserting the required ten units without any exchanging at all; and exchanging from the tens column in the second number. All these can be observed in any primary classroom where the method of decomposition is taught as a meaningless procedure, a recipe to be learnt by rote, without the emphasis on understanding that has been promoted in this chapter.

In an interesting comparison of American (English-speaking) and Chinese children doing calculations, Sun and Zhang (2001) draw attention to the advantage that Chinese children have in the way their words for numbers consistently match the base-ten numeration system – so that, for example, the word for 13 is the equivalent of 'ten-three'. This makes the intermediate steps that turn up in a formal subtraction (such as 13 – 8) much easier, because the 13 is seen more readily as a ten and a three. There is also in the Chinese approach a much greater emphasis on fluency with single-digit mental additions and related subtraction facts before learning a formal subtraction algorithm for larger numbers.

Fiori and Zuccheri (2005) looked at what caused errors in subtraction for two groups of Italian primary school children, those who had been taught the usual algorithm and those who had been taught equal addition (which they refer to as the Austrian Method). They found that for those using the usual subtraction algorithm it was the process of exchange that caused the biggest problems, especially when 'borrowing from 0'. For those using equal addition, zeros were less of a problem; and errors were primarily associated with larger numbers. Errors were categorized according to 4 types: (a) application of the borrowing technique; (b) not borrowing (exchanging) when you should; (c) incorrect recall of numerical facts; (d) oversights lacking mathematical logic (such as adding instead of subtracting). For those children employing the usual algorithm, over four times as many errors were of type (a) than any other type.

LEARNING RESOURCES

Access activities for your **lesson plans** at: https://study.sagepub.com/haylock7e

Before trying the self-assessment questions below, you should complete the **self-assessment questions** for this chapter at: https://study.sagepub.com/haylock7e

9.1: Using the same kind of explanation with counters as that given in this chapter (white, red and orange to represent ones, tens and hundreds), work through the addition of 208, 156 and 97.

9.2: Practise the explanation of the process of subtraction by decomposition using counters and appropriate examples, such as 623 – 471.

9.3: Practise the explanation of the process of decomposition using base-ten blocks to represent units, tens, hundreds and thousands, with examples with zero in the first number, such as 2006 – 438.

9.4: Find the answer to 2006 – 438 using the constant difference method.

FURTHER PRACTICE

FROM THE STUDENT WORKBOOK

Questions 9.01–9.10: Checking understanding (written methods for addition and subtraction)

Questions 9.11–9.17: Reasoning and problem solving (written methods for addition and subtraction)

Questions 9.18–9.26: Learning and teaching (written methods for addition and subtraction)

GLOSSARY OF KEY TERMS INTRODUCED IN CHAPTER 9

Column addition and **column subtraction** Ways of setting out an addition or subtraction calculation in which the ones, tens, hundreds and thousands (and so on) in the numbers in the calculation are arranged in columns.

Exchange In calculations, the replacement of ten ones by a ten, ten tens by a hundred, ten hundreds by a thousand and so on, and vice versa.

Carrying (one) In an addition calculation, the process of replacing ten in one column by one in the column to the left; for example, 10 tens are replaced by 1 hundred, which is then *carried* to the hundreds column.

Decomposition (subtraction) The column method of subtraction which uses the principle of exchange to overcome the difficulty caused when the digit to be subtracted in a particular column is less than the one it is being subtracted from; for example, if there are not sufficient tens in the top number to do the subtraction in that column, one of the hundreds is exchanged for 10 tens.

Equal additions A formal procedure for doing subtraction calculations, based on the idea of adding 10 or 100 or 1000 (and so on) to both numbers, thus keeping the difference the same; the method is no longer taught in British primary schools because decomposition is easier to understand in terms of the manipulation of base-ten materials.

Constant difference method An informal, ad hoc method for doing subtraction calculations, based on the idea that the difference between the two numbers does not change if you add the same number to both or subtract the same number from both.

MULTIPLICATION AND DIVISION STRUCTURES

IN THIS CHAPTER, THERE ARE EXPLANATIONS OF

- two different structures of real-life problems modelled by multiplication;
- the contexts in which children will meet these multiplication structures;
- the commutative law of multiplication;
- the idea of a rectangular array associated with multiplication;
- three different structures of real-life problems modelled by division;
- the contexts in which children will meet these division structures.

READ THIS CHAPTER'S CURRICULUM LINKS AT: HTTPS://STUDY.SAGEPUB.COM/HAYLOCK7E

WHAT ARE THE DIFFERENT KINDS OF MULTIPLICATION SITUATIONS?

This chapter is not about doing multiplication and division calculations, but focuses on understanding the mathematical structures of these operations. Chapters 10–12 follow the pattern used in Chapters 7–9 for addition and subtraction. First, in this chapter, we identify the range of situations that children have to learn to connect with the operations of multiplication and division. So, here we are concerned essentially with step 1 of the modelling process introduced in Chapter 5 (see Figure 5.2): setting up the mathematical model corresponding to a given

situation. Then in the following two chapters we will consider step 2 of the modelling process – the mental and written methods for doing multiplication and division calculations.

We can identify at least two categories of situation that have a structure that corresponds to the mathematical operation represented by the symbol for multiplication. These two structures, which are essentially extensions of the two structures of addition discussed in Chapter 7, are:

- the repeated aggregation structure;
- the scaling structure.

As well as being the basis for their understanding of the operation itself, experience of these different structures associated with multiplication is essential for developing children's confidence in a range of mental and written calculation methods. For example, a key mental strategy for multiplication makes use of doubling, which is based on the scaling structure.

WHAT IS THE REPEATED AGGREGATION STRUCTURE FOR MULTIPLICATION?

Repeated aggregation (or repeated addition) is the elementary idea that multiplication means 'so many sets of' or 'so many lots of'. If I have '10 sets of 3 counters' then the question, 'how many counters altogether?', is associated with the multiplication, 3 × 10 (see Figure 10.1). This structure is simply an extension of the aggregation structure of addition, with, for example, the repeated addition, 3 + 3 + 3 + 3 + 3 + 3 + 3 + 3 + 3 + 3, becoming the multiplication, 3 × 10. The result of this multiplication, 30, is called the **product** of 3 and 10.

Figure 10.1   Multiplication as repeated aggregation, 3 × 10

WHAT IS THE SCALING STRUCTURE FOR MULTIPLICATION?

The **scaling** structure is a rather more difficult idea. It is an extension of the augmentation structure of addition. In that structure, addition means increasing a quantity by a certain amount. With multiplication, we also increase a quantity, but we increase it by a *scale factor*. So, multiplication by 10 would be interpreted in this structure as scaling a quantity by a factor of 10, as illustrated in Figure 10.2.

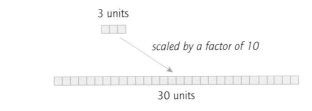

3 units

scaled by a factor of 10

30 units

Figure 10.2 Multiplication as scaling, 3 × 10

It should be mentioned here that multiplication by a number less than 1 would correspond to a scaling that *reduces* the size of the quantity, not increases it. For example, scaling 3 by a factor of 0.5 would reduce it to 1.5, corresponding to the multiplication, 3 × 0.5 = 1.5.

I'M NOT SURE WHETHER 3 × 5 MEANS '3 SETS OF 5' OR '5 SETS OF 3'

Which of the pictures in Figure 10.3 would we connect with 3 × 5? If I were to be really pedantic, I suppose I would have to say that, strictly speaking, 3 × 5 means '5 sets of 3'. You start with 3, and multiply it by 5. That is, you reproduce the three, five times in all, as illustrated in Figure 10.3(a). But I would prefer to let the meaning of the symbol be determined by how it is used. It seems to me that, in practice, people use the symbols 3 × 5 to mean both '3 sets of 5' and '5 sets of 3' – in other words, both the examples shown in Figures 10.3(a) and 10.3(b). I am happy therefore to let 3 × 5 refer to both (a) and (b). And the same goes for 5 × 3. One symbol having more than one meaning is something we have to learn to live with in mathematics – as well as being a feature that makes mathematical symbols so powerful in their application.

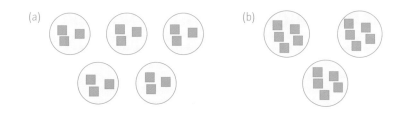

Figure 10.3   (a) 5 sets of 3; (b) 3 sets of 5

The underlying mathematical principle here is what is called the **commutative law of multiplication**. This refers to the fact that when you are multiplying two numbers together, the order in which you write them down does not make any difference. We have already seen in Chapter 7 that addition also has this property of commutativity. We recognize this commutative property formally by the following generalization, which is true whatever the numbers a and b: $b \times a = a \times b$.

There are two important points to note about the commutative property in relation to multiplication. First, it is important to realize that division does *not* have this property. For example, 10 ÷ 5 is not equal to 5 ÷ 10. Second, use of the commutative property enables us to simplify some calculations. For example, many of us would evaluate '5 lots of 14' by changing the question to the equivalent, '14 lots of 5' – because fives are much easier to handle than fourteens. Grasping the principle of commutativity also cuts down significantly the number of different results we have to memorize from the multiplication tables: for example, if I know the 9-times table, because it has such a strong pattern built into it, then I also know nine 7s (the same as seven 9s), nine 8s (the same as eight 9s) and so on.

> **LEARNING AND TEACHING POINT**
>
> There is no need to make a fuss about whether $a \times b$ means 'a lots of b things' or 'b lots of a things'. Instead, work hard with children to establish as soon as possible the commutative principle in multiplication: that $a \times b$ and $b \times a$ are equal.

> **LEARNING AND TEACHING POINT**
>
> Teach children how to use the commutative principle to help in mastering the multiplication tables. For example, if they know seven 5s then they know five 7s.

IS THERE A PICTURE THAT CAN USEFULLY BE CONNECTED WITH THE MULTIPLICATION SYMBOL?

It is interesting to note, first, that the commutative property of multiplication is by no means obvious. Other than by counting the numbers in each picture, we would not immediately recognize that (a) and (b) in Figure 10.3 have the same number of counters. So this picture is not especially helpful. But there is one very significant picture of multiplication that does make this commutative property obvious. This is the association of multiplication with the image of a **rectangular array**. Figure 10.4 shows some examples of rectangular arrays that correspond to 3 × 5 (or 5 × 3). This is the image of multiplication that we should carry round in our heads, particularly when we want to talk to children about multiplication and to illustrate our discussions with diagrams. This picture really does make the commutative property transparently obvious. We can actually see that 3 sets of 5 and 5 sets of 3 come to the same thing, because the array can be thought of as 3 columns of 5 or 5 rows of 3.

Figure 10.4   Examples of rectangular arrays for 3 × 5

There are other good reasons for strongly associating this image of a rectangular array with multiplication. For example, this idea leads on naturally to the use of multiplication for determining the area of a rectangle. In example (c) in Figure 10.4, 3 × 5 gives the number of square units in the rectangle and therefore determines its area. We can extend this idea to develop an effective method for multiplying together numbers with two or three digits (see Chapter 12).

APART FROM 'SO MANY SETS OF SO MANY', ARE THERE OTHER CONTEXTS IN WHICH CHILDREN MEET MULTIPLICATION IN THE REPEATED AGGREGATION STRUCTURE?

The repeated aggregation structure of multiplication applies to what are sometimes referred to as 'correspondence problems', in which *n* objects are connected to each of *m* objects. For example, if each of 28 children in a class requires 6 exercise books, then the total number of exercise books required is 28 × 6. Similarly, if there are 8 coaches on a train and each coach can seat 48 people, then the total number of seats available is 8 × 48. Note particularly the use of the word *each* in these examples; it is an important word in expressing mathematical statements of this kind unambiguously.

Similarly, any situation in which we aggregate a certain number of portions of a given quantity – such as mass, liquid volume, length and time – provides an application of this multiplication structure: for example, finding the total mileage for 42 journeys of 38 miles each (42 × 38); finding the total volume of drink required to fill 32 glasses if each holds 225 ml (32 × 225); finding the total time required for 12 events each lasting 25 minutes (12 × 25).

But, not surprisingly, an important context for multiplication will be shopping, particularly where we have to find the cost of a number of items given the *unit cost*. This again involves the concept of 'each'. For example, we might need to find the cost of 25 cans of drink at 39p *each* (25 × 39). Another important word here is **per**. For example, we might purchase 25 tickets at £3.50 *per* ticket (25 × 3.50). In both cases, we have to associate the

> **LEARNING AND TEACHING POINT**
>
> Use rectangular arrays frequently to illustrate and to support your explanations about multiplication, particularly for reinforcing the commutative principle. Get the children to draw an array (say 3 rows of 5) on a white board, and then to rotate the whiteboard through ninety degrees (to become 5 rows of 3). Figure 10.4(c) is a particularly important diagram for 3 multiplied by 5.

> **LEARNING AND TEACHING POINT**
>
> Help children to use the important words *each* and *per* with confidence and accuracy in describing multiplication situations. Provide plenty of experience in associating the practical problems about *unit cost* and *cost per unit of measurement* with the corresponding multiplications.

language and the structure of the example with the operation of multiplication. Then there are important situations where we encounter repeated aggregation in the context of *cost per unit of measurement*. For example, if we purchase 28 litres of petrol at a cost of £1.75 per litre, we should recognize that a multiplication (28 × 1.75) is required to determine the total cost, although in practice the petrol pump will do it for us. (For multiplication with decimal numbers, see Chapter 17.) Likewise, we should connect multiplication with situations such as finding the cost of so many metres of material given the cost per metre, or someone's earnings for so many hours of work given the rate of pay per hour and so on.

When multiplication is used in real situations like any of those above, the number or quantity being multiplied is sometimes called the **multiplicand** and the number it is being multiplied by is called the **multiplier**. For example, in 'Spinach costs 65p a bag, how much for 3 bags?' the 65 is the multiplicand and the 3 is the multiplier.

WHAT ARE SOME OF THE CONTEXTS IN WHICH CHILDREN WILL MEET MULTIPLICATION IN THE SCALING STRUCTURE?

Most obviously, this structure is associated with scale models and scale drawings. For example, if a scale model is built using a scale factor of 1 to 10, then each linear measurement in the actual object is 10 times the corresponding measurement in the model. Similarly, if we have made a plan of the classroom using a scale factor of 1 to 100 and the width of the whiteboard in the drawing is 2 cm, then the width of the actual whiteboard will be 200 cm (2 × 100).

This is also the multiplication structure that lies behind the idea of a **pro rata increase**. For example, if we all get a 13% increase in our salary, then all our salaries get multiplied by the same scale factor, namely 1.13 (see Chapter 18 for an explanation of percentage increases). The simplest experience of this structure is when we talk about *doubling* or *trebling* a given quantity: this is simply increasing the given quantity by applying the scale factors of 2 or 3 respectively.

Then we also sometimes use this multiplication structure to express a comparison between two numbers or amounts, where we make statements using phrases such as 'so many times as much (or as many)' or 'so many times bigger (longer, heavier and so on)'. For example, 225 × 3 would be the calculation corresponding to this situation: John earns £225 a week, but his brother earns three times as much; how much does his brother earn per week? The phrase *three times as much* has to be connected with multiplication by 3.

> **LEARNING AND TEACHING POINT**
>
> Division is not just 'sharing'. 'Equal sharing between' is only one of the structures for division. Teachers should not overemphasize the language and imagery of *sharing* at the expense of the other important language and imagery that are associated with division, particularly division as the inverse of multiplication.

WHAT ARE THE DIFFERENT KINDS OF DIVISION SITUATIONS?

There is a wide range of situations in which we have to learn to recognize that the appropriate operation is division. These situations can be categorized into at least the following three structures:

- the equal-sharing-between structure;
- the inverse-of-multiplication structure;
- the ratio structure.

As with the other operations of addition, subtraction and multiplication, it can help us to connect these mathematical structures with the operation of division if we ask ourselves the question: 'What is the calculation I would enter on a calculator in order to solve this problem?' In each case, the answer will involve using the division key on the calculator. As we saw earlier, particularly with subtraction in Chapter 7, one of the difficulties in understanding the meanings of the symbols we use in mathematics is that one symbol can have a number of strikingly different meanings. This is certainly the case with the division symbol.

> **LEARNING AND TEACHING POINT**
>
> The key phrases to use in problems with the equal-sharing-between structure of division are *shared equally between* and *how many (or how much) each?*

WHAT IS THE EQUAL-SHARING-BETWEEN STRUCTURE FOR DIVISION?

The **equal-sharing-between** structure refers to a situation in which a quantity is shared out equally into a given number of portions and we are asked to determine how many or how much there is in each portion. For example, 20 marbles might be *shared equally between* 4 children in a game, as shown in Figure 10.5. The calculation to be entered on a calculator to correspond to this situation is 20 ÷ 4. This is the structure that teachers most naturally connect with division and which is strongly associated with the language of 'sharing' and 'how many (or how much) each?'

> **LEARNING AND TEACHING POINT**
>
> Young children will benefit from plenty of practical experience of 'equal sharing between' and sorting sets of objects into equal subsets, and the associated informal language, before they meet the formal language and the symbol for division.

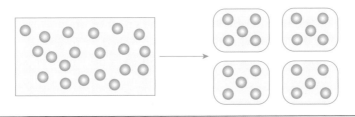

Figure 10.5   Division as equal sharing between (20 ÷ 4)

WHAT IS THE INVERSE-OF-MULTIPLICATION STRUCTURE FOR DIVISION?

The **inverse-of-multiplication** structure of division interprets 20 ÷ 4 in a completely different way, as shown in Figure 10.6. Now the question being asked is: 'How many groups of 4 marbles are there in the set of 20 marbles?'

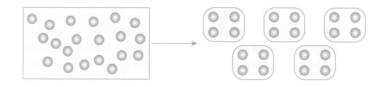

Figure 10.6   Division as the inverse of multiplication (20 ÷ 4)

Figures 10.5 and 10.6 are equally valid interpretations of the division, 20 ÷ 4, even though they are answering two different questions:

Figure 10.5: share 20 equally into 4 groups. How many in each group? (equal sharing between)

Figure 10.6: sort 20 into groups of 4. How many groups? (inverse of multiplication)

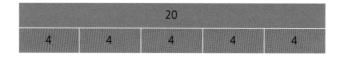

Figure 10.7   'How many 4s in 20?' illustrated with bar-modelling

Figure 10.7 shows a bar-modelling diagram for illustrating how many groups of 4 make 20. This provides a very clear image of how the five 4s are equivalent to the 20. The top bar representing 20 might be set out as two tens.

The phrase 'inverse of multiplication' underlines the idea that division and multiplication are inverse processes – just as addition and subtraction were seen to be inverse processes in Chapter 7. This means, for example, that since 6 × 9 is equal to 54, then 54 ÷ 9 must be 6. The division by 9 'undoes' the effect of multiplying by 9. Hence, to solve a problem of the form 'what must *A* be multiplied by to give *B*?' we divide *B* by *A*. For example, how many tickets costing £1.50 each do I need to sell to raise £90? The calculation that would be entered on a calculator to solve this problem is the division, 90 ÷ 1.50.

The actual problems that occur in practice which have this inverse-of-multiplication structure can be further subdivided. First, there are problems that incorporate the notion of **repeated subtraction** from a given quantity, such as 'how many sets of 4 can I get from a set of 20?' So the process of sharing out the 20 marbles in the above example can be thought of as *repeatedly subtracting* sets of 4 marbles from the set of 20 until there are none left, counting the number of sets as you do this.

Second, there are those problems that incorporate the idea of **repeated addition** to reach a target, such as 'how many sets of 4 do you need to get a set of 20?'. For example, the question, 'how many groups of 4 marbles are there in a set of 20 marbles?' could mean, in practical terms, *repeatedly adding* sets of 4 marbles until the target of 20 is achieved, counting the number of sets required as you do this.

LEARNING AND TEACHING POINT

The image of a rectangular array, as used in representing multiplication, is very helpful in connecting the equal sharing and the inverse-of-multiplication structures of division. For example, the array in Figure 10.4(c) can be used to show 15 ÷ 3 as both '15 squares are shared equally between 3 columns, how many in each column?' and 'how many rows of 3 squares in this rectangular array of 15 squares?'

WHAT IS THE RATIO STRUCTURE FOR DIVISION?

The **ratio** structure for division refers to situations where we use division to compare two quantities. In Chapter 7, we saw that in a situation where a comparison has to be made between two numbers, two sums of money or two measurements of some kind, one way to make the comparison is by subtraction, focusing on the *difference* between the quantities. For example, if A earns £300 a week and B earns £900 a week, one way of comparing them is to state that B earns £600 more than A, or A earns £600 less than B. The 600 is the result of the subtraction, 900 − 300. But we could also compare A's and B's earnings by looking at their *ratio*, stating, for example, that B earns *three times more* than A. The three here is now the result of the division, 900 ÷ 300.

This process is simply the inverse of the scaling structure of multiplication described above, since

LEARNING AND TEACHING POINT

Division is the operation needed to compare two quantities by their ratio. The key language in introducing this division structure is *how many times greater (larger, longer, higher, taller, heavier ...)?* For example, children might compare how much school time they spend in mathematics lessons in a week (say, 200 minutes) with the time spent doing physical activity (say, 50 minutes) and calculate that mathematics is 4 times longer than physical activity.

what we are doing here is finding the scale factor by which one quantity must be increased in order to match the other. In the above example, the question would be, by what factor must 300 be multiplied to give 900? Or, how many times bigger than 300 is 900? Here is another example. The longest word in this paragraph (multiplication) has 14 letters; one of the shortest words (by) has 2 letters. Comparing them by ratio, we might say that 'multiplication' has 7 times more letters than 'by'. This corresponds to $14 \div 2 = 7$. The underlying idea here is the sophisticated notion of ratio: see the discussion of rational numbers in Chapter 6 and further explanation of ratio in Chapter 15.

WHAT ABOUT REMAINDERS?

The reader may have noticed that all the examples of division used in this chapter so far have had whole-number answers and no **remainders**. In practice, most division questions do not work out nicely like this, of course. For example, a teacher may want to put the 32 children in her class into six equal groups for some activity. The mathematical model is '$32 \div 6 = 5$, remainder 2'. This means that sharing 32 equally into 6 groups gives 5 in each group, with 2 remaining not in a group. Alternatively, using the inverse-of-multiplication structure, the teacher's plan might be to put the 32 children into groups of 6. The mathematical model is again '$32 \div 6 = 5$, remainder 2', but now the interpretation is that there can be 5 groups of 6 children, with once again 2 children not in a group. These examples also highlight the importance of the *context* for a calculation; a teacher is unlikely to consider leaving two children out as a professional solution to these problems.

If a division calculation works out exactly without a remainder, then we say that the dividend is **divisible** by the divisor. For example, because '$32 \div 8$' is exactly 4, with no remainder, we say that 32 is divisible by 8.

Remainders do not occur where division is used to model situations with the ratio structure. For example, if we were calculating how many times greater than a journey of 6 km is a journey of 32 km, the division calculation is once again $32 \div 6$, but the solution '5, remainder 2' would not make any sense in this context. The solution would have to be something like 'about 5.33 times greater' (rounded to two decimal places). In practice, this aspect of division often involves a decimal answer to the division calculation (see Chapter 17). For example, to compare something costing £125 with another thing costing £130, we might do the division $130 \div 125 = 1.04$ and make the observation that the second is '1.04 times more expensive than the first'.

WHAT ARE SOME OF THE CONTEXTS IN WHICH CHILDREN WILL MEET DIVISION IN THE EQUAL-SHARING-BETWEEN STRUCTURE?

At first sight, we might think that sharing is a very familiar experience for children: sharing sweets, sharing pencils, sharing books, sharing toys and so on. But this idea of sharing a set into subsets corresponds to division only under certain conditions. First, the set must be shared into *equal* subsets, which is certainly not always the case in children's experience of sharing. Second, it is important to note that the language is *sharing between* rather than *sharing with*. Children's normal experience is to share sets of things *with* a number of friends. Division requires sharing between a number of people. The division 12 ÷ 3 does not correspond to 'I have 12 marbles and I share them *with* my 3 friends.' The situation required is: 'Share 12 marbles equally *between* 3 people.' This is a somewhat artificial process and may not be encountered as often in the children's experience as we might imagine. So, sharing does not always correspond to division: it must be not just 'sharing', nor just 'sharing equally', but 'sharing equally between'.

In the context of measurement, it is not difficult to come up with imaginary situations where we might share a given quantity into a number of equal portions. Cutting up a 750-cm length of wood into 6 equal lengths, or pouring out 750 ml of wine equally into 6 glasses, or sharing out 750 g of chocolate equally between 6 children, for example, all correspond to the division 750 ÷ 6. But these do feel like situations contrived for mathematics lessons rather than genuine problems (see Boaler, 2020, chapter 5, 'Learning without Reality').

> **LEARNING AND TEACHING POINT**
>
> As far as possible, avoid talking about 'sharing' when explaining division. Instead, talk about 'sharing equally between'. Each word in this phrase is crucial to children's understanding of the equal-sharing-between structure.

> **LEARNING AND TEACHING POINT**
>
> Children should experience all the division structures in a range of practical and relevant contexts, including especially shopping, rates of pay and the many kinds of problems associated with the word *per*, such as *price per unit*.

The context of money does, however, provide some of the more natural examples in real life for this structure of the division operation. For example, a group of people might share a prize in a lottery, or share a bill in a restaurant: in both cases, it is likely that we would *share equally between* the people in the group. An important class of everyday situations is where items are sold in multiple packs: a familiar requirement is to want to know the price per item. So, for example, if a shop is selling a pack of 8 nectarines for £3.20, the cost per nectarine, in pence, is found by the division $320 \div 8$ (= 40). We can think to ourselves that a set of 320 pence is being shared equally between the 8 nectarines, so that each one is allocated 40p.

This then extends naturally to the idea of price per unit of measurement. For example, to find the cost per pint of a 6-pint bottle of milk costing 240p, we should recognize that this situation corresponds to the division, $240 \div 6$. It is as though a set of 240 pence is being shared out equally between the 6 pints, giving 40p for each pint. Once again, the word 'per', meaning 'for each', plays an important part in our understanding of this kind of situation.

This idea of 'per' turns up in numerous other situations in everyday life: for example, when calculating how many or how much we get per pound (£), when finding miles per litre, when determining how much someone earns per hour, an average speed in miles per hour, the number of words typed per minute and so on: all these situations correspond to the operation of division using the equal-sharing-between structure.

WHAT ARE SOME OF THE CONTEXTS IN WHICH CHILDREN WILL MEET DIVISION IN THE INVERSE-OF-MULTIPLICATION STRUCTURE?

There are many practical situations in which a set is to be sorted into subsets of a given size and the question to be answered is, 'how many subsets are there?' For example, the head of a school with 240 children may wish to organize them into classes of 30 children. How many classes do we need? This is modelled by $240 \div 30$. In other words, how many 30s make 240? Then the teacher with a class of 30 children may wish to organize them into groups of 5 children and asks: how many groups? This is modelled by $30 \div 5$. In other words, how many 5s in 30?

Once again, the structure extends quite naturally into the context of money. A familiar question is: How many of these can I afford? This kind of question incorporates the idea of *repeated subtraction from a given quantity*. For example, how many items costing £6 each can I buy with £150? The question is basically 'how many 6s can I get out of 150?' We could imagine repeatedly spending (subtracting) £6 until all the £150 is used up. Similar situations occur in the context of measurement. For instance, the question, 'how many 150-ml servings of wine from a 750-ml bottle?' is an example of the inverse-of-multiplication structure of division, in the context of liquid volume and capacity, corresponding to the calculation, 750 ÷ 150. In other words, how many 150s make 750? Again, the notion of repeated subtraction from a given quantity is evident here. We can imagine repeatedly pouring out (subtracting) 150-ml servings, until the 750 ml is used up.

Then there are problems in the context of money and measurement that ask the question: How many do we need? This kind of question incorporates the idea of *repeated addition to reach a target*. For example, how many items priced at £6 each must I sell to raise £150? We could imagine repeatedly adding £6 to our takings until we reach the target of £150. In spite of the language used, the problem is modelled by the division, 150 ÷ 6.

The word 'per' turns up again in this division structure. For example, if we know that the price per kg of potatoes is £1.25 (125p), then we might find ourselves asking a question like, 'how many kilograms can I get for £10 (1000p)?' Similarly, if I save £1.25 (125p) per week, I might ask the question, 'how long will it take me to save £10?' Or, if the price of petrol is 125p per litre, the question might be, 'how many litres can I get for £10?' Each of these is, of course, an instance of 'how many 125s make 1000?', so they are again examples of the inverse-of-multiplication structure, corresponding in these cases to the division, 1000 ÷ 125.

Exactly the same mathematical structure occurs in finding the time for a journey given the average speed. For example, the question, 'how long will it take me to drive 1000 miles, if I average 50 miles per hour?', is equivalent to the question, 'how many 50s make 1000?', and, hence, using the inverse-of-multiplication structure, to the division, 1000 ÷ 50.

WHAT ABOUT SITUATIONS USING THE RATIO DIVISION STRUCTURE?

Many primary school children can learn to recognize the need to use division to compare two quantities by ratio. Situations where comparisons could be made between numbers in sets, between amounts of money or between measurements of various kinds, are readily available. The problem is, however, that, in practice, unless the questions are contrived carefully, the answers tend to be quite difficult to interpret. It's easy enough to deal with, say, comparing two children's journeys to school of 10 minutes and 30 minutes, and, using division (30 ÷ 10 = 3), making the statement that one child's journey is three times longer than the other. But it's a huge step from interpreting a statement like that with whole numbers to making sense of, say, comparing the heights of two children, at 125 cm and 145 cm, using a calculator to do the division (145 ÷ 125 = 1.16) and concluding that one child is 1.16 times taller than the other.

> ### LEARNING AND TEACHING POINT
>
> Primary children can be introduced to the idea of using division to find the ratio between two quantities in order to compare them, but the results can be difficult to interpret if they are not whole numbers.

RESEARCH FOCUS: WORD PROBLEMS, MULTIPLICATION AND DIVISION

There have been many studies of children's abilities to interpret word problems with multiplication and division structures. A helpful summary is provided by Verschaffel, Greer and De Corte (2007: 582–9). In one study, for example, De Corte, Verschaffel and Van Coillie (1988) investigated how well children aged 10 to 11 years could recognize multiplication structures in a series of word problems, involving whole numbers and decimals less than 1. They found that children were very successful (around 98%) in recognizing the operation required to be multiplication when the multiplier was a whole number, regardless of whether the multiplicand was a whole number or a decimal less than 1. But the facility dropped dramatically (to 32%) when the multiplicand was a whole number but the multiplier a decimal less than 1 (for example, 'Milk costs 90p a litre, how much for 0.8 litre?'). The explanation for this is twofold. First, the idea of multiplication as repeated addition does not fit easily with a multiplier less than 1. Second, children seem to have the (mistaken) idea that multiplication always makes things bigger, whereas in this case the product is smaller than the multiplicand.

The word problems children have to deal with are generally generated by teachers, so an important question is to what extent teachers themselves can construct such problems with various structures and in different contexts. Rizvi (2004) investigated the ability of prospective teachers in an Australian university to construct word problems requiring division. Prior to instruction, the correct word problems they constructed all used an equal-sharing structure and none were set in measurement contexts involving non-countable quantities (such as litres and kilograms). Even after instruction, the prospective teachers used only a very limited range from the structures and contexts for division outlined in this chapter. See also the Research Focus for Chapter 16.

⊏⊐ LEARNING RESOURCES

Access activities for your **lesson plans** at: https://study.sagepub.com/haylock7e

Before trying the self-assessment questions below, you should complete the **self-assessment questions** for this chapter at: https://study.sagepub.com/haylock7e

10.1: Give two problems associated with the multiplication 29 × 12, one using the idea of '29 lots of 12' and the other using '12 lots of 29'.

10.2: Make up a problem using the repeated aggregation structure and the word *per*, in the context of shopping, corresponding to the multiplication, 12 × 25.

10.3: A box of yoghurts consists of 4 rows of 6 cartons arranged in a rectangular array. How would you use this as an example to illustrate the commutative property of multiplication?

10.4: A model of an aeroplane is built on a scale of 1 to 25. Make up a question about the model and the actual aeroplane, using the scaling structure.

10.5: The real car is 300 cm long and the model car is 15 cm long: how many times longer is the real car? (That is, what is the scale factor?) What is the calculation to be entered on a calculator to answer this? Of what division structure is this question an example?

10.6: How many months do I need to save up £300 if I save £12 a month? What is the calculation to be entered on a calculator to answer this? Of what division structure is this question an example?

10.7: Make up a problem that corresponds to the division, 60 ÷ 4, using the equal-sharing structure in the context of shopping.

10.8: Make up a problem that corresponds to the division, 60 ÷ 4, using the inverse-of-multiplication structure in the context of shopping.

10.9: Make up a problem using the ratio structure in the context of salaries. Use a calculator to answer your own problem.

FURTHER PRACTICE

Access the website material for Knowledge Check 3: The commutative laws at:
https://study.sagepub.com/haylock7e

FROM THE STUDENT WORKBOOK

Questions 10.01–10.09: Checking understanding (multiplication and division structures)

Questions 10.10–10.24: Reasoning and problem solving (multiplication and division structures)

Questions 10.25–10.33: Learning and teaching (multiplication and division structures)

GLOSSARY OF KEY TERMS INTRODUCED IN CHAPTER 10

Repeated aggregation   The process modelled by multiplication related to the idea of 'so many sets of so many'; also called repeated addition.

Product   The result of a multiplication; for example, the product of 37 and 27 is 999.

Scaling of quantity   The process modelled by multiplication in which a given quantity is increased by a scale factor; doubling and trebling are examples of scaling by factors of 2 and 3, respectively. Scaling by a factor less than 1 (for example, halving) reduces the size of the quantity.

Commutative law of multiplication   The principle that the order of two numbers in a multiplication calculation makes no difference; for example, $5 \times 7 = 7 \times 5$. In symbols, the commutative law of multiplication states that, whatever the numbers a and b, $a \times b = b \times a$.

Rectangular array   A set of objects or shapes arranged in rows and columns, in the shape of a rectangle; for example, 7 rows of 5 counters, or a 7 by 5 grid of squares. Rectangular arrays are important images to be associated with multiplication.

Per   An important word in many multiplication and division situations, meaning 'for each'; used, for example, in problems about cost per unit of measurement.

Multiplicand   A number or quantity that is to be multiplied.

Multiplier   The number by which a multiplicand is multiplied.

Pro rata increase   An increase applied, for example, to salaries or prices, in which each amount is increased by the same scale factor.

Equal sharing between   The process modelled by division in which a set of items or a given quantity is shared equally between a number of individuals; the key phrase in this structure of division is 'shared equally between'.

Inverse of multiplication   The process modelled by division in which the question is 'how many groups of a given number are there in a given set?' For example, 'how many 4s make 20?' corresponds to $20 \div 4$.

Repeated subtraction (division)   One of the ways of experiencing the inverse-of-multiplication structure of division by repeatedly subtracting a quantity from a given amount; for example, 'how many times can £6 be taken away from £24 until there is nothing left?' is connected with $24 \div 6$.

Repeated addition (division)   One of the ways of experiencing the inverse-of-multiplication structure of division by repeatedly adding a quantity to reach a given target; for example, 'how many payments of £6 are required to make a total of £24?' is connected with $24 \div 6$.

Ratio   The inverse of the scaling structure of multiplication, where division is used to compare two quantities; for example, the ratio of £36 to £12 is 3; this is represented by the division $36 \div 12 = 3$; the result tells you that £36 is 3 times £12.

Remainder   In a division situation that does not work out exactly, the surplus number after an equal sharing or grouping has been completed; for example, $45 \div 7 = 6$, remainder 3. This could mean 45 shared equally between 7 gives 6 each, with 3 not shared out; or it could mean that 6 subsets of 7 can be made from a set of 42, with 3 not in a subset.

Divisible   If a number a can be divided exactly by a number b, without a remainder, then a is said to be divisible by b. For example, 21 is divisible by 7.

MENTAL STRATEGIES FOR MULTIPLICATION AND DIVISION

IN THIS CHAPTER, THERE ARE EXPLANATIONS OF

- the associative and distributive laws of multiplication;
- the distributive laws of division;
- quotient, dividend and divisor;
- how these laws are used in multiplication and division calculation strategies;
- some prerequisite skills for being efficient in mental multiplication and division calculations;
- how factors can be used to simplify multiplications;
- how doubling can be used as an ad hoc approach to multiplication;
- the use of ad hoc additions and subtractions in multiplication and division;
- the constant ratio method for a division calculation.

READ THIS CHAPTER'S CURRICULUM LINKS AT: HTTPS://STUDY.SAGEPUB.COM/HAYLOCK7E

WHAT ARE THE ASSOCIATIVE AND DISTRIBUTIVE LAWS OF MULTIPLICATION?

It should be noted that my aim in this chapter is to increase the reader's own facility in informal multiplications and divisions, as well as to explain how some of the fundamental laws for combining numbers can be exploited to simplify some challenging mental and informal calculations. We have already met the commutative laws, first in Chapter 7, where commutativity was explained as a fundamental property of addition, and then in Chapter 10, where the corresponding commutative law of multiplication was explained in detail (see Figure 10.4). In Chapter 8, the associative law of addition was explained. We now meet the **associative law of multiplication**. There are also two further important laws of multiplication, called the **distributive laws of multiplication**.

For completeness, here are all four of the fundamental laws of multiplication, written formally as algebraic generalizations:

Commutative law of multiplication:	$a \times b = b \times a$
Associative law of multiplication:	$(a \times b) \times c = a \times (b \times c)$
Distributive law, multiplication over addition:	$(a + b) \times c = (a \times c) + (b \times c)$
Distributive law of multiplication over subtraction:	$(a - b) \times c = (a \times c) - (b \times c)$

These are true whatever numbers are chosen for a, b and c.

Written down baldly, as they are above, the four fundamental laws of multiplication look a bit daunting and obscure. But, rather like Monsieur Jourdain in *Le Bourgeois Gentilhomme*, who discovered to his delight that he had been speaking prose for more than 40 years without knowing it, readers can be assured that they probably use one or another of these laws unconsciously every time they undertake a multiplication calculation.

HOW ARE THESE LAWS USED IN MULTIPLICATION CALCULATIONS?

The commutative law allows you to choose which of the two numbers in a multiplication question should be the multiplicand and which the multiplier (see definitions in the glossary at the end of Chapter 10). Take as an example the calculation of 5×28. First, I might prefer to think of this as 28 fives, rather than five 28s, simply because I am better at my 5-times table than I am at my 28-times table. It is the commutative law that allows me to switch the order of two numbers in a multiplication freely like this: $5 \times 28 = 28 \times 5$.

Now to work out 28 × 5, I could think of the 28 as 14 × 2, choose to do 2 × 5 first (to get 10) and then multiply this by 14 (that is, 14 × 10 = 140). What I am using here is the associative law of multiplication. I am 'associating' the 2 with the 5, rather than with the 14, in order to make the calculation easier: (14 × 2) × 5 = 14 × (2 × 5).

An alternative approach to calculating 28 × 5 would be to split the 28 into 20 + 8 and then to multiply the 20 and 8 separately by the 5, to get 100 and 40, which add up to 140. This is using the distributive law. The multiplication by 5 is being 'distributed' across the addition of 20 and 8. Written down formally, the first step of this strategy might look like this:

$$(20 + 8) \times 5 = (20 \times 5) + (8 \times 5)$$

Finally, we could have chosen to think of the 28 as 30 – 2 and then to 'distribute' the multiplication by 5 across this subtraction: 30 × 5 is 150, 2 × 5 is 10, so the answer is 150 – 10, which is 140. This is using the second of the distributive laws of multiplication:

$$(30 - 2) \times 5 = (30 \times 5) - (2 \times 5)$$

The commutative and associative laws of multiplication together give us the freedom to rearrange two or more numbers in a multiplication in any order we wish. For example:

$$
\begin{aligned}
(5 \times 3) \times 8 &= 5 \times (3 \times 8) && \text{(using the associative law of multiplication)} \\
&= 5 \times (8 \times 3) && \text{(using the commutative law of multiplication)} \\
&= (5 \times 8) \times 3 && \text{(using the associative law of multiplication)} \\
&= (8 \times 5) \times 3 && \text{(using the commutative law of multiplication)} \\
&= 40 \times 3 = 120
\end{aligned}
$$

This means that we can write down 5 × 3 × 8 without any brackets, recognizing that we can multiply the numbers together in any order we like.

The distributive laws give us the option to deal with a complicated multiplication in easy stages, breaking up the numbers into easier components. We shall see below that these four laws are really all we need to become very efficient at multiplication with informal mental strategies.

WHAT ARE QUOTIENTS, DIVIDENDS AND DIVISORS?

These are three technical words that are used in the context of division calculations. They correspond to the words used in multiplication, introduced in the previous chapter: product, multiplicand and multiplier. In division, a **dividend** is divided by a **divisor** and the result is the **quotient**. For example, when 56 is divided by 8 (that is, 56 ÷ 8 = 7) the first number in the division, that which is to be divided (in this example, 56), is called the *dividend*. The number by which it is divided (in this example, 8) is called the *divisor*. The result, 7, is the *quotient*.

ARE THERE ANY FUNDAMENTAL LAWS OF DIVISION?

We have already noted in Chapter 10 that division (like subtraction) is not commutative. We should note here that division is also not associative (again like subtraction). For example, (24 ÷ 6) ÷ 2 is not equal to 24 ÷ (6 ÷ 2). Dealing with the divisions in the brackets first, in the first of these we get the answer 2, whereas in the second we get the answer 8.

But, like multiplication, division can be *distributed* across addition and subtraction. There are the following two **distributive laws of division**:

Distributive laws of division: $(a + b) \div c = (a \div c) + (b \div c)$
$(a - b) \div c = (a \div c) - (b \div c)$

whatever numbers are chosen for a, b and c (provided c does not equal 0).

The proviso about c not being equal to zero is because division by zero is not possible. Informally, the first of these laws means that if the dividend can be thought of as the sum of two numbers ($a + b$), then you can divide each of a and b separately by the divisor (c) and then add up the results. And the second means that if the dividend can be thought of as the difference between two numbers ($a - b$), then you can divide each of a and b separately by the divisor (c) and then find the difference between the results.

> **LEARNING AND TEACHING POINT**
>
> To explain why division by zero is not possible, use the inverse-of-multiplication structure of division. For example: '7 ÷ 0' could mean 'how many sets containing zero hedgehogs must I put on the table to get a set of 7 hedgehogs?' Also discuss with children what happens when you enter, say, 7 ÷ 0 on a calculator.

Again, the reader should recognize these as the basis of some very familiar procedures in the way we handle division calculations. We often employ this principle to simplify division questions. For example, since 45 = 30 + 15, you can split 45 ÷ 3 into two easier divisions: 30 ÷ 3 and 15 ÷ 3, giving us 10 and 5, which total 15. Written formally, we see that it is the first of the distributive laws of division being used here:

$$(30 + 15) ÷ 3 = (30 ÷ 3) + (15 ÷ 3)$$

The trick here was to transform the 45 into 30 + 15, being aware that with a divisor of 3, numbers like 30 and 15 are easily dealt with. In using the distributive law in an ad hoc approach, we should always look for numbers that are easy to handle with the particular divisor. So, to take another example, when tackling 143 ÷ 11, I might make use of the fact that 99 is a good number to have around when dividing by 11, and think of 143 as 99 + 44 as follows:

$$143 ÷ 11 = (99 + 44) ÷ 11$$
$$(99 + 44) ÷ 11 = (99 ÷ 11) + (44 ÷ 11) \text{ (distributing division across addition)}$$
$$= 9 + 4 = 13$$

Of course, there is no need to set out the working as formally as I have done here. It will probably be done mentally with a few numbers written down to keep track of where you are. Here is an example where we might use the same trick but with subtraction, for calculating 162 ÷ 9. The trick here is to spot that 162 is not too far from 180, which is an easy number to divide by 9. So we think of 162 as 180 − 18:

$$162 ÷ 9 = (180 − 18) ÷ 9$$
$$(180 − 18) ÷ 9 = (180 ÷ 9) − (18 ÷ 9) \text{ (distributing division across subtraction)}$$
$$= 20 − 2 = 18$$

Again, there is no need to set out the working in full like this, unless, like me, you are trying to explain your strategy to someone else.

WHAT ARE THE PREREQUISITE SKILLS FOR BEING EFFICIENT AT MENTAL STRATEGIES FOR MULTIPLICATION AND DIVISION?

The first prerequisite is that you know thoroughly and can recall easily all the results in the multiplication tables up to 10 × 10. There is, of course, value in knowing by heart results for multiplication tables beyond 10. For example, it is useful in many practical situations involving measurements to know the 25-times table. Pharmacists find it useful to know their 28-times table, because many tablets come in packs of 28.

Historically, the 12-times table was significant in Britain for calculations with inches/feet and pennies/shillings, so there may still be the rather arbitrary expectation that children should learn their multiplication tables up to 12 × 12. In this case, children would have to be able to recall all the results shown in Figure 11.1.

If children struggle in memorizing some of these multiplication facts then help them to develop strategies for working out what they do not know from what they do know. For example, if they are not sure of the product of 6 and 8, start with 6 × 2 = 12, double it to get 6 × 4 = 24 and double it again to get 6 × 8 = 48. Then, if they cannot remember the product of 7 and 8, but they now know 6 × 8 = 48, they can get 7 × 8 = 56 by adding on another 8. Or

LEARNING AND TEACHING POINT

To develop mastery of multiplication facts, encourage children to find and articulate relationships and patterns in the table of multiplication results shown in Figure 11.1. For example, how is each entry in the table related to the numbers above it and below it? And to the numbers to the left and right of it? And to its mirror image in the main diagonal? Which number appears most often in the table? What would we get if we added to the table a row and a column for zero? How are the numbers in the 4-column related to the numbers in the 8-column?

×	**1**	**2**	**3**	**4**	**5**	**6**	**7**	**8**	**9**	**10**	**11**	**12**
1	1	2	3	4	5	6	7	8	9	10	11	12
2	2	4	6	8	10	12	14	16	18	20	22	24
3	3	6	9	12	15	18	21	24	27	30	33	36
4	4	8	12	16	20	24	28	32	36	40	44	48
5	5	10	15	20	25	30	35	40	45	50	55	60
6	6	12	18	24	30	36	42	48	54	60	66	72
7	7	14	21	28	35	42	49	56	63	70	77	84
8	8	16	24	32	40	48	56	64	72	80	88	96
9	9	18	27	36	45	54	63	72	81	90	99	108
10	10	20	30	40	50	60	70	80	90	100	110	120
11	11	22	33	44	55	66	77	88	99	110	121	132
12	12	24	36	48	60	72	84	96	108	120	132	144

Figure 11.1 Multiplication results up to 12 × 12

they could use $5 \times 8 = 40$ and $2 \times 8 = 16$ to deduce that $7 \times 8 = 40 + 16$. As suggested in our discussion of learning to learn mathematics with understanding in Chapter 3, this is all about developing mastery by encouraging learners to make connections.

The second prerequisite is that you should be able to derive from any one of these results a whole series of results for multiplications involving multiples of 10 and 100. For example, knowing $7 \times 8 = 56$, we should be able to deduce the following:

$$70 \times 8 \quad = \quad 560$$
$$7 \times 80 \quad = \quad 560$$
$$70 \times 80 \quad = \quad 5600$$
$$7 \times 800 \quad = \quad 5600$$
$$700 \times 8 \quad = \quad 5600$$
$$70 \times 800 \quad = 56{,}000$$
$$700 \times 80 \quad = 56{,}000$$

It helps enormously just to notice that the total number of zeros written in the numbers on the left-hand side of each of these results is the same as the extra number of zeros written after the 56 on the right-hand side. To help understand this, using the example 700×80, we can think of the 700 as 7×100 and the 80 as 8×10. Then the whole calculation becomes: $7 \times 100 \times 8 \times 10$. Using the freedom granted to us by the commutative and associative laws of multiplication to rearrange this how we like, we can think of it as $(7 \times 8) \times (100 \times 10)$, which leads to $56 \times 1000 = 56{,}000$.

The third prerequisite is that you should be able to recognize all the division results that are simply the inverses of any of the above results. For example:

$$56 \div 8 \quad = 7$$
$$56 \div 7 \quad = 8$$
$$560 \div 8 \quad = 70$$
$$5600 \div 70 \quad = 80$$
$$56{,}000 \div 800 = 70$$

Once you have these skills in place, you will be able to use a wide range of strategies, such as:

- the use of factors as an ad hoc approach to multiplication;
- the use of doubling as an ad hoc approach to multiplication;
- ad hoc additions and subtractions in multiplication;
- ad hoc additions and subtractions in division;
- the constant ratio method for division.

HOW CAN FACTORS BE USED AS AN AD HOC APPROACH TO MULTIPLICATION?

When we think of a number such as 6 as 3×2 we are splitting it up into what are called **factors**. A factor of any natural number is a natural number by which it can be divided exactly without any remainder; factors are discussed in more detail in Chapter 13. As we have already seen above, we can use the associative law to enable us to multiply by each factor in turn, in order to simplify a product.

For example, one way to tackle 37×6 would be to think of the 6 as 3×2:

$$37 \times 6 = 37 \times (3 \times 2)$$
$$= (37 \times 3) \times 2 \text{ (using the associative law)}$$
$$= 111 \times 2 = 222$$

This strategy is particularly effective when there are numbers ending in 5 around, since they are especially easy to multiply by 2 or 4. For example, to calculate 26×15, I would jump at the opportunity to make use of the fact that $2 \times 15 = 30$. To achieve this, we split the 26 into factors as 13×2:

$$26 \times 15 = (13 \times 2) \times 15$$
$$= 13 \times (2 \times 15) \text{ (using the associative law)}$$
$$= 13 \times 30 = 390$$

Similarly, when I see 25 in a multiplication, such as 25×32, I immediately look to see if I can use the fact that $25 \times 4 = 100$. In this example, we can get the 4 we are looking for by splitting the 32 into factors, writing it as 4×8:

$$25 \times 32 = 25 \times (4 \times 8)$$
$$= (25 \times 4) \times 8 \text{ (using the associative law)}$$
$$= 100 \times 8 = 800$$

Again, it should be stressed that I am writing these out in detail only to explain the mathematical basis of what will essentially be a mental process. Figure 11.2 shows some examples of an 11-year-old using factors to assist with some multiplications, writing down her method in a way that shows her thinking to someone else:

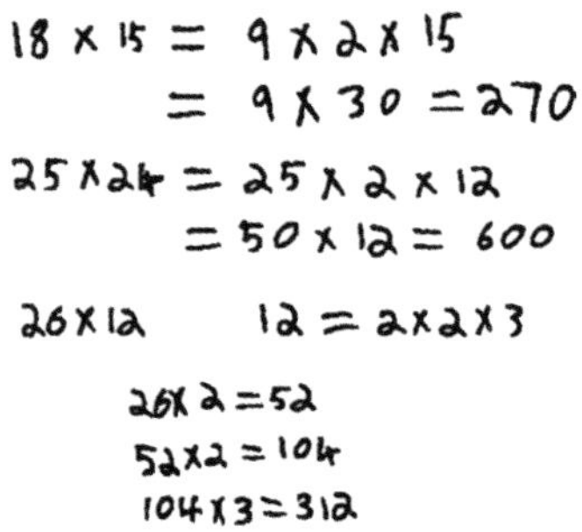

Figure 11.2   An 11-year-old using factors in multiplication

HOW CAN DOUBLING BE USED AS AN AD HOC APPROACH TO MULTIPLICATION?

We have previously noted (in Chapter 8) that we can make use of our confidence in doubling numbers in informal approaches to additions and subtractions. We can also deal with *any* multiplication with whole numbers simply by a process of doubling! Approaches to multiplication based on doubling have in fact been around much longer than has what people regard as the 'traditional' method for long multiplication. The trick here is to note that any number can be obtained by adding together some of the following numbers (called the powers of 2): 1, 2, 4, 8, 16, 32, 64 … and so on. For example, 23 = 16 + 4 + 2 + 1. This means that we can multiply a number by 23 (using the distributive law) in bits, multiplying by 16, 4, 2 and 1, and adding the results. For example, to calculate 26 × 23 the working might look like this:

LEARNING AND TEACHING POINT

Explore with children the way in which factors can sometimes be used to simplify a multiplication, especially with multiples of 2, 5 and 10.

LEARNING AND TEACHING POINT

Value and encourage informal, ad hoc methods of tackling multiplications and divisions that build on children's personal confidence with number and number relationships.

First, by repeatedly doubling the 26 and then selecting the results for the multipliers that add to 23 (16, 4, 2 and 1):

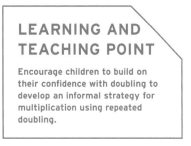

$$26 \times \ 1 = 26$$
$$26 \times \ 2 = 52$$
$$26 \times \ 4 = 104$$
$$26 \times \ 8 = 208$$
$$26 \times 16 = 416$$
$$\text{So, } 26 \times 23 = 416 + 104 + 52 + 26 = 598$$

Again, I have provided more detail than would be necessary for an informal calculation, as illustrated in Figure 11.3 in which an 11-year-old uses this method in a shopping context.

$$
\begin{array}{r l}
1 & £17 \\
2 & £34 \\
4 & £68 \\
8 & £136 \\
\hline
13 & £221
\end{array}
$$

Figure 11.3 Using doubling to find the cost of 13 items at £17 each

WATCH THE PROBLEM SOLVED! VIDEO AT: HTTPS://STUDY.SAGEPUB.COM/HAYLOCK7E

HOW CAN YOU USE AD HOC ADDITIONS AND SUBTRACTIONS IN MULTIPLICATION?

As with all calculations, we can always make use of those number facts and relationships with which we are confident to find our own individual approaches that make sense to us. Again, there is great value in encouraging children, even in primary school, to build on their growing confidence with number to develop their own approaches and to share different approaches to the same calculation. The distributive law gives us the freedom to break up a number in a multiplication calculation in any way we like, using an ad hoc combination of additions and subtractions of whatever numbers are easiest to handle.

For instance, here are two ad hoc ways of evaluating 26×34. First, by breaking up the 26 into $10 + 10 + 2 + 2 + 2$, on the basis that I am confident in multiplying by 10 and by 2, we can transform 26×34 into $(10 \times 34) + (10 \times 34) + (2 \times 34) + (2 \times 34) + (2 \times 34)$, as follows:

$$10 \times 34 = 340$$
$$10 \times 34 = 340$$
$$2 \times 34 = 68$$
$$2 \times 34 = 68$$
$$2 \times 34 = 68$$
$$\text{So, } 26 \times 34 = 884$$

Visually, this method is represented by the diagram in Figure 11.4. The trick in this kind of approach is to make sure you only ever multiply by easy numbers, like 1, 2, 5 and 10.

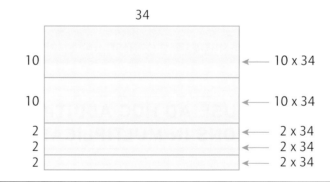

Figure 11.4   An ad hoc approach to 26×34 (diagram not to scale)

A second ad hoc approach to this calculation would be to think of the 34 as $10 + 10 + 10 + 5 - 1$, so that 26×34 becomes $(26 \times 10) + (26 \times 10) + (26 \times 10) + (26 \times 5) - (26 \times 1)$, as follows:

$$26 \times 10 = 260$$
$$26 \times 10 = 260$$
$$26 \times 10 = 260$$
$$26 \times 5 = 130$$

So, $26 \times 35 = 910$ (adding, to get thirty-five 26s)

$$26 \times 1 = 26$$ (one 26 to be subtracted)

So, $26 \times 34 = 910 - 26 = 884$

AND HOW CAN YOU USE AD HOC ADDITIONS AND SUBTRACTIONS IN DIVISION?

You can sometimes make a division much simpler by writing it as the sum or difference of numbers that are easier to divide by the given divisor. For example, if you have your wits about you, given $608 \div 32$, you might spot that it would have been much nicer if the question had been $640 \div 32$ (answer 20). So, we could deal with this division as follows:

$$608 \div 32 = (640 - 32) \div 32$$
$$= (640 \div 32) - (32 \div 32)$$
$$= 20 - 1 = 19$$

In practice, the context of the problem that gave rise to the calculation will often suggest an appropriate ad hoc approach. For example, in Figure 11.5 a child is calculating how many classes of 32 children would be needed for a school of 608. The child's approach is to build up to the given total of 608, by an ad hoc process of addition, using first 10 classes, then another 5, then a further 2 and another 2. Formally, the child is breaking the 608 up into $320 + 160 + 64 + 64$ and distributing the division by 32 across this addition as follows:

Figure 11.5 Ad hoc addition used to solve a division problem

$$608 \div 32 = (320 + 160 + 64 + 64) \div 32$$
$$= (320 \div 32) + (160 \div 32) + (64 \div 32) + (64 \div 32)$$
$$= 10 + 5 + 2 + 2 = 19$$

The child's own way of writing down the thinking involved is clear and is, of course, perfectly acceptable. Many children are very successful in dealing with division calculations by this kind of ad hoc addition, building up to the given target. In the next chapter, I will explain how an approach based on ad hoc subtraction can be used to develop an efficient written method for division calculations.

WHAT IS THE CONSTANT RATIO METHOD FOR DIVISION?

We have here a parallel with the constant difference method for subtraction explained in Chapter 9. We make use of this important principle: that we do not change the answer to a division calculation if we multiply both the numbers by the same thing. To understand the **constant ratio method for division**, think of the division in terms of the ratio structure (introduced in Chapter 10): if both quantities are scaled by the same factor, then their ratio does not change – just as when you add the same thing to two numbers, their difference does not change. For instance, all these divisions give the same result as $6 \div 2$ (= 3), because in each case the two numbers 6 and 2 have been multiplied by the same scale factor:

$60 \div 20$ (multiply both numbers by 10)
$12 \div 4$ (multiply both numbers by 2)
$30 \div 10$ (multiply both numbers by 5)
$6000 \div 2000$ (multiply both numbers by 1000)

This principle can be illustrated in the context of money. For example, imagine that we are comparing two prices, £3.75 and £1.25, by dividing one by the other to find the ratio. If we change the prices into pence (multiplying both numbers by 100), the ratio stays the same: $3.75 \div 1.25 = 375 \div 125$.

The division $75 \div 5$ can be used to demonstrate the application of this principle. Multiply both numbers by 2 and the question becomes $150 \div 10$. So the answer is clearly 15. This approach is particularly useful when it comes to divisions involving decimals, as we shall see in Chapter 17. For example, to handle $6 \div 1.5$, which looks tricky, we could multiply both numbers by 2, to give $12 \div 3$, which is easy!

We can also use the reverse principle: that we do not change the answer to a division calculation if we *divide* both numbers by the same thing. I might use this principle if I was dealing with, for example, $648 \div 24$, as follows:

$648 \div 24$ is the same as $324 \div 12$ (dividing both numbers by 2)
$324 \div 12$ is the same as $108 \div 4$ (dividing both numbers by 3)
$108 \div 4$ is the same as $54 \div 2$ (dividing both numbers by 2), which is 27

These two principles can be combined, as in the following example, where I might spot that multiplying both numbers by 2 would turn $225 \div 15$ into an easier question:

$$225 \div 15 = 450 \div 30 \quad \text{(multiplying both numbers by 2)}$$
$$= 45 \div 3 \quad \text{(dividing both numbers by 10)}$$
$$= 15$$

Although this constant ratio method is mathematically sound, the reader should be warned that it could lead you astray if you are dealing with a division that does not work out exactly and you wish to give the answer *with a remainder*. For example, if the question was $48 \div 5$ and you doubled both numbers, you would produce the equivalent ratio of $96 \div 10$. The exact value of this quotient (9.6) is indeed the correct answer to $48 \div 5$. However, $96 \div 10$ expressed as '9, remainder 6' is *not* the correct result for $48 \div 5$, because the remainder has been doubled as well (the correct result is 9, remainder 3).

In Chapter 15, we shall use this principle (that you do not change the ratio if you multiply or divide both numbers by the same thing) in explaining the idea of equivalent fractions; and there's more to come on the key idea of ratio in Chapter 18.

> **LEARNING AND TEACHING POINT**
>
> Make explicit the principle that you do not change the ratio if you multiply or divide two numbers by the same thing. Before children are able to give their answers as fractions or decimals, use only examples where the division works out exactly without a remainder.

RESEARCH FOCUS: UNDERSTANDING THE DISTRIBUTIVE LAW

Squire, Davies and Bryant (2004) studied the ability of children aged 9 and 10 years to use the commutative and distributive laws in multiplication questions. The children showed a much better than expected understanding of commutativity, but a very poor grasp of the distributive law. A question would be preceded by a cue, designed to prompt the application of the distributive law. For example, for the question, 'How many squares in a bar of chocolate 27 squares long and 21 squares wide?', the following cue would be given: 'In a bar of chocolate that is 26 squares long and 21 squares wide there are altogether 546 squares.' The most common response was to select the answer 547 from the available options. The children tended therefore to think that adding one to one of the numbers in the multiplication added one to the answer. This is a major possible misunderstanding on which to focus explicitly in teaching mental strategies for multiplication.

LEARNING RESOURCES

Access activities for your **lesson plans** at: https://study.sagepub.com/haylock7e

Before trying the self-assessment questions below, you should complete the **self-assessment questions** for this chapter at: https://study.sagepub.com/haylock7e

11.1: How would the commutative law of multiplication help in calculating mentally the number of children in 25 groups of 16?

11.2: Shozna calculated 25×24 by thinking of it as $25 \times (4 \times 6)$ and then using the associative law of multiplication. Complete Shozna's calculation.

11.3: Sam calculated 25×24 by thinking of it as $25 \times (20 + 4)$ and then using the distributive law. Complete Sam's calculation.

11.4: Bev calculated 22×38 by thinking of it as $22 \times (40 - 2)$ and then using the distributive law. Complete Bev's calculation.

11.5: Deduce eight other multiplication results, involving 4, 40, 400, 9, 90 and 900, from the result $4 \times 9 = 36$.

11.6: Use an ad hoc method, based on factors, to find 48×25.

11.7: Use the fact that $26 = 2 + 8 + 16$ and the doubling strategy to find 26×103.

11.8: Use the fact that $26 = 10 + 10 + 2 + 2 + 2$ to find 26×103.

11.9: (a) Use the distributive law and the fact that $154 = 88 + 66$ to find the answer to $154 \div 22$.

   (b) Now do this again using the fact that $154 = 220 - 66$.

11.10. Find $483 \div 21$ by the ad hoc addition of groups of 21, building up to the total of 483.

11.11. Find $385 \div 55$ by using the constant ratio method.

FURTHER PRACTICE

Access the website material for:

- Knowledge check 4: The associative laws
- Knowledge check 5: The distributive laws
- Knowledge check 6: Mental calculations, multiplication strategies
- Knowledge check 7: Mental calculations, division strategies

at: https://study.sagepub.com/haylock7e

FROM THE STUDENT WORKBOOK

Questions 11.01–11.10: Checking understanding (mental strategies for multiplication and division)

Questions 11.11–11.21: Reasoning and problem solving (mental strategies for multiplication and division)

Questions 11.22–11.38: Learning and teaching (mental strategies for multiplication and division)

GLOSSARY OF KEY TERMS INTRODUCED IN CHAPTER 11

Associative law of multiplication   The principle that if there are three numbers to be multiplied together, it makes no difference whether you start by multiplying the first and second, or by multiplying the second and third. In symbols, this law states that, for any three numbers a, b and c, $(a \times b) \times c = a \times (b \times c)$.

Distributive laws of multiplication   The laws that allow you to distribute a multiplication across an addition or across a subtraction. For example, 28×4 can be split up into 25×4 add 3×4, or 30×4 subtract 2×4. Formally, the laws state that for any numbers, a, b and c, then $(a + b) \times c = (a \times c) + (b \times c)$ and $(a - b) \times c = (a \times c) - (b \times c)$.

Quotient, dividend and divisor   Technical names for the three numbers involved in a division calculation; for example, in $999 \div 37 = 27$, the 999 is the dividend, 37 is the divisor and 27 is the quotient.

Distributive laws of division   The laws that allow you to distribute a division across an addition or across a subtraction. For example, $92 \div 4$ can be split up into $80 \div 4$ add $12 \div 4$, or $100 \div 4$ subtract $8 \div 4$. Formally, the laws state that for any numbers, a, b and c (provided c is not zero), then $(a + b) \div c = (a \div c) + (b \div c)$ and $(a - b) \div c = (a \div c) - (b \div c)$.

Factor   A natural number by which a given natural number can be divided exactly, without a remainder; for example, 7 is a factor of 28.

Constant ratio method for division   A method for simplifying a division calculation by multiplying the dividend and the divisor by the same thing, or by dividing them by the same thing, thus keeping the ratio the same. For example, $180 \div 15$ can be simplified to $360 \div 30$, by doubling both numbers; and $360 \div 30$ can then be simplified to $36 \div 3$, by dividing both numbers by 10.

WRITTEN METHODS FOR MULTIPLICATION AND DIVISION

IN THIS CHAPTER, THERE ARE EXPLANATIONS OF

- short multiplication;
- the areas method for multiplication;
- the grid method for multiplication;
- the long multiplication algorithm;
- short division;
- the ad hoc subtraction method of doing division calculations;
- the algorithm known as long division.

READ THIS CHAPTER'S CURRICULUM LINKS AT: HTTPS://STUDY.SAGEPUB.COM/HAYLOCK7E

WHAT IS SHORT MULTIPLICATION?

In this chapter, I shall explain various written methods for doing multiplication and division calculations, using only positive whole numbers. Calculations involving decimals are considered in Chapters 16 and 17.

Short multiplication refers to a formal way of writing out a multiplication of a number with two or more digits by a single-digit number. Figure 12.1 shows one way in which this might be done for the calculation 38 × 4. In essence, all that we do here is to use the distributive law explained in Chapter 11, breaking down the 38 into 30 + 8, and then multiplying the 8 by 4 and the 30 by 4 and adding the results. The calculation is written down in vertical format with digits representing tens and ones (units) lined up in columns, as in Figure 12.1(a). Then, as shown in Figure 12.1(b), the 8 in the units column is multiplied by 4, giving 32. The 2 in this 32 is written in the answer in the units column and the 3 tens are carried over to the next column. This 'carry 3' is indicated by the 3 written in the tens position below the answer line. Finally, in Figure 12.1(c) we multiply the 3 tens by 4, giving 12 tens; add on the 3 tens carried to give a total of 15 tens, which is recorded as shown in the answer, 152.

$$
\begin{array}{c}
\text{(a)} \quad 3\,8 \\
\times \quad \underline{4} \\
\\
\end{array}
\qquad
\begin{array}{c}
\text{(b)} \quad 3\,8 \\
\times \quad \underline{4} \\
2 \\
3 \\
\end{array}
\qquad
\begin{array}{c}
\text{(c)} \quad 3\,8 \\
\times \quad \underline{4} \\
\underline{1\,5\,2} \\
3 \\
\end{array}
$$

Figure 12.1   Short multiplication for 38 × 4

The procedure extends to the multiplication of numbers with any number of digits by a single-digit number. Three examples are given in Figure 12.2. The commentary for 416 × 8 to accompany Figure 12.2(a) would be something like this: '6 multiplied by 8 is 48; that's 8 in the units column and carry 4 into the tens column; 1 in the tens column multiplied by 8 is 8, add the 4 carried to give 12; that's 2 in the tens column and carry 1 into the hundreds column; 4 in the hundreds column multiplied by 8 is 32, add the 1 carried to give 33; write this in the answer line, giving the result 3328.'

$$
\begin{array}{c}
\text{(a)} \quad 4\,1\,6 \\
\times \quad \underline{8} \\
3\,3\,2\,8 \\
1\,4 \\
\end{array}
\qquad
\begin{array}{c}
\text{(b)} \quad 4\,3\,2\,9 \\
\times \quad \underline{6} \\
2\,5\,9\,7\,4 \\
1\,1\,5 \\
\end{array}
\qquad
\begin{array}{c}
\text{(c)} \quad 6\,2\,0\,9 \\
\times \quad \underline{9} \\
5\,5\,8\,8\,1 \\
1\,8 \\
\end{array}
$$

Figure 12.2   Short multiplication for (a) 416 × 8, (b) 4329 × 6, (c) 6209 × 9

 WATCH THE PROBLEM SOLVED! VIDEO AT: HTTPS://STUDY.SAGEPUB.COM/HAYLOCK7E

Once the learner is confident in recalling their multiplication tables, then, because they are not expending much cognitive effort in the individual multiplications involved in this process, they will be able to remember the numbers being carried at each stage and add them into the next column without writing them down. Being able to do this is a great bonus when it comes to long multiplication (see later in this chapter), which otherwise can get very messy with little digits all over the place.

HOW MIGHT MULTIPLICATION OF TWO 2-DIGIT NUMBERS BE INTRODUCED?

An introductory method for multiplying two 2-digit numbers – and, in fact, an effective method for multiplication that extends to numbers with 3 digits – is an approach based on the idea of splitting up the two numbers being multiplied into tens and units (ones). So the 26 becomes 20 + 6 and the 34 becomes 30 + 4. Then we have four multiplications to do: the 20 by the 30, the 20 by the 4, the 6 by the 30 and the 6 by the 4. To make sense of this, we can visualize the multiplication as a question of finding the number of counters in a rectangular array of 26 by 34 (see Chapter 10), as shown in Figure 12.3. Thinking of the 26 and the 34 as 20 + 6 and 30 + 4 respectively suggests that we can split the array up into four separate rectangular arrays of counters, representing 20 × 30, 20 × 4, 6 × 30 and 6 × 4.

Drawing pictures with hundreds of counters in them, like Figure 12.3, helps to explain the method, but it is clearly very tedious. A more efficient picture, therefore, uses the idea, suggested in Chapter 10, that we can extend the notion of a rectangular array into that of the area of a rectangle. We can then explain 26 × 34 very simply using the diagram shown in Figure 12.4. The answer to the multiplication is obtained by working out the areas of the four separate rectangles and adding them up. This can be called the **areas method** for multiplication.

> ### LEARNING AND TEACHING POINT
>
> To help children master the short multiplication process with understanding, rather than just learn it as a recipe, give them the opportunity to work through the calculation with base-ten Dienes blocks or place value counters. Help them to make the connections between what they do with the blocks (or counters) and the recording in symbols. For example, for 38 × 4, using blocks, they put out four sets of 'three longs (tens) and eight units (ones)'. They combine all the units (32 in total) and exchange 30 of these for three longs (tens), leaving two units. They then have 3 × 4 = 12 tens, plus the 3 exchanged, giving a total of 15 tens. Ten of these ten-blocks are then exchanged for a flat (a hundred-block). The result is 1 hundred block, 5 ten-blocks and 2 units: that is 152. While 1p coins are still in circulation, this same process can be carried out with 1p, 10p and £1 coins (or tokens).

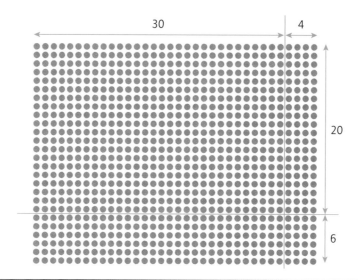

Figure 12.3 Visualizing 26 × 34 as a rectangular array

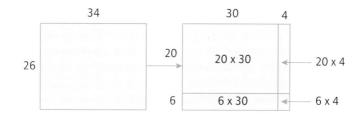

Figure 12.4 Visualizing 26 × 34 as four areas (not to scale)

The actual calculation can be written out as follows:

$$
\begin{array}{r}
20 \times 30 = 600 \\
6 \times 30 = 180 \\
20 \times 4 = 80 \\
\underline{6 \times 4 = 24} \\
884
\end{array}
$$

Once you have grasped this approach, it becomes unnecessary to draw the rectangles every time. The steps in the calculation can be set out in a grid, as shown below. Because of this, many teachers call this the **grid method**:

> **LEARNING AND TEACHING POINT**
>
> Introduce primary school children to written methods for multiplying 2-digit numbers by using the method based on the areas of four rectangles, splitting each of the two numbers into tens and units (as shown in Figure 12.4). Once they are fluent with this, they will not need to draw the rectangle and can record the steps in a grid.

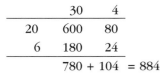

	30	4
20	600	80
6	180	24

$$780 + 104 = 884$$

▶ WATCH THE PROBLEM SOLVED! VIDEO AT: HTTPS://STUDY.SAGEPUB.COM/HAYLOCK7E

HOW IS THIS METHOD USED WITH THREE-DIGIT NUMBERS?

The method extends quite easily to multiplications involving three-digit numbers. Figure 12.5 shows a rough sketch that might be used to visualize 348 × 25. (The rectangles in the diagram are not drawn to scale.)

Clearly, in this example, there are now six bits to deal with separately, and then to be added. This leads to the calculation being set out, for example, as follows, with the six products in the right-hand column being summed to give 348 × 25 = 8700.

LEARNING AND TEACHING POINT

Children in primary schools will be able to extend the areas approach to multiply together a three-digit number and a two-digit number (as shown in Figure 12.5), using six areas – and then to record the steps using the grid format.

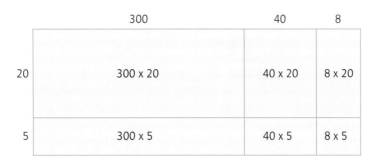

	300	40	8
20	300 x 20	40 x 20	8 x 20
5	300 x 5	40 x 5	8 x 5

Figure 12.5   The areas method applied to 348 × 25

$$300 \times 20 = 6000$$
$$40 \times 20 = 800$$
$$8 \times 20 = 160$$
$$300 \times 5 = 1500$$
$$40 \times 5 = 200$$
$$8 \times 5 = 40$$
$$8700$$

Using the grid format, the steps in this calculation could be set out like this:

	300	40	8
20	6000	800	160
5	1500	200	40

7500 + 1000 + 200 = 8700

HOW DO YOU MAKE SENSE OF LONG MULTIPLICATION?

The standard algorithm for multiplying together two numbers with two or more digits is usually called **long multiplication**. Figure 12.6 shows how the method might be set out for calculating 26 × 34.

```
    2 6
  × 3 4
  ─────
  7 8 0  ←──── this is 26 × 30
  1 0 4  ←──── this is 26 × 4
  ─────
  8 8 4  ←──── this is 26 × 34
  ─────
```

Figure 12.6 Long multiplication, 26 × 34

When we did the multiplication 26 × 34 in Figure 12.4 above, using the four areas method, the calculation was broken down into four relatively straightforward multiplications. The long multiplication method, however, breaks the calculation down into just two (more difficult) multiplications. It is therefore a much more condensed method and really requires considerable confidence in short multiplication, as explained at the beginning of this chapter.

Long multiplication is also based on the distributive law for multiplication distributed over addition (see Chapter 11). Effectively, what happens is that one of the numbers is broken up into the sum of its tens and units, and the multiplication by the other number is distributed across these. Applying this principle to 26 × 34, we can think of it as 26 × (30 + 4) and get the answer by multiplying the 26 first by the 30 and then by the 4, and then adding the results. This is precisely what goes on in long multiplication, as shown in Figure 12.6. But the problem with the method is that multiplications like 26 × 30 and 26 × 4 might be quite tricky to do mentally in the middle of a series of steps required for a longer calculation. Note that in Figure 12.6 a choice has been made to deal with the multiplication by 30 first and then the multiplication by 4. It makes no difference which of these you do first.

> **LEARNING AND TEACHING POINT**
>
> Note again the importance of children being thoroughly confident in multiplication with multiples of 10 and 100 (for example, 300 × 20) as a prerequisite for going on to multiply two- and three-digit numbers. Keep reinforcing these basic skills.

> **LEARNING AND TEACHING POINT**
>
> Resist the temptation to teach long multiplication as a routine to be learnt simply by rote; make frequent reference to the place values of the digits involved at each stage. Illustrate the process with base-ten Dienes blocks or place-value counters. Explain how and why 'putting down a zero' in the units place when multiplying by the tens automatically turns a multiplication by 4, for example, into a multiplication by 40. (See Figure 12.7 (b) and (c).)

	(a) 74	(b) 74	(c) 74	(d) 74
	× 43	× 43	× 43	× 43
	222	222	222	222
		0	2960	2960
				3182

Figure 12.7 The steps involved in long multiplication, 74 × 43

Figure 12.7 shows the steps involved in calculating 74 × 43, doing the multiplication of 74 by 3 first (to give 222), and then 74 by 40 second (to give 2960). In practice, the multiplication of 78 by 40 is done by first writing down a zero, as in Figure 12.7(b), and then just multiplying 74 by 4, as in (c). The zero has the effect of multiplying the result of 74 × 4 by 10, hence producing 74 × 40, as required. In Figure 12.7(d), the results of the two multiplications are added to get the overall answer.

Note that when the 74 in Figure 12.7 is multiplied by the 3 and then by the 4 these are two short multiplications (see earlier in this chapter) that involve 'carrying'. There are various solutions to the problem of where to write the little digits that indicate what is being carried, some of which can be very messy and confusing when recorded in a child's handwriting (or mine, to be honest!). The best way is to be so fluent in short multiplication that you do not have to write them down at all, as in Figure 12.7. If I do have to write down

> **LEARNING AND TEACHING POINT**
>
> Prerequisites for being able to master long multiplication with some degree of understanding are: (a) application of the distributive law for multiplication by a 2-digit number (for example, multiplication by 37 is equivalent to multiplication by 30 added to multiplication by 7); (b) fluency and accuracy in short multiplication; and (c) for the final step, fluency and accuracy in column addition.

the digits being carried, my personal approach is just to jot them down somewhere well away from the formal calculation and cross them out when I have added them in.

Given the availability of calculators and other technology in the real world, it really is unnecessary for children in primary school to do multiplications harder than that in Figure 12.7 by formal written methods. However, when a national curriculum and the school inspectorate demand that children in primary school should be able to multiply a 3-digit or 4-digit number by a 2-digit number using long multiplication, teachers will have to teach children how to do this. Figure 12.8 provides two examples. In Figure 12.8(a), the calculation 632 × 47 is done in the two multiplications: 632 × 7 in the first line and 632 × 40 in the second. Note that 632 × 40 is done by first writing down a zero in the units column, which effectively multiplies by 10, and then multiplying 632 by 4 (= 2528). In Figure 12.8(b), the calculation 3784 × 65 is done in the two multiplications: 3784 × 5 in the first line and 3784 × 60 in the second. Again, note that in the second line the calculation of 3784 × 60 is done by writing a zero in the units position to multiply by 10, and then multiplying 3784 by 6 (= 22,704).

```
(a)      632          (b)      3784
       ×  47                 ×    65
        4424                 18920
       25280                227040
       29704                245960
```

Figure 12.8 Further examples of long multiplication

▶ **WATCH THE PROBLEM SOLVED! VIDEO AT: HTTPS://STUDY.SAGEPUB.COM/HAYLOCK7E**

WHAT IS SHORT DIVISION?

Short division is a standard algorithm often used for divisions with a single-digit number as the divisor, such as $75 \div 5$, $438 \div 6$ and $944 \div 4$. If you are confident with, say, the 12-times table or the 25-times table, it could also be used for divisions such as $1956 \div 12$ and $620 \div 25$.

One of the standard ways of recording this process is shown in Figure 12.9 for the calculation of $75 \div 5$. It is based on the idea of distributing the division by 5 across the addition of the 7 tens and the 5 ones in the number 75. First, you divide the 7 (tens) by 5. This gives the result 1 (ten) remainder 2. The 1 is written in the tens position in the answer above the line. The remainder, 2 (tens), is then exchanged for 20 ones. This exchange is indicated by the little 2 written in front of the 5. There are now 25 ones to be divided by 5. This gives the result 5 (exactly), which is written above the line in the units position. The 15 written above the line (1 ten and 5 ones) is therefore the answer to $75 \div 5$.

$$\begin{array}{r} 1\,5 \\ \hline 5\,|\,7\,{}^{2}5 \end{array}$$

Figure 12.9 Short division, $75 \div 5$

Figure 12.10 shows the calculation of $438 \div 6$, using short division. The dividend (the number being divided, 438) and the divisor (6) are set out as shown in Figure 12.10(a). The result of the division appears above the line, as shown in Figure 12.10(b). The commentary to accompany this process would be along these lines: 'Our number has 4 hundreds, 3 tens and 8 ones. We start with the hundreds. There are only 4 hundreds, not enough to divide by 6. So, we mentally exchange the 4 hundreds for 40 tens and now think of our number as 43 tens and 8 ones. So we are now looking at '43 (tens) to be divided by 6'. The answer is 7 (tens) with 1 (ten) remaining. Write the 7 in the tens position in the answer. Exchange the remainder 1 (ten) for 10 ones, and indicate this with a little 1 written in front of the 8 ones. We now have a total of 18 ones to be divided by the 6, which gives 3 (ones). Write this 3 in the units position in the answer. So, $438 \div 6 = 73$.

$$\text{(a)} \quad \begin{array}{r} \hline 6\,|\,4\,3\,8 \end{array} \qquad \text{(b)} \quad \begin{array}{r} 7\,3 \\ \hline 6\,|\,4\,3\,{}^{1}8 \end{array}$$

Figure 12.10 Short division, $438 \div 6$

Figure 12.11 shows how children might be helped to follow this process by covering up the digits not yet being used with a card, which is then used to reveal gradually the hundreds, tens and ones in the number being divided. Figure 12.11(a) shows 438 ÷ 6 set out as in Figure 12.10, but with H, T, U headings to indicate hundreds, tens and units (ones). In (b) a card is used to cover the tens and units. This makes clear that we are looking just at the 4 hundreds: not enough to divide by 6. So, move the card along, as shown in (c), and it is now clear that we have 43 tens to be divided by 6. This gives us 7 with 1 to carry, as recorded in (d). The card is then removed, to reveal that we have 18 ones to be divided by 6, as in (e), giving the 3 in the answer above the line, as shown in (f).

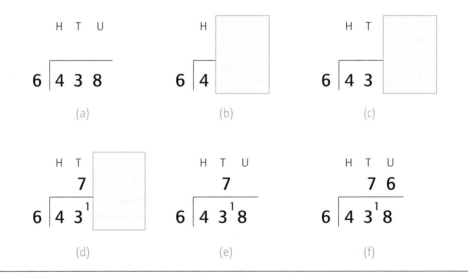

Figure 12.11   Using a card in short division

Figure 12.12 shows some further examples of the short division method, including some with a remainder in the answer. The calculation of 1956 ÷ 12 done in this way assumes confidence with the 12-times table: the first step is to divide 19 (hundreds) by 12, giving the result 1 (hundred) and remainder 7. The 7 is exchanged for 70 in the next position, as indicated by the little 7, giving a total of 75 (tens). You then need to know that this is 6 twelves, remainder 3 and so on. Similarly, to get started with 620 ÷ 25 you need to know that 62 divided by 25 is 2 remainder 12; the 12 is exchanged for 120 in the units position, giving 120 to be divided by 25. This gives 4, remainder 20.

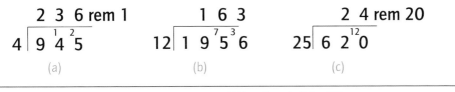

(a)   (b)   (c)

Figure 12.12 Further short divisions: 945 ÷ 4, 1956 ÷ 12 and 620 ÷ 25

WATCH THE PROBLEM SOLVED! VIDEO AT: HTTPS://STUDY.SAGEPUB.COM/HAYLOCK7E

Note that, in these examples, the interpretation of the remainder depends on the context in which the division calculation arose. For example, if 620 ÷ 25 is the calculation used to find how many groups of 25 children there are in a school of 620 children, then the result tells us that there are 24 complete groups, and the remainder of 20 represents the 20 children not included in these 24 groups. But if 620 ÷ 25 is the calculation used to find how much money each child gets if 620p is shared equally between 25 children, then the remainder represents the surplus of 20 pence not allocated in the equal sharing process. We have done the same calculation, with the divisor 25 representing 25 children in both examples: but in one situation the remainder represents children and in the other case it represents pence! In Chapter 17 we shall look at how the result of a division calculation with a remainder, such as 945 ÷ 4 and 620 ÷ 25, can be expressed as a decimal number, when the context is appropriate.

WHAT IS THE AD HOC SUBTRACTION METHOD FOR DIVISION?

The examples in Figure 12.12 show that making sense of the steps in the short division process can be quite challenging. Providing a commentary to match what you do with symbols is far from straightforward and there is always the temptation to drill learners in the steps involved and to accept rote learning. This is even more the case with long division, as we shall see later in this chapter. Any teacher committed to children learning how to learn mathematics (see Chapter 3) will be reluctant to go down this route.

By contrast, the **ad hoc subtraction** approach (sometimes called 'chunking') allows for a variety of ways of doing a particular division, depending on the numbers involved and the individual's personal confidence with multiplication. It is a method that is much easier to understand and it builds on the mental and informal approaches to division discussed in Chapter 11. It works very well with problems up to the level of difficulty of dividing a three-digit number by a two-digit number – which, in our technological age where calculators are freely available on every child's mobile phone, really ought to be sufficient.

The method uses the inverse-of-multiplication structure for division and the idea of repeated subtraction, explained in Chapter 10. I will take as an example the division, $648 \div 24$. Hence, the question is interpreted as: how many 24s make 648? (Not 'share 648 between 24'.) We approach this step-by-step, using whatever multiplications we are confident with, repeatedly subtracting from the dividend various multiples of the divisor, in an ad hoc manner, as shown in Figure 12.13.

(a)		$648 \div 24$	(b)		$648 \div 24$	(c)		$648 \div 24$	(d)		$648 \div 24$
10	240		10	240		10	240		10	240	
	408			408			408			408	
			10	240		10	240		10	240	
				168			168			168	
						2	48		2	48	
							120			120	
									2	48	
										72	
									2	48	
										24	
									1	24	
									27	0	

Figure 12.13   Ad hoc subtraction approach for $648 \div 24$

For example, we should know easily that ten 24s make 240. We subtract this 240 from the 648. That leaves us with 408 to find. At this stage, our working might be as shown in (a) in Figure 12.13. Try another ten 24s. That's a further 240, leaving us with 168, as shown in (b). We do not have enough for a further ten 24s, so we might try two 24s (or one 24, if we prefer it, or whatever we are confident to do mentally). This gives us the situation shown in (c). And so we proceed, until we have used up all the 648, as shown in (d). Totting up the numbers of 24s we've used down the left-hand side gives us the answer to the calculation: $10 + 10 + 2 + 2 + 2 + 1 = 27$.

Someone with greater confidence with multiplication might get to the result more quickly, as shown in Figure 12.14. Here I have gone straight in with twenty 24s (that is, 20 × 24 = 480), followed this up with five 24s (that is, 5 × 24 = 120) and finished off with two 24s (that is, 2 × 24 = 48). The beauty of this method is that each individual can approach the calculation in a way that suits them and draws on their personal level of confidence with number relationships. We should note, however, that you would not get far with this method for division by a two-digit number if you were not fluent in subtraction. And it also helps to be really at ease with multiplying by at least 1, 2, 5, 10, 20 and 50.

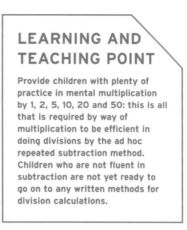

LEARNING AND TEACHING POINT

Provide children with plenty of practice in mental multiplication by 1, 2, 5, 10, 20 and 50: this is all that is required by way of multiplication to be efficient in doing divisions by the ad hoc repeated subtraction method. Children who are not fluent in subtraction are not yet ready to go on to any written methods for division calculations.

The reader should compare this method with the reverse process of ad hoc addition of multiples of the divisor building up to the given dividend, explained in the previous chapter (see Figure 11.5).

```
        |  648 ÷ 24
    20  |  480
        | ─────
        |  168
     5  |  120
        | ─────
        |   48
     2  |   48
    ─── | ─────
    27  |    0
```

Figure 12.14 Finding 648 ÷ 24 with fewer steps

This method can be introduced with simpler examples, of course, such as those in Figure 12.15, which show how someone might approach the divisions (a) 75 ÷ 5 and (b) 438 ÷ 6. In (a) the person doing the division has subtracted ten 5s from 75 and then five 5s, giving a total of 15 as the result of the division. In (b) they have subtracted fifty 6s from 438, then twenty 6s and then three 6s, giving a total of 73 as the result.

```
(a)         |  75 ÷ 5        (b)          |  438 ÷ 6
       10   |  50                    50   |  300
            | ────                        | ─────
        5   |  25                         |  138
       ───  | ────                   20   |  120
       15   |   0                         | ─────
                                          |   18
                                     3    |   18
                                    ───   | ─────
                                    73    |    0
```

Figure 12.15 Introducing ad hoc subtraction with division by a single digit

DOES THIS METHOD WORK WHEN THERE'S A REMAINDER?

Yes. Take as an example 437 ÷ 18 (Figure 12.16). This time I have started with twenty 18s, because I can do that in my head easily (20 × 18 = 360), then followed it up with two 18s and a further two 18s. At this stage, I am left with 5. Since this is less than the divisor, 18, I can go no further. The answer to the question is therefore '24, remainder 5'. The meaning of this answer will, of course, depend on the actual practical situation that gave rise to the calculation.

$$
\begin{array}{r|l}
 & 437 \div 18 \\
20 & 360 \\
\hline
 & 77 \\
2 & 36 \\
\hline
 & 41 \\
2 & 36 \\
\hline
24 & \text{rem } 5 \\
\end{array}
$$

Figure 12.16   An example with a remainder, 437 ÷ 18

 WATCH THE PROBLEM SOLVED! VIDEO AT: HTTPS://STUDY.SAGEPUB.COM/HAYLOCK7E

WHAT ABOUT LONG DIVISION?

The conventional algorithm for division, usually known as **long division**, can involve some tricky multiplications and is, to say the least, not easy to make sense of. Some would argue that the method of long division is more efficient than any other method, provided the steps are carried out in the prescribed manner. My response would be that if you want efficiency then use a calculator!

Figure 12.17 illustrates the method for 648 ÷ 24. I'll talk you through this. The first step is to ask, how many 24s are there in 64? The answer to this question is 2, which is written above the 4 in 648. You then write the product of 2 and 24 (48) under the 64 and subtract, giving 16. The 8 in the 648 is then brought down and written next to the 16, making 168 in this row. You then ask, how many 24s are there in 168? The answer to this is 7, which is written above the 8 in 648. You then write the product of 7 and 24 (168) under the 168 and subtract it, giving zero. So 648 ÷ 24 = 27, with no remainder. That's what you do, but does it make much sense? It really is quite difficult to explain what is going on here. What are you actually doing when you divide the 24 into 64? What does the 48 mean? Why do you have to write various numbers where you write them? The challenge for the primary school teacher who is required to teach this procedure is to get children to master this method without just relying on drill and on rote learning (rather than learning with understanding) – which, as we have seen in Chapter 3, undermines the priority we would give to children learning how to learn mathematics.

$$
\begin{array}{r}
2\,7 \\
24\,\overline{\smash{)}6\,4\,8} \\
4\,8 \\
\hline
1\,6\,8 \\
1\,6\,8 \\
\hline
0
\end{array}
$$

Figure 12.17   Long division, 648 ÷ 24

WATCH THE PROBLEM SOLVED! VIDEO AT: HTTPS://STUDY.SAGEPUB.COM/HAYLOCK7E

If you are aiming for children to master this traditional process for long division, then it helps to introduce the format with examples of dividing by a single digit, such as those used for short division earlier in the chapter (see Figures 12.9 and 12.10), and also to extend the use of a card to reveal hundreds, tens and units (see Figure 12.11). An example is shown in Figure 12.18, where 9396 is divided by 7. The calculation is set out as in (a), with headings to indicate thousands, hundreds, tens and units. In (b) the 9 (thousands) are divided by 7, giving 1 in the thousand position in the answer. The 7 (thousands) that have now been used are subtracted from the 9 to give 2 thousands still to be used. The card is then moved to reveal the hundreds, as in (c). The 3 hundreds in the 9396 are then brought down to join the unused 2 thousands, giving a total of 23 hundreds. These are now divided by the 7, as shown in (d), giving 3 in the hundreds position in the answer. This 3 multiplied by 7 gives 21 hundreds used up, which,

> ### LEARNING AND TEACHING POINT
>
> If you have to teach long division, introduce the format with examples that could be done by short division. Use a card to reveal and deal step by step with the thousands, hundreds, tens and ones. To help the children remember the four things to be done at each stage (Divide, Multiply, Subtract, Bring down), devise an appropriate mnemonic. For example: Do MacDonalds Sell Burgers?

when subtracted from the 23, leaves 2 hundreds still to be dealt with. The card is moved, as in (e), to reveal the tens. The 9 tens are brought down to join the 2, giving 29 (tens). This 29 divided by 7 gives 4, which uses up 28 (4 × 7) of them, leaving 1 ten still to be divided. The card is then removed to reveal the units, and the calculation completed, as shown in (f). The final 16 (units) divided by 7 gives 2 with remainder 2. So, the overall result is 9396 ÷ 7 = 1342, remainder 2.

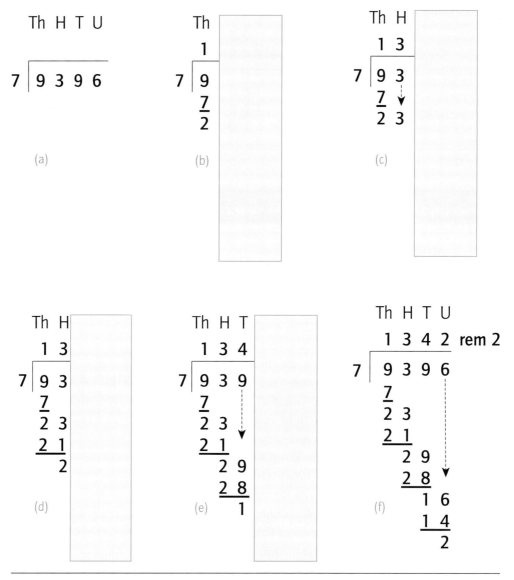

Figure 12.18   Calculating 9396 ÷ 7 using the long division format

RESEARCH FOCUS: DIVISION CALCULATION METHODS

A study of over 500 children in England and the Netherlands (Anghileri, 2001) revealed the impact on children's performances of the different standard methods for division that had been taught in primary schools in the two countries. The Dutch approach had been to progress from repeated subtraction of the divisor to more and more efficient subtraction of multiples of the divisor, using essentially the ad hoc subtraction procedure proposed in this chapter (see Figures 12.13, 12.14, 12.15 and 12.16). The traditional short-division algorithm (see Figures 12.9 and 12.10) was taught much more in English schools and took precedence over informal methods. This had the clear disadvantages of being inflexible and disguising the number relationships involved. It was clear that the Dutch approach proved to be more successful because it had the flexibility for individuals to use the knowledge of multiplication facts with which they were confident. It also gave children access to division by two-digit numbers much earlier.

Another interesting study related to written division calculations is that by Lee (2007), who worked with a small group of children in the USA, giving them plenty of time to investigate the meaning of the long division algorithm. While noting some initial difficulties, the class episodes in this article provide examples of internalization that highlight the active role of the learner in transforming concrete representations into an abstract algorithm. Factors that encouraged students to be deeply engaged in making sense of the long division algorithm included meaningful tasks and dynamic class discussions.

LEARNING RESOURCES

Access activities for your **lesson plans** at: https://study.sagepub.com/haylock7e

Before trying the self-assessment questions below, you should complete the **self-assessment questions** for this chapter at: https://study.sagepub.com/haylock7e

12.1: Use the method of Figure 12.4 to find 42 × 37.

12.2: Use the method of Figure 12.5 to find 345 × 17.

12.3: Write down the values of: 7 × 3, 7 × 5, 7 × 10. Now find 126 ÷ 7, using the method of ad hoc subtraction.

12.4: Write down the values of: 23 × 2, 23 × 5, 23 × 10, 23 × 20. Now find 851 ÷ 23, using the method of ad hoc subtraction.

12.5: Write down the values of: 8 × 5, 8 × 10, 8 × 50. Now find 529 ÷ 8, using the method of ad hoc subtraction.

12.6: Using long multiplication for 432 × 57, the calculation is broken down into two multiplications. What are they?

12.7: Make sure you can calculate: (a) 64 × 7, using short multiplication; (b) 209 × 88, using long multiplication; (c) 184 ÷ 8, using short division; (d) 1089 ÷ 15, using long division, with a remainder.

FURTHER PRACTICE

Access the website material for

- Knowledge check 8: More multiplication strategies
- Knowledge check 9: More division strategies

at: https://study.sagepub.com/haylock7e

FROM THE STUDENT WORKBOOK

Questions 12.01–12.10: Checking understanding (written methods for multiplication and division)

Questions 12.11–12.21: Reasoning and problem solving (written methods for multiplication and division)

Questions 12.22–12.28: Learning and teaching (written methods for multiplication and division)

GLOSSARY OF KEY TERMS INTRODUCED IN CHAPTER 12

Short multiplication   A formal column method for setting out a multiplication by a single-digit number; the hundreds, tens and ones in the multiplicand are multiplied separately by the multiplier and the results added.

Areas method for multiplication   A more expanded approach to multiplication of two numbers, in which the two numbers are interpreted as the sides of a rectangle and their product is the area. In this procedure, 426 × 37 is calculated as the sum of six areas: 400 × 30, 400 × 7, 20 × 30, 20 × 7, 6 × 30 and 6 × 7.

Grid method for multiplication   An alternative way of recording the steps in the areas method, without actually drawing the rectangle.

Long multiplication   A condensed and formal written algorithm for multiplying two numbers, based on the distributive law for multiplication. In this procedure, for example, 426 × 37 is calculated in two steps: 426 × 30 and 426 × 7.

Short division   A compact standard algorithm for a division calculation involving a single-digit divisor. The divisor is divided into each digit in turn, working from left to right, with any remainders being transferred to the next column.

Ad hoc subtraction   An acceptable written method for doing division calculations, in which ad hoc multiples of the divisor are subtracted from the dividend until no more can be subtracted. The method is easier to understand than long division and learners can operate at their own level of confidence with number relationships.

Long division   A condensed and formal written method for division by two-digit numbers (and larger). The procedure is difficult to understand and involves some tricky multiplications; learners have to be able to recall accurately a complicated sequence of steps.

SUGGESTIONS FOR FURTHER READING FOR SECTION C

1   Chapter 2 of Haylock and Cockburn (2017) is about understanding number and counting. We outline the mathematical development of number, through natural numbers, integers and rational numbers to real numbers. Chapter 7 is on understanding place value. Chapter 3 is on understanding addition and subtraction and Chapter 4 is about understanding multiplication and division. These chapters will provide the opportunity to look again at the structures of the four operations, specifically from the perspective of teaching young children.

2   Chapters 2, 3 and 4 in Hansen (2020) provide interesting analyses of children's errors and misconceptions in understanding number and in handling number operations.

3   Chapter 3 of Nunes and Bryant (1996) is on understanding numeration systems. The authors of this classic study provide an insightful analysis of young children's understanding of numeration systems. Chapters 6 and 7 of this book are recommended for readers who want to dig more deeply into children's understanding of the ideas of addition, subtraction, multiplication and division.

4   Chapter 6, 'Highlighting the learning process', written by Littler and Jirotková, in Cockburn and Littler (2008), explores some of the difficulties children encounter with place value.

5   There is a useful section on mental calculation strategies for addition and subtraction in Chapter 3 of English, R. (2013). Chapter 4 of this book shows how informal methods for calculation can help to develop formal written methods. Chapter 5 is a helpful guide to traditional written methods for doing calculations.

6   Murphy (2011), in an article comparing the use of the empty number line in primary schools in England and the Netherlands, reveals some significant differences in the ways teachers in the two countries teach children strategies for addition and subtraction.

7   Search out the classic study of the unorthodox calculation skills of Brazilian street children in Nunes et al. (1993). This book may make you reassess the balance between teaching formal calculation methods and encouraging children to develop mental and informal approaches that make sense to them.

8   Robinson's chapter on 'Learning effective strategies for mental calculation', in Koshy and Murray (2011), provides practical guidance on teaching mental strategies for all four operations.

9   Guidance on teaching calculations is available on the NCETM website (NCETM, 2015).

10 Clarke (2005), in Australia, reviews the reasons given for an emphasis on teaching written algorithms for calculations in primary school, identifying the potential detrimental effects on the development of mental strategies and number sense; while, also in Australia, Hartnett (2015) reports on the apparent success of a project in a primary school that taught no formal written methods at all.

11 For an extensive review of research into children's learning of number concepts and all four operations with whole numbers, see Verschaffel, Greer and De Corte (2007).

12 Barmby et al. (2009) examine whether the array representation can support children's understanding and reasoning in multiplication. They adopt a 'representational-reasoning' model of understanding, where understanding is seen as connections being made between mental representations of concepts, with reasoning linking together the different parts of the learner's understanding.

13 Check out Ofsted's Research review of mathematics (Ofsted, 2021) and the detailed response to this by the primary mathematics group of the Association of Teachers of Mathematics and the Mathematical Association (ATM/MA, 2021).

SECTION D

FURTHER NUMBER CONCEPTS AND SKILLS

13	Natural Numbers: Some Key Concepts	221
14	Integers: Positive and Negative	242
15	Fractions and Ratios	250
16	Decimal Numbers and Rounding	271
17	Calculations with Decimals	293
18	Proportionality and Percentages	313

WATCH THE SECTION OPENER VIDEO AT: HTTPS://STUDY.SAGEPUB.COM/HAYLOCK7E

NATURAL NUMBERS: SOME KEY CONCEPTS

IN THIS CHAPTER, THERE ARE EXPLANATIONS OF

- multiples, including lowest common multiple;
- some ways of spotting multiples of various numbers;
- digital sums and digital roots;
- factors, including highest common factor;
- the transitive property of multiples and factors;
- prime numbers and composite (rectangular) numbers;
- square numbers and some of their properties;
- cube numbers;
- square roots and cube roots;
- the relationship between sequences of geometric and numerical patterns;
- triangle numbers.

READ THIS CHAPTER'S CURRICULUM LINKS AT: HTTPS://STUDY.SAGEPUB.COM/HAYLOCK7E

WHAT DO I NEED TO KNOW ABOUT MULTIPLES?

It should be made clear at the start that in this chapter we are focusing here only on *natural numbers* (see Chapter 6). These are the numbers we use for counting: 1, 2, 3, 4, 5, 6 and so on, continuing ad infinitum. So when you read the word 'number' in this chapter, it refers to one of these. Hence, when we say 'number' we are excluding zero, negative numbers and anything other than positive whole numbers. It is also the case that some of the material in this chapter goes beyond the level you might expect to teach in a primary school, but I include it in the hope that I can encourage the reader to be fascinated by numbers, their properties and patterns.

We have already met the concept of 'multiple' in Chapter 4 (see the glossary at the end of that chapter). The multiples of any given (natural) number are obtained by multiplying the number in turn by each of the natural numbers. For example:

- multiples of 3 are: 3, 6, 9, 12, 15, 18, 21, 24, 27, 30, 33, 36, 39, 42, 45 and so on;
- multiples of 7 are: 7, 14, 21, 28, 35, 42, 49, 56, 63, 70, 77, 84, 91 and so on;
- multiples of 37 are: 37, 74, 111, 148, 185, 222, 259, 296, 333, 370 and so on.

Note that to say that *b* is a *multiple* of *a* is equivalent to saying that *b* is **divisible** by *a*. For example, 28 is a multiple of 7, so 28 is divisible by 7. This means that 28 can be divided by 7 without any remainder.

The mathematical relationship 'is a multiple of' applied to numbers possesses what is called the **transitive property**. Formally, this means that if *A* is a multiple of *B* and *B* is a multiple of *C*, then it follows that *A* is a multiple of *C*. This is illustrated in general terms in Figure 13.1(a). Figure 13.l(b) gives an example: any number that is a multiple of 6, such as 24, must also be a multiple of 3, because 6 itself is a multiple of 3. Applying this principle, we can deduce that all multiples of 6 are multiples of 3 (but not vice versa). Similarly, all multiples of 28 must be multiples of 7, because 28 is itself a multiple of 7.

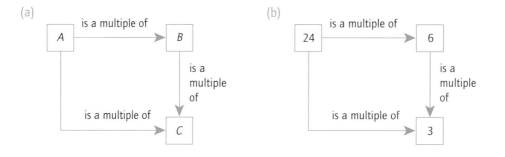

Figure 13.1   The transitive property of multiples

Being able to recognize multiples and having an awareness of some of the patterns and relationships within them help to develop a high level of confidence and pleasure in working with numbers. For example, this pattern in the multiples of 37 may appeal to some readers:

$3 \times 37 = 111$

$6 \times 37 = 222$

$9 \times 37 = 333$

$12 \times 37 = 444$

$15 \times 37 = 555$

$18 \times 37 = 666$

The pattern in the multiples of 9 up to 90, with the tens digit increasing by one and the units digit decreasing by one each time, is a useful aid for learning the 9-times table:

$1 \times 9 = 09$

$2 \times 9 = 18$

$3 \times 9 = 27$

$4 \times 9 = 36$

$5 \times 9 = 45$

$6 \times 9 = 54$

$7 \times 9 = 63$

$8 \times 9 = 72$

$9 \times 9 = 81$

$10 \times 9 = 90$

There are a number of ways of spotting certain multiples. These are sometimes called **rules of divisibility**. For example, you will surely be able to tell at a glance that all these numbers are multiples of ten: 20, 450, 980, 7620. That is using a very obvious rule, namely, that all multiples of 10 end with the digit zero and all numbers ending in zero are divisible by 10. Similarly, you probably know that all multiples of two (the even numbers) have 0, 2, 4, 6 or 8 as their final digit; and that all multiples of five end in 0 or 5.

There is a simple way to spot whether a number greater than 100 is a multiple of 4. Since 100 is a multiple of 4, then any multiple of 100 is a multiple of 4. So, given, for example, 4528, we can think of it as 4500 + 28. We know that the 4500 is a multiple of 4, because it's a multiple of 100. So all we need to decide is whether the 28 is a multiple of 4, which it is. Hence, if you have a number with three or more digits, you need only look at the last two digits to determine whether or not you are dealing with a multiple of 4.

WATCH THE PROBLEM SOLVED! VIDEO AT: HTTPS://STUDY.SAGEPUB.COM/HAYLOCK7E

Would you spot immediately that all these are multiples of nine: 18; 72; 315; 567; 4986? If so, you may be making use of the **digital sum** for each number. This is the number you get if you add up the digits in the given number. If you then add up the digits in the digital sum, and keep going with this process of adding the digits until a single-digit answer is obtained, the number you get is called the **digital root**. For example, 4986 has a digital sum of 27. This is itself a multiple of nine! This is true for any multiple of 9: the digital sum is itself a multiple of 9. If you then add up the digits of this digital sum (2 + 7) you get the single-digit number, 9, which is therefore the digital root. Fascinatingly, the digital root of a multiple of 9 is always 9.

Here is a summary of some useful tricks for spotting various multiples:

- Every natural number is a multiple of 1.
- All even numbers (numbers ending in 0, 2, 4, 6 or 8) are multiples of 2.
- A number that has a digital sum that is a multiple of 3 is itself a multiple of 3.
- The digital root of a multiple of 3 is always 3, 6 or 9.
- If the last two digits of a number give a multiple of 4, then it is a multiple of 4.
- Any number ending in 0 or 5 is a multiple of 5.
- Multiples of 6 are multiples of both 3 and 2. So, any even number with a digital root of 3, 6 or 9 must be a multiple of 6.
- If the last three digits of a number give a multiple of 8, then the number is a multiple of 8.
- A number that has a digital sum that is a multiple of 9 is itself a multiple of 9.
- The digital root of a multiple of 9 is always 9.
- Any number ending in 0 is a multiple of 10.

WHAT IS A 'LOWEST COMMON MULTIPLE'?

If we list all the multiples of each of two numbers, then inevitably there will be some multiples common to the two sets. For example, with 6 and 10, we obtain the following sets of multiples:

- Multiples of 6: 6, 12, 18, 24, 30, 36, 42, 48, 54, 60, 66, …
- Multiples of 10: 10, 20, 30, 40, 50, 60, 70, …

The numbers common to the two sets are the 'common multiples' of 6 and 10: 30, 60, 90, 120 and so on. The smallest of these (30) is known as the **lowest common multiple**. The lowest common multiple of 6 and 10 is therefore the smallest number that can be split up into groups of 6 and into groups of 10. This concept occurs in a number of practical situations. For example, a class of 30 children is the smallest class-size that can be organized into groups of 6 children for mathematics and teams of 10 for games. Or, if I have to feed one plant every six days and another plant every ten days, then the lowest common multiple indicates the first day on which both plants have to be fed, namely, the thirtieth day. Or, if I can only buy a certain kind of biscuit in packets of 10 and I want to share all the biscuits equally between 6 people, the number of biscuits I buy must be a multiple of both 6 and 10; so the smallest number I can purchase is the lowest common multiple, which is 30 biscuits. In Chapter 15, we shall see that the lowest common multiple of the bottom numbers can be useful when comparing two fractions or when doing additions and subtractions with fractions.

WHAT IS A FACTOR?

The concept of *factor* (see glossary for Chapter 11) is the inverse of multiple. So, if *A* is a multiple of *B* then *B* is a factor of *A*. For example, 24 is a multiple of 6, so 6 is a factor of 24, as illustrated in Figure 13.2. Colloquially, we say '6 goes into 24'. This means the same thing as '6 is a factor of 24'.

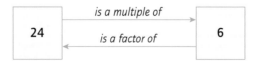

Figure 13.2   If *A* is a multiple of *B* then *B* is a factor of *A*

So the factors of 24 are all those natural numbers by which 24 can be divided exactly: 1, 2, 3, 4, 6, 8, 12 and 24. Notice that 1 and 24 are included as factors of 24. Of course, 1 is a factor of all numbers and every number is a factor of itself. The ability to recognize quickly all the factors of a given number is very useful and is another indication of an individual's confidence and familiarity with numbers. For example, with a set of 24 people we should know instantly that they can be put into groups of 2, 3, 4, 6, 8 or 12. This makes a number like 24, with lots of factors, much more useful for many practical purposes than a number like 23, which has no factors other than 1 and itself.

The idea of a *rectangular array*, introduced in Chapter 10, provides a good illustration of the concept of a factor. Figure 13.3 shows all the different rectangular arrays possible with a set of 24 crosses. The dimensions of these arrays are all the possible **factor pairs** for 24: 1 and 24; 2 and 12; 3 and 8; 4 and 6.

X X X X X X X X X X X X X X X X X X X X X X X X   X X X X
1 by 24   X X X X
                                                     X X X X
                           X X X X X X X X   X X X X
X X X X X X X X X X X X   X X X X X X X X   X X X X
X X X X X X X X X X X X   X X X X X X X X   X X X X
2 by 12   3 by 8   4 by 6

Figure 13.3 Factors of 24 shown in rectangular arrays

If, for some reason, you want to know if 23 is a factor of 1955, then the simplest way is to use a calculator to divide 1955 by 23: if the answer is a whole number then 23 is a factor, otherwise it is not. Using a calculator, I get 1955 ÷ 23 = 85, so 23 *is* a factor of 1955. Is 15 a factor of 1955? Using a calculator, I get 1955 ÷ 15 = 130.33333, so 15 is *not* a factor of 1955.

The mathematical relationship, 'is a factor of', also possesses the transitive property, as illustrated in Figure 13.4. So, for example, any factor of 12 must be a factor of 24, because 12 is a factor of 24.

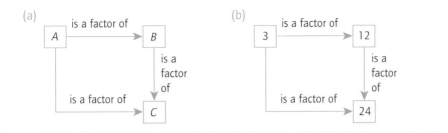

Figure 13.4   The transitive property of factors

WHAT ABOUT THE 'HIGHEST COMMON FACTOR'?

This is a similar idea to the lowest common multiple. If we list all the factors of two numbers, the two sets of factors may have some numbers in common. (Since 1 is a factor of all natural numbers, they must at least have this number in common!) The largest of these common factors is called the **highest common factor** (or greatest common factor). For example, with 24 and 30 we have the following two sets of factors:

- Factors of 24: 1, 2, 3, 4, 6, 8, 12, 24.
- Factors of 30: 1, 2, 3, 5, 6, 10, 15, 30.

The factors in common are 1, 2, 3 and 6. So the highest common factor is 6. This concept occurs in a number of practical situations. For example, imagine that two classes in a year group have 24 and 30 children respectively, and that we wish to divide them up into a number of groups, with each class of children shared equally between the groups. The number of groups must clearly be a factor of both 24 and 30. The largest possible number of groups is therefore 6, since this is the highest common factor. Or, imagine that some teachers in a primary school require 40 exercise books a term and others require 32. It would be convenient to store the exercise books in packs of 8, so that some teachers can pick up 5 packs and others 4 packs. Why 8? Because 8 is the highest common factor of 40 and 32. We shall meet highest common factor again in Chapter 15 where it has an important role in calculations with fractions.

WHAT IS A PRIME NUMBER?

Any number that has precisely two factors, and no more than two, is called a **prime number**. This is the strict mathematical definition. In practice, we think of a prime number as a number that cannot be divided exactly by any number apart from 1 and itself. So, for example, 7 is a prime number because it has precisely two factors, namely 1 and 7. But 10 is not a prime number because it has four factors, namely 1, 2, 5 and 10. A number, such as 10, with more than two factors is sometimes called a **composite number**, or, because it can be arranged as a rectangular array with more than one row (see Figure 13.5), a *rectangular number*. Prime numbers cannot be arranged as rectangular arrays, other than with a single row. The first twenty prime numbers are: 2, 3, 5, 7, 11, 13, 17, 19, 23, 29, 31, 37, 41, 43, 47, 53, 59, 61, 67 and 71.

(a) X X X X X   (b)

X X X X X   X X X X X X X

10   7

Figure 13.5   (a) 10 is a composite number; (b) 7 is a prime number

WATCH THE PROBLEM SOLVED! VIDEO AT: HTTPS://STUDY.SAGEPUB.COM/HAYLOCK7E

Notice that according to the strict mathematical definition, the number 1 is *not* a prime number since, uniquely, it has only one factor (itself). So 1 is the only number that is neither prime nor composite. The exclusion of 1 from the set of prime numbers often puzzles students of mathematics. The reason is related to the most important property of prime numbers: given any composite number whatsoever, there is only one combination of prime numbers that, multiplied together, gives the number.

I will illustrate this with the number 24. This number can be obtained by multiplying together various combinations of numbers, such as: 2×12, $2 \times 2 \times 6$, $1 \times 2 \times 3 \times 4$ and so on. If, however, we stipulate that only *prime* numbers can be used, there is only one combination that will produce 24: namely, $2 \times 2 \times 2 \times 3$. This is called the **prime factorization** of 24. This is such a powerful property of prime numbers that it would be a pity to mess it up by allowing 1 to be a prime number! If we did, then the prime factorization of 24, for example, would not be unique, because we could also get to 24 with $1 \times 2 \times 2 \times 2 \times 3$, or $1 \times 1 \times 2 \times 2 \times 2 \times 3$ and so on. The theorem that states that there is only one prime factorization of any composite number is considered to be so important in number theory that it is actually called the fundamental theorem of arithmetic. So perhaps we had better not undermine it by calling 1 a prime number!

The study of primes is a fascinating branch of number theory. Nowadays, computers are employed to search for very large numbers that are prime. At the time of writing, to my knowledge, the largest known prime number is '$2^{82,589,933} - 1$'. This means 82,589,933 twos multiplied together, minus 1: this produces a number with nearly 25 million digits. To get some sense of how big this number is, just to write this number down would require 15 books the size of this one! By the time you read this, there may well be a new 'largest known prime number', since searching for larger and larger primes continues to be encouraged by the offer of a prize for finding the first prime number with more than 100 million digits! However, when they find one this big, it will definitely not be the largest prime number, because, as the Greek mathematician, Euclid, proved, as long ago as 300 BC, there is no largest prime number. The really annoying facet of prime numbers is that there is no pattern or formula that will generate the complete set of prime numbers.

WHAT IS THE POINT OF LEARNING ABOUT MULTIPLES, FACTORS AND PRIMES?

I recall once, in response to this question, asking a student whether she felt differently about the numbers 47 and 48. I was surprised to be told that she did not! To me, 48 seems such a friendly number, flexible and amenable. If I have 48 in a group, there are so many ways I can reorganize them: 6 sets of 8, 3 sets of 16, 4 sets of 12 and so on. By contrast, 47 is such an awkward number! The difference, of course, is that 47 is prime, but 48 has lots of factors. This is all part of what is sometimes called 'having a feel for number'. Our confidence in responding to numbers in the everyday situations where they occur will be improved enormously by having this kind of feel for numbers; by being aware of the significant relationships between them; and by recognizing at a glance which properties they possess and which they do not. And the more we are aware of properties like multiples, factors and primes, the more we learn to delight in the pattern and fascination of number. Being able to spot at a glance which car registration numbers are multiples of 11, or how many of the hymn numbers in church on Sunday morning are prime, might be diverting but is of no immediate practical use. However, it all leads to yet greater confidence when we have to respond to numerical situations that do matter. Whether the reader is convinced or not by this argument will, no doubt, be evident in the level of enthusiasm with which they tackle the self-assessment questions at the end of this chapter.

> **LEARNING AND TEACHING POINT**
>
> Encourage children to be fascinated by number and patterns in number. By exploring the properties of multiples, factors and prime numbers, build up children's feel for number and therefore their confidence in responding to numerical situations.

WHY ARE SOME NUMBERS CALLED SQUARES? A SQUARE IS A SHAPE, ISN'T IT?

Connecting pictures with number concepts helps to build up our understanding and confidence. We have seen earlier in this chapter that a prime number, like 7, can be represented as a rectangle of dots only with one row, whereas a composite (rectangular) number, like 10, can be shown as a rectangular array with more than one row (see Figure 13.5).

> **LEARNING AND TEACHING POINT**
>
> Use square arrays of dots and square grids to explain square numbers. Connect the square of a number with the area of a square grid, given by counting the number of square units in the grid.

Some rectangles have *equal* sides: these are the rectangles that are called **squares** (see Chapter 25). So numbers, such as 1, 4, 9, 16, 25 and so on, which can be represented by square arrays, as shown in Figure 13.6, are called **square numbers**. If we use an array of small squares, called **square units**, as in Figure 13.6(b), rather than just dots, as in Figure 13.6(a), then the number of squares in the array also corresponds to the total area (see Chapter 22). For example, the area of the 5 by 5 square grid is 25 square units. Square numbers are also composite (rectangular) numbers, just as squares are rectangles. (Strictly speaking, there is one exception: the number 1 is considered a square number but, as we saw earlier, it is neither prime nor composite.)

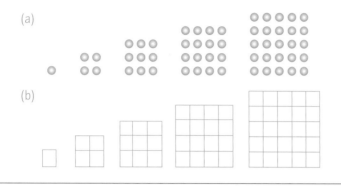

Figure 13.6 Pictures of square numbers: (a) using dots; (b) using square grids

This explains the geometric idea of square numbers. The arithmetic idea that corresponds to this is that a square number is any number that is obtained by multiplying a (whole) number by itself. The number 16 is represented in Figure 13.6 by 4 rows of 4 dots, or 4 rows of 4 squares, which, of course, corresponds to the multiplication 4 × 4. Likewise, the representation of 25 as a square corresponds to 5 × 5. There is a special mathematical notation that can be used as shorthand for writing 5 × 5. This is 5^2, which means simply that there are two fives to be multiplied together. This is read as 'five **squared**' or 'five to the power two' (see the discussion about powers of 10 in Chapter 6).

The square numbers can be obtained easily using a basic, non-scientific calculator. For example, to find 6 squared, just enter: 6, ×, =. This sequence of keys multiplies 6 by itself. (This works on most basic calculators, but may be different on some. A scientific calculator would have a specific squaring function.) This is the numerical pattern for the set of square numbers:

$1 \times 1 = 1^2 = 1$

$2 \times 2 = 2^2 = 4$

$3 \times 3 = 3^2 = 9$

$4 \times 4 = 4^2 = 16$

$5 \times 5 = 5^2 = 25$

$6 \times 6 = 6^2 = 36$

$7 \times 7 = 7^2 = 49$

$8 \times 8 = 8^2 = 64$

$9 \times 9 = 9^2 = 81$ and so on.

WHAT ARE SOME INTERESTING THINGS ABOUT SQUARE NUMBERS?

First, here is an example of a pattern in square numbers, revealed when you connect the numbers with the picture of a square array. I start with the question: can you make a 4-by-4 square using strips of 3 units? Figure 13.7(a) shows that you can nearly do it with 5 strips. You just need 1 more unit in the bottom right-hand corner. So the picture here shows that 4^2 is equal to 5 lots of 3 plus 1. So what about making a 5-by-5 square using strips of 4 units? Figure 13.7(b) shows that this can be nearly done with 6 strips of 4, but once again you need 1 more unit in the corner. So this picture shows that 5^2 is equal to 6 lots of 4 plus 1.

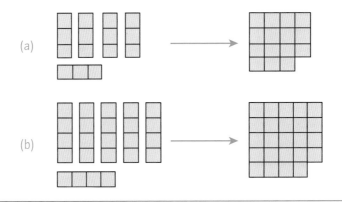

(a)

(b)

Figure 13.7   Pictures showing: (a) $4^2 = (5 \times 3) + 1$; (b) $5^2 = (6 \times 4) + 1$

A pattern is emerging. Check this out with a 6-by-6 square made from strips of 5 units, and a 7-by-7 square made from strips of 6 units. You should easily be able to use the visual images to confirm this numerical pattern:

$2^2 = (3 \times 1) + 1$

$3^2 = (4 \times 2) + 1$

$4^2 = (5 \times 3) + 1$

$5^2 = (6 \times 4) + 1$

$6^2 = (7 \times 5) + 1$ and so on.

So, you have momentarily forgotten what 7^2 is equal to? Well, it's just 8 times 6, plus 1 (48 + 1 = 49). You can't remember 11^2? Well, that would be 12 times 10, plus 1 (120 + 1 = 121). You want to work out 21^2? Easy! That's 22 multiplied by 20, plus 1; which is 440 + 1 = 441.

What about 1^2? Does that fit the pattern, or is it a 'special case'? (See Chapter 4.) This is left for the reader to consider in self-assessment question 13.10 at the end of this chapter.

In Chapter 4 we identified another interesting property of square numbers that could be explored by older children in primary school. Most positive integers have an even number of factors. This is because factors tend to come in pairs. For example, 24 has four factor pairs, giving a total of 8 factors: 1 and 24; 2 and 12; 3 and 8; 4 and 6 (each pair multiplying together to make 24). Similarly, 15 has four factors: 1 and 15; 3 and 5. And, 60 has twelve factors: 1 and 60; 2 and 30; 3 and 20; 4 and 15; 5 and 12; 6 and 10. But what about a square number? A little exploration reveals that square numbers always have an *odd* number of factors! For example, 36, which is a square number, has 9 factors: 1 and 36; 2 and 18; 3 and 12; 4 and 9; and 6. The 6 at the end of

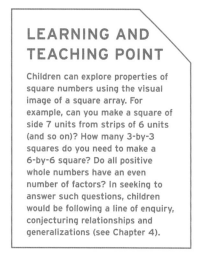

LEARNING AND TEACHING POINT

Children can explore properties of square numbers using the visual image of a square array. For example, can you make a square of side 7 units from strips of 6 units (and so on)? How many 3-by-3 squares do you need to make a 6-by-6 square? Do all positive whole numbers have an even number of factors? In seeking to answer such questions, children would be following a line of enquiry, conjecturing relationships and generalizations (see Chapter 4).

this list is multiplied by itself to give 36, so it's on its own in the list of factors, not one of a factor pair. In the same way, every square number has an odd number of factors. For example, 121 has 3 factors (1, 11 and 121); and 64 has 7 factors (1, 2, 4, 8, 16, 32 and 64).

WHAT ARE CUBE NUMBERS?

Just as some numbers can be represented by square arrays, there are those, such as 1, 8 and 27, that can be represented by arrangements in the shape of a **cube**. These **cube numbers** will turn up in exploring the volumes of cubes with older children. Figure 13.8

shows how the first three cube numbers are constructed from small cubes, called **cubic units**. The first three cube numbers are made from 1 cubic unit, from 8 cubic units and from 27 cubic units, respectively. The 27, for example, is produced by 3 layers of cubes, with 3 rows of 3 cubes in each layer, and is therefore equal to 3 × 3 × 3. The number of cubic units in the whole construction corresponds to the total volume of the cube. For example, the volume of the 3-by-3-by-3 cube is 27 cubic units. It is difficult, of course, to represent these cubic constructions in a two-dimensional picture, so readers are encouraged to build various-sized cubes from cubic units and to generate these cube numbers for themselves.

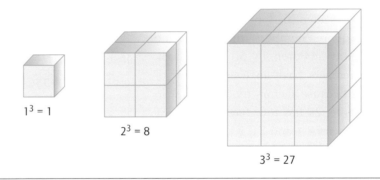

$1^3 = 1$

$2^3 = 8$

$3^3 = 27$

Figure 13.8 Examples of cube numbers

The same kind of notation is used for cube numbers as for square numbers: 3 × 3 × 3 is abbreviated to 3^3 and read as 'three **cubed**' or 'three to the power of three' or 'three to the power three'. Again, it is easy to generate cube numbers using a simple calculator; for example, to obtain 6 cubed, simply enter: 6, ×, =, =. By analogy with the square numbers above, we can construct the following pattern for the cube numbers:

$1 \times 1 \times 1 = 1^3 = 1$

$2 \times 2 \times 2 = 2^3 = 8$

$3 \times 3 \times 3 = 3^3 = 27$

$4 \times 4 \times 4 = 4^3 = 64$

$5 \times 5 \times 5 = 5^3 = 125$

$6 \times 6 \times 6 = 6^3 = 216$ and so on.

> ## LEARNING AND TEACHING POINT
>
> Cube numbers can be explored by older and more able children in primary school. Get them to construct cubes from cubic units. Connect the cube of a number with the volume of the cube (see Chapter 22), given by counting the number of cubic units used to construct it.

WHAT ARE SQUARE ROOTS AND CUBE ROOTS?

What is the length of the side of a square that has an area of 729 square units? Another way of asking the same question is: Which number, when multiplied by itself, gives 729? Or, which number has a square equal to 729? The answer (27) is called the **square root** of 729. (Strictly speaking, 27 is the *positive* square root of 729; you can also get 729 by multiplying the negative number, –27, by itself, but we are not concerned with negative numbers in this chapter.) The mathematical abbreviation for 'the (positive) square root of' is: $\sqrt{}$. So we could write, for example, $\sqrt{729} = 27$. Finding a square root is the *inverse* process of finding a square. This means that one process undoes the effect of the other, for example:

$2^2 = 4$ so $\sqrt{4} = 2$

$3^2 = 9$ so $\sqrt{9} = 3$

$4^2 = 16$ so $\sqrt{16} = 4$

$27^2 = 729$ so $\sqrt{729} = 27$

The recurrent mathematical idea of inverse was introduced in Chapter 7, where we saw addition and subtraction as inverse processes, and in Chapter 10 where we saw multiplication and division as inverse processes.

The idea of a **cube root** follows the same logic. In geometric terms, the question would be: what is the length of the side of a cube with a total volume of 729 cubic units? Or, in arithmetic terms, what number has a cube equal to 729? The answer (9) is called the cube root of 729. As with square roots, we can think in terms of an *inverse* process: finding the cube root is the inverse process of finding the cube. For example, the cube of 14 is 2744, so the cube root of 2744 is 14. The symbol for a cube root is: $\sqrt[3]{}$. So, for example, we could write: $\sqrt[3]{2744} = 14$.

> ### LEARNING AND TEACHING POINT
>
> Squaring and finding a square root are interesting examples of inverse processes to discuss with older primary school children. Inverse processes (like addition and subtraction, or multiplication and division) are those where one process undoes the effect of the other (see Chapters 7 and 10).

Unlike multiples, factors, primes and composite numbers, the concepts of squares, cubes, square roots and cube roots can be extended beyond just natural numbers and applied particularly to decimal numbers (see Chapter 17).

ARE THERE OTHER SETS OF GEOMETRIC SHAPES THAT CORRESPOND TO SETS OF NUMBERS?

Almost any sequence of geometric shapes or patterns, such as those shown in Figure 13.9, can be used to generate a corresponding set of numbers. Exploring these kinds of sequences, trying to relate the geometric and numerical patterns, can produce some intriguing mathematics. For example, the reader might consider why the sequence of patterns in Figure 13.9(a) generates the odd numbers: 1, 3, 5, 7, 9 and so on; and why the sequence in Figure 13.9(b) generates the multiples of three: 3, 6, 9, 12 and so on.

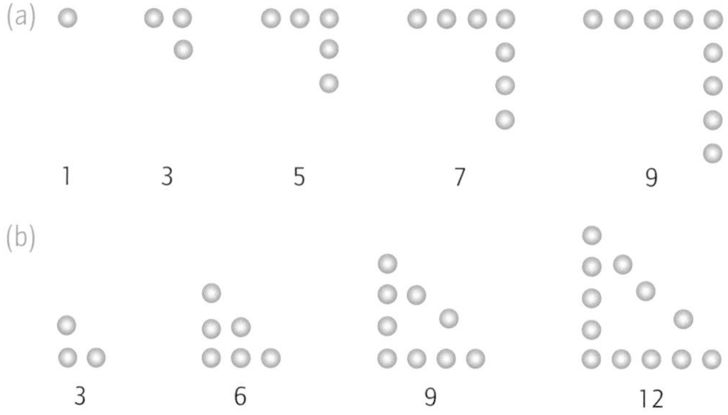

Figure 13.9 Geometric patterns generating sets of numbers

Some of these patterns turn out to be particularly interesting and are given special names. For example, you may come across the so-called **triangle numbers**. These are the numbers that correspond to the particular pattern of triangles of dots shown in Figure 13.10: 1, 3, 6, 10, 15 and so on. Notice that you get the second triangle by adding two dots to the first; the third by adding three dots to the second; the fourth by adding four dots to the third; and so on. The geometric arrangements of dots show that these triangle numbers have the following numerical pattern:

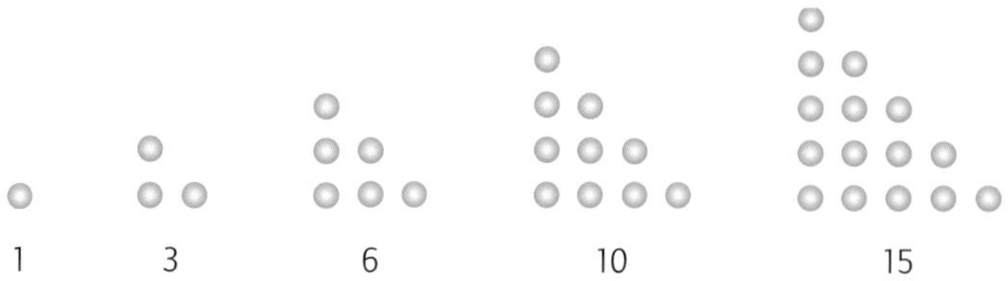

Figure 13.10 Triangle numbers

$1 = 1$

$3 = 1 + 2$

$6 = 1 + 2 + 3$

$10 = 1 + 2 + 3 + 4$

$15 = 1 + 2 + 3 + 4 + 5$ and so on.

A general formula for triangle numbers is given in Chapter 19. Self-assessment questions 13.10 and 13.11 below provide two examples of explaining patterns in number sequences by thinking about the geometric patterns to which they correspond.

> ## LEARNING AND TEACHING POINT
>
> Give children opportunities to investigate the relationships between sequences of geometric patterns and numerical sequences. The kind of thinking involved is an introduction to algebraic reasoning, involving the recognition and articulation of generalizations.

RESEARCH FOCUS: AWARENESS OF MATHEMATICAL PATTERN AND STRUCTURE

In this chapter, I have highlighted the significance of children connecting numerical and spatial patterns in their mathematical development. Mulligan and Mitchelmore (2009), working with children aged 5–6 years, have identified a construct they call 'Awareness of Mathematical Pattern and Structure' (AMPS), which they claim can be reliably measured and which they have found to be a strong indicator of general mathematical understanding. The children were given a range of tasks that required them to identify, visualize, represent or replicate elements of pattern and structure, such as: identifying multiples on a number track; completing a partially drawn 3-by-4 rectangular grid; drawing a triangular pattern of six dots and then extending it. The children who showed the most developed level of AMPS produced representations of mathematical patterns that correctly integrated numerical and spatial structural features. The researchers provide evidence that a focus on patterning can lead to a significant improvement in mathematical achievement, and propose that an exploration of pattern and structure should be the core of our teaching of mathematics in primary school.

In this respect, a paper by Fluellen (2008) makes intriguing reading. Working with children aged 4 and 5 years (in a US kindergarten), using story and games as starting points, Fluellen enabled them to explore number, pattern and relationships in simple arrays, using objects and drawings. One of the games involved a story about a magic pot found in a garden that doubled everything that was put into it. The children were able to show what happened to an input of 5 pieces of chocolate by, for example, making a 2 × 5 array of pieces of chocolate on a chessboard. In the next story, there was another pot that squared everything, so if the child put in 3 pieces of chocolate, 9 came out. Even children as young as these were able to engage with square numbers in this way, independently creating arrays for squares of natural numbers from 1 to 5. Examples are cited of children discussing the pattern of growth in the square numbers, connecting what the magic pot did and their square arrays by writing 5 × 5 = 25, and showing mathematical memory based on generalizations.

LEARNING RESOURCES

Access activities for your **lesson plans** at: https://study.sagepub.com/haylock7e

Before trying the self-assessment questions below, you should complete the **self-assessment questions** for this chapter at: https://study.sagepub.com/haylock7e

13.1: Drawing on the ideas of this chapter, find at least one interesting thing to say about each of the numbers from 20 to 29.

13.2: Continue the pattern shown earlier in this chapter for certain multiples of 37. When does the pattern break down?

13.3: Use some of the methods for spotting multiples to decide whether the following numbers are multiples of 2, 3, 4, 5, 6, 8 or 9: (a) 2652; (b) 6570; and (c) 2401.

13.4: These numbers are all multiples of 11: (a) 561; (b) 594; (c) 418; (d) 979; and (e) 330. Add up the two outside digits and compare the total with the middle digit. Do this with a few more three-digit multiples of 11. Can you state a rule?

13.5: What is the smallest number of people that can be split up equally into groups of 8 and groups of 12? What mathematical concept is used in solving this problem?

13.6: Find all the factors of: (a) 95; (b) 96; and (c) 97. Which of these two numbers would be most flexible as a year-group size for breaking up into smaller-sized groups for various activities?

13.7: List all the factors of 48 and 80. What is the highest common factor of 48 and 80?

13.8: Starting with 1, add 4, add 2, add 4, add 2, and continue this sequence until you pass 60. How many of the answers are prime?

13.9: Look again at the pattern revealed in Figure 13.7. Does a 1-by-1 square fit this pattern or is it a special case?

13.10: Look at the sequence of square numbers: 1, 4, 9, 16, 25, 36 and so on. Find the differences between successive numbers in this sequence. What do you notice? Can you explain this in terms of patterns of dots, as shown in Figure 13.6(a)?

13.11: List all the triangle numbers less than 100. Find the sums of successive pairs of triangle numbers, for example, 1 + 3 = 4, 3 + 6 = 9, 6 + 10 = 16 and so on. What do you notice about the answers? Can you explain this numerical pattern by reference to the geometric patterns for triangle numbers shown in Figure 13.10?

FURTHER PRACTICE

FROM THE STUDENT WORKBOOK

Questions 13.01–13.24: Checking understanding (natural numbers: some key concepts)

Questions 13.25–13.40: Reasoning and problem solving (natural numbers: some key concepts)

Questions 13.41–13.54: Learning and teaching (natural numbers: some key concepts)

GLOSSARY OF KEY TERMS INTRODUCED IN CHAPTER 13

Divisible   The natural number n is divisible by the natural number m if n can be divided by m exactly, without any remainder. This is equivalent to saying that n is a multiple of m.

Rules of divisibility   The various ways of spotting that a (natural) number is divisible by a particular number (and therefore a multiple of it). For example, any natural number ending in 0 or 5 is divisible by 5.

Transitive property   A property that any given mathematical relationship may or may not possess; the property is that if A is related to B and B is related to C then it always follows that A is related to C. Each of the relationships 'is a factor of' and 'is a multiple of' is transitive.

Digital sum   The sum of all the digits in a given natural number; for example, the digital sum of 8937 is 27 (8 + 9 + 3 + 7).

Digital root   The result of finding the digital sum of the digital sum of a natural number repeatedly until a single-digit answer is obtained; for example, 8937 has a digital sum of 27 (because 8 + 9 + 3 + 7 = 27) and therefore a digital root of 9 (because 2 + 7 = 9).

Lowest common multiple   For two (or more) natural numbers, the smallest number that is a multiple of both (or all) of them.

Factor pair   Two factors of a number whose product is equal to the number; for example, factor pairs for 20 are 1 and 20, 2 and 10, 4 and 5.

Highest common factor   For two (or more) natural numbers, the highest number that is a factor of both (or all) of them; for example, the highest common factor of 24 and 36 is 12. For 'factor', see glossary for Chapter 11.

Prime number   A natural number that has precisely two factors (namely, 1 and itself). The first ten prime numbers are 2, 3, 5, 7, 11, 13, 17, 19, 23 and 29.

Composite (rectangular) number   A natural number that has more than two factors. A composite number can be illustrated as a rectangular array with more than one row; for example, 21 is a composite number (with factors 1, 3, 7 and 21) and can be arranged as 3 rows of 7. All non-prime numbers except 1 are composite.

Prime factorization   Writing a given natural number as the product of prime numbers; for example, the prime factorization of 63 is 3 × 3 × 7. Each composite number has a unique prime factorization.

Square (shape)   A rectangle (see Chapter 25) with all four sides equal in length.

Square number   A number that can be represented as a square array; a number that is obtained by multiplying a whole number by itself. Square numbers are 1, 4, 9, 16, 25, 36, 49, 64 ...

Square unit   A square shape used as a measure of area; for example, a square made up of 5 rows of 5 square units has an area of 25 square units.

Squared   'To the power of 2'; for example, 'five squared' is written 5^2 and is equal to 5 × 5.

Cube (shape)   A solid shape with six square faces and all its edges equal in length.

Cube number   A number that can be represented as an arrangement of cubic units in the shape of a cube; a number that is obtained by multiplying a whole number by itself and by itself again. Cube numbers are 1, 8, 27, 64, 125, 216 …

Cubic unit   A cube shape used as a measure of volume; for example, a cube made up of 5 layers of 5 rows of 5 cubic units has a volume of 125 cubic units.

Cubed   'To the power of 3'; for example, 'five cubed' is written 5^3 and is equal to $5 \times 5 \times 5$.

Square root   The (positive) square root of a given number is the positive number which when squared gives that number; for example, because $5^2 = 25$, the (positive) square root of 25 is 5. In symbols, this is written $\sqrt{25} = 5$.

Cube root   The cube root of a given number is the number which when cubed gives that number; for example, because $5^3 = 125$, the cube root of 125 is 5. In symbols, this is written $\sqrt[3]{125} = 5$.

Triangle numbers   Numbers that can be arranged as triangles of dots in the way shown in Figure 13.10. The set of triangle numbers is 1, 3, 6, 10, 15, 21, 28 and so on. The eighth triangle number, for example, is the sum of the natural numbers from 1 to 8.

INTEGERS: POSITIVE AND NEGATIVE

IN THIS CHAPTER, THERE ARE EXPLANATIONS OF

- how to make sense of negative numbers;
- situations in the contexts of temperatures and bank balances that are modelled by the addition and subtraction of positive and negative numbers.

READ THIS CHAPTER'S CURRICULUM LINKS AT: HTTPS://STUDY.SAGEPUB.COM/HAYLOCK7E

HOW CAN WE MAKE SENSE OF NEGATIVE NUMBERS?

Integers – positive and negative whole numbers and zero – were introduced in Chapter 6. Many people have difficulty with the concept of a negative number, mainly because we overemphasize the idea that a number represents a set of things. There is also some confusion caused by the same symbol being used for the operation of subtraction (a 'minus' sign) and for a negative number. So we might read '3 – 7' as '3 subtract 7' or '3 minus 7', but the result of this subtraction (–4) is read as 'negative 4'.

The concept of a negative number is not difficult if we make strong connections between the number line and the ordinal aspect of number (numbers as labels for putting things in order). The number line, either drawn left to right (see Figures 6.2 and 6.3 in Chapter 6), or drawn vertically with positive numbers going up from zero and negative numbers going down, is the most straightforward image for us to associate with positive and negative integers. There are some other contexts that also help us to make sense of negative numbers.

The most familiar is probably the context of temperature. Quite young children can grasp the idea of the temperature falling below zero, associating this with feeling cold and icy roads, and are often familiar with the use of negative numbers to describe this. We can also associate positive and negative integers with levels in, say, a multi-storey car park or department store, with, for example, 1 being the first floor, 0 being ground level, –1 being one floor below ground level, –2 being two floors below ground level and so on. We have a department store locally that has the buttons in the lift labelled in this way. Similarly, the specification of heights of locations above and below sea level provides another application for positive and negative numbers. For some people, the context of bank balances is one where negative numbers make real, if painful, sense. For example, being overdrawn by £5 at the bank can be represented by the negative number, –5.

LEARNING AND TEACHING POINT

Children can have fun generating the integers on a basic handheld or a simple online calculator. To produce the positive integers, just enter: +, 1, =, =, =, =, ... and continue pressing the equals sign as many times as you wish. To produce the negative integers, just enter: –, 1, =, =, =, =, ... The reader should be warned, however, that various calculators have different ways of displaying negative numbers and that this may not work on more sophisticated calculators.

Finally, in football league tables we find another application of positive and negative integers. If two teams have the same number of points, their order in the table is determined by their goal difference, which is 'the number of goals-for subtract the number of goals-against'. We should note that this is an unusual use of the word 'difference'. Usually, the difference between two numbers is given as a positive number. The difference between 23 and 28 is the same as the difference between 28 and 23, namely 5. But, in the context of football, a team with 28 goals-for and 23 goals-against has a goal difference of +5. And a team with 23 goals-for and 28 goals-against has a goal difference of –5. Many children will find this a relevant and realistic context for experiencing the process of putting in order a set of positive and negative numbers.

> **LEARNING AND TEACHING POINT**
>
> Use familiar contexts such as temperatures, multi-storey buildings, heights above and below sea level, bank balances and 'goal difference' to give meaning to positive and negative numbers.

HOW DO YOU EXPLAIN ADDITION INVOLVING NEGATIVE INTEGERS?

Many primary school teachers and trainee teachers tell me that they found negative numbers something of a mystery when they encountered them at school. The difficulty in making sense of the ways in which we manipulate positive and negative integers is that we really need to use different images to support different operations. With addition, we would need contexts and problems that help us to make sense of such calculations as these:

Example 1: $10 + (-2)$

Example 2: $2 + (-7)$

Example 3: $(-3) + 4$

Example 4: $(-5) + (-3)$

Note that for clarity I am choosing to write the negative numbers in brackets when in a calculation. If using a superscript for the negative sign (see Chapter 6) Example 1 above could also be written as $10 + {}^-2$. So let us try to understand the calculations in Examples 1–4 using bank balances. We can interpret the addition as follows: the first number represents your starting balance; the second number represents either a credit (a positive number) or a debit (a negative number). So, with this interpretation, each of the examples 1–4 above can be seen as a mathematical model (see Chapter 5) for a real-life situation, as follows.

Example 1: We start with £10 and add a debit of £2. The result is a balance of £8. The corresponding mathematical model is: $10 + (-2) = 8$.

Example 2: We start with a balance of £2 and add a debit of £7. The result is a balance of £5 overdrawn. The corresponding mathematical model is: $2 + (-7) = -5$.

Example 3: We start with a balance of £3 overdrawn and add a credit of £4. The result is a balance of £1. The corresponding mathematical model is: $(-3) + 4 = 1$.

Example 4: We start with a balance of £5 overdrawn and add a debit of £3. The result is a balance of £8 overdrawn. The corresponding mathematical model is: $(-5) + (-3) = -8$.

We could also interpret these collections of symbols in the context of temperatures, with the first number being a starting temperature and the second being either a rise or a fall in temperature, or, on the number line, which, of course, is just like the scale on a thermometer, with the first number being the starting point and the second number a move in either the positive or the negative direction. The reader is invited to construct problems of this kind in self-assessment question 14.2.

WHAT ABOUT SUBTRACTION INVOLVING NEGATIVE INTEGERS?

The key to making sense of subtraction with positive and negative integers is to get out of our heads the idea that subtraction means 'take away'. The take-away structure applies only to positive numbers: you cannot 'take away' a negative number. The calculation $6 - (-3)$, for example, cannot model a problem about having a set of six things and 'taking away negative three things': the words here are just nonsense. We make sense of subtracting with negative numbers by drawing on situations that incorporate some of the other structures for subtraction, notably the comparison or the inverse-of-addition structures (see Chapter 7).

LEARNING AND TEACHING POINT

Use additions with positive and negative integers to model simple questions about temperatures falling and rising, with the first number representing a starting temperature and the second a rise or fall of so many degrees. Use parallel examples about bank balances, credits and debits. Never talk about 'taking away' a negative number. This language is meaningless and just adds to the confusion.

The kinds of calculations to which we need to give meaning, through experience in context, would include examples where the first number in the subtraction is greater than the second, such as:

Example 5: 6 − (−3)
Example 6: (−3) − (−8)

We will look at these in the context of temperatures, interpreting the subtractions in terms of the comparison structure. In this structure, the subtraction $a − b$ is asking us to compare a with b. It models a question of the form, 'How much greater is a than b?' So the question in the context of temperature would be to find *how much higher* is the first temperature than the second. In my experience, surprisingly young children can answer questions like, 'How much higher is a temperature of +16 degrees inside than a temperature of −2 degrees outside?', although they would not necessarily record this formally in symbols. Such questions become really straightforward when we connect this language with the picture of the number line, as shown in Figure 14.1.

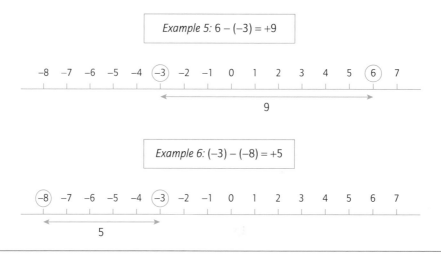

Figure 14.1   Subtractions with integers: interpreted as comparison

Example 5: Using the comparison structure, we could be looking at the difference between a temperature of 6 degrees at noon and a temperature of −3 degrees at midnight. The number line diagram for Example 5 in Figure 14.1 makes it clear that the temperature of 6 degrees is 9 degrees higher than that of −3. The corresponding mathematical model is: 6 − (−3) = +9.

Example 6: Again, using the idea of subtraction as comparison, we might be comparing a temperature of −3 degrees at noon with a temperature of −8 degrees at midnight, as shown in the number line diagram for Example 6 in Figure 14.1. It is clear from the diagram that the noon temperature is 5 degrees higher than the midnight temperature. The corresponding mathematical model is: $(-3) - (-8) = +5$.

This idea of comparison makes a lot of sense when we are using a subtraction to model a situation in which the first temperature is *higher* than the second. Primary school children can experience this idea of subtraction with negative numbers: comparing two numbers, one or both of which might be negative, to find the difference. In the context of temperatures, this will always correspond to a subtraction in which the first number is the higher of the two temperatures. We could just as easily interpret the subtractions in Examples 5 and 6 in terms of comparing bank balances. In Example 5, $(6 - (-3))$ corresponds to comparing a bank balance of £6 in credit with one that is £3 overdrawn. See self-assessment question 14.3.

WATCH THE PROBLEM SOLVED! VIDEO AT: HTTPS://STUDY.SAGEPUB.COM/HAYLOCK7E

I hope that these illustrations make it clear that we do not need a nonsense rule like, 'two minuses make a plus'. Apart from anything else, to be just a little pedantic, in a question such as $6 - (-3)$, remember that the first '−' is a minus sign, indicating subtraction, and the second is a negative sign, indicating a negative number. If we simply interpret the subtraction as 'compare the first number with the second and find the difference' and put these questions into contexts such as temperatures and bank balances, then there is some chance of actually understanding what is going on.

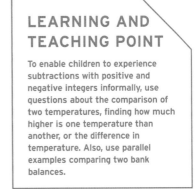

LEARNING AND TEACHING POINT

To enable children to experience subtractions with positive and negative integers informally, use questions about the comparison of two temperatures, finding how much higher is one temperature than another, or the difference in temperature. Also, use parallel examples comparing two bank balances.

However, having explained all this, I should say that the context that gives rise to the problem will very often suggest that the actual calculation which you do makes no use of negative numbers whatsoever. For example, if the problem had been to find the difference in height between two points, one 184 metres above sea level and the other 78 metres below sea level, then the image formed in my mind by the context leads me simply to add 184 and 78.

My conclusion therefore is that we rarely need to use calculations with negative numbers to solve real-life problems; but we do need the real-life problems to help to explain the way we manipulate positive and negative numbers when we are doing abstract mathematical calculations!

RESEARCH FOCUS: METAPHORS FOR NEGATIVE NUMBERS

Kilhamn (2008) identifies four aspects of understanding negative numbers: the idea that a negative number has both magnitude and direction; proficiency in calculations; the meaning of the minus/negative sign; and the use of models and metaphorical reasoning when dealing with calculations involving negative numbers. A key research question is the extent to which the metaphors that are used to promote understanding of negative numbers, such as temperature scales and number lines, actually support the reasoning required to do calculations presented in purely symbolic form. Kilhamn found that relying on a single metaphor created difficulties when the model is insufficient to make sense of all types of calculation. The problem with subtraction of integers on the number line is that different interpretations have to be applied to different calculations in order for them to make any sense. Kilhamn concludes that it is important therefore that learners are aware of the limitations of any metaphor used to make sense of positive and negative integers. A study with 150 primary school children in Turkey by Altiparmak and Ozdogan (2010) looked at the impact of teaching negative numbers through context and conceptual animations. They report a positive impact on progress when comparing their conceptual approach based on understanding with traditional approaches based on the memorization of rules and algorithms.

LEARNING RESOURCES

Access activities for your **lesson plans** at: https://study.sagepub.com/haylock7e

Before trying the self-assessment questions below, you should complete the **self-assessment questions** for this chapter at: https://study.sagepub.com/haylock7e

14.1: In a football league table, Arsenal, Blackburn and Chelsea (A, B and C) all have the same number of points. A has 18 goals for and 22 against, B has 32 for and 29 against, C has 25 for and 30 against. Work out the goal differences and put the teams in order in the table.

14.2: Make up situations about temperatures that are modelled by the additions: (a) 4 + (−12); and (b) (−6) + 10. Give the answers to these additions.

14.3: Give situations about bank balances that are modelled by the subtractions: (a) 20 − (−5); (b) (−10) − (−15); and (c) (−10) − 20. Give the answers to these subtractions.

14.4: Yesterday I was overdrawn at the bank by £187.85. Someone has since made an online payment into my account and this morning I am £458.64 in credit. Model the difference between the two balances with a subtraction. Use a calculator to find out how much was the online payment.

FURTHER PRACTICE

FROM THE STUDENT WORKBOOK

Questions 14.01–14.17: Checking understanding (integers, positive and negative)

Questions 14.18–14.27: Reasoning and problem solving (integers, positive and negative)

Questions 14.28–14.36: Learning and teaching (integers, positive and negative)

FRACTIONS AND RATIOS

IN THIS CHAPTER, THERE ARE EXPLANATIONS OF

- five different meanings of the fraction notation: a proportion of a unit, a point on a number line, a proportion of a set, a division and a ratio;

- numerator and denominator;

- proper fractions, improper fractions and mixed numbers;

- the important idea of equivalent fractions;

- equivalent ratios and their use in scale drawings and maps;

- simplifying fractions and ratios by cancelling;

- how to compare two simple fractions;

- interpreting the remainder in a division calculation as a fraction;

- how to add and subtract simple fractions;

- how to find a simple fraction of a quantity;

- how to multiply two simple fractions;

- various examples of other calculations involving fractions.

READ THIS CHAPTER'S CURRICULUM LINKS AT: HTTPS://STUDY.SAGEPUB.COM/HAYLOCK7E

I THINK OF A FRACTION AS REPRESENTING A PART OF A WHOLE. IS THERE ANY MORE TO IT THAN THAT?

In Chapter 6 we introduced the set of rational numbers. **Fraction** notation is one of the ways in which we can represent rational numbers. But what precisely is the meaning of a fraction such as $^3/_8$? Once again, we encounter the special difficulty presented by mathematical symbols: that one symbol (or collection of symbols) has to be connected to a number of different kinds of situation in the real world. The mathematical notation used for a fraction might, in fact, be used in at least five different ways:

- to represent a proportion of a whole or of a unit;
- to represent a point on a line;
- to represent a proportion of a set;
- to model a division problem;
- as a ratio.

Note that for ease of printing and reading, a fraction in this book is printed as $^3/_8$, rather than $\frac{3}{8}$, which may be the more usual way in handwritten work.

> ## LEARNING AND TEACHING POINT
>
> Ask children to discuss at home with members of their family whether they can think of any situations where they actually use fractions. Share with the rest of the class any examples that they come up with. There may not be many examples other than halves and quarters (particularly in the context of telling the time).

HOW DOES A FRACTION REPRESENT A PROPORTION OF A UNIT?

We should note first that the word **proportion** is used in two different ways in mathematics. Here I am using it to refer to a share or a part or a portion of something. (In Chapter 18 it will be used in the other sense, as a particular kind of relationship between two variables.) Consider, for example, the fraction three-eighths, which in symbols is $^3/_8$. The commonest interpretation of these symbols is illustrated by the diagrams in Figure 15.1. I find the most useful everyday examples are chocolate bars (rectangles) and pizzas (a mathematical pizza is, of course, a perfect circle). One item (sometimes called the 'whole'), such as a bar of chocolate or a pizza, is somehow subdivided into eight equal sections, called 'eighths', and three of these, 'three-eighths', are then selected.

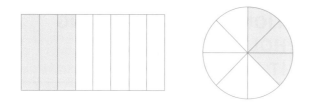

Figure 15.1 The shaded sections are three-eighths of the whole shape

Note that the word 'whole' does sound the same as 'hole'; this can be confusing for children in some situations. I recall one teacher tearing a sheet of paper into four quarters and then, in the course of her explanation, asking a child to show her 'the whole'; not surprisingly, the child kept pointing to the space in the middle. It is also not uncommon colloquially to hear someone talk about 'a whole half', as in 'I ate a whole half of a pizza'. Because of all this, I prefer to talk about 'fractions of a unit'.

HOW DOES A FRACTION REPRESENT A POINT ON A LINE?

Like all real numbers, fractions can be understood as points on a number line. This is actually just an extension of the idea of a fraction as a part of a unit. In Figure 15.2, the section of the number line between 0 and 1, which is 1 unit in length, has been divided up into eight equal sections. Each of these sections is one-eighth of a unit. So, the points from 0 to 1 can be labelled as 0, $\frac{1}{8}$, $\frac{2}{8}$, $\frac{3}{8}$, $\frac{4}{8}$, $\frac{5}{8}$, $\frac{6}{8}$, $\frac{7}{8}$ and 1. Notice also that each step marked along this section of the number line can be thought of as an eighth: so $\frac{3}{8}$ would also be represented by the step from 0 to the point $\frac{3}{8}$; or, indeed, by the step from $\frac{2}{8}$ to $\frac{5}{8}$ and so on. This image of a step along a number line is helpful when making sense of the addition and subtraction of fractions.

Figure 15.2 Eighths of a unit represented as points on a number line from 0 to 1

HOW DOES A FRACTION REPRESENT A PROPORTION OF A SET?

The idea of the fraction $^3/_8$ as meaning 3 parts selected from 8 parts of a unit can then be extended to a proportion of a set. This is the way it is used in situations where a set of items is sub-divided into eight equal subsets and three of these subsets are selected. For example, the set of 40 dots in Figure 15.3(a) has been subdivided into eight equal subsets (of 5 dots each) in Figure 15.3(b). The 15 dots selected in Figure 15.3(c) can therefore be described as three-eighths of the set of 40. So, the diagram shows that one-eighth of 40 is 5 and three-eighths of 40 is 15.

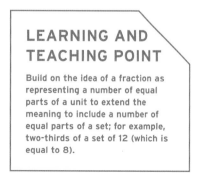

Figure 15.3   Three-eighths of a set of 40

HOW DOES A FRACTION REPRESENT A DIVISION?

The fraction $^3/_8$ can also be used to represent the division of 3 by 8, thinking of division as 'equal sharing between' (see Chapter 10). It might, for example, represent the result of sharing three bars of chocolate equally between eight people. Notice the marked difference here: in Figure 15.1, it was one bar of chocolate that was being subdivided; now we are talking about cutting up three bars. The actual process we would have to go through to solve this problem practically is not immediately obvious.

One way of doing it is to lay the three bars side by side, as shown in Figure 15.4, and then to slice through all three bars simultaneously with a knife, cutting each bar into eight equal pieces. The pieces then form themselves nicely into eight equal portions. Figure 15.4 shows that each of the eight people gets the equivalent of three-eighths of a whole bar of chocolate. (If you are doing this with pizzas, you have to place them one on top of the other, rather than side by side, but otherwise the process is the same.)

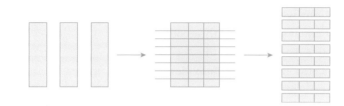

Figure 15.4   Three shared between eight

So, what we see here is, first, that the symbols $^3/_8$ can mean 'divide 3 units by 8' and, second, that the result of doing this division is 'three-eighths of a unit'. Hence, the symbols $^3/_8$ represent both an *instruction* to perform an operation and the *result* of performing it! We often need the idea that the fraction p/q means 'p divided by q' in order to handle fractions on a calculator. Simply by entering $p \div q$ we can express the fraction as a decimal. This also allows us to use fraction notation as an alternative to the division sign (÷).

The division sign itself is, of course, a representation of a fraction, with dots representing the numerator and the denominator. The division sign is used less and less beyond primary school mathematics, so that, for example, '450 divided by 25' will often be written as 450/25. This is certainly the case in algebra, where division ($x \div y$) is almost always indicated by fraction notation ($^x/_y$).

HOW DOES A FRACTION REPRESENT A RATIO?

We have seen in Chapter 10 that one of the categories of problems modelled by division is where two quantities are compared by means of ratio. So, because the symbols $^3/_8$ can mean 'three divided by eight', we can extend the meanings of the symbols to include 'the ratio of three to eight'. This is sometimes written as 3:8. For example, in Figure 15.5(a), when comparing the set of circles with the set of squares, we could say that 'the ratio of circles to squares is three to eight'. This means that for every three circles there are eight squares. Arranging the squares and circles, as shown in Figure 15.5(b), shows this to be the case. The reason why we also use the fraction notation ($^3/_8$) to represent the ratio (3:8) is simply that another way of expressing the comparison between the two sets is to say that the number of circles is three-eighths of the number of squares. The reader may recall from Chapter 6 that *rational* numbers are given that name because they can be expressed as the *ratio* of two integers. So, the principle that any fraction can be understood as a ratio is a really fundamental idea – indeed, mathematically, this is probably the most important meaning of a fraction.

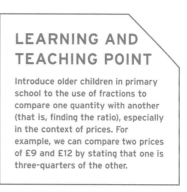

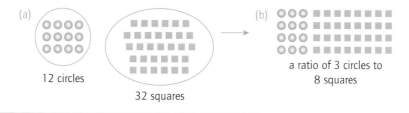

(a) 12 circles

32 squares

(b) a ratio of 3 circles to 8 squares

Figure 15.5 A ratio of three to eight

WHAT ABOUT NUMERATORS, DENOMINATORS, VULGAR FRACTIONS, PROPER AND IMPROPER FRACTIONS, AND MIXED NUMBERS?

The term 'fraction' is used colloquially nowadays to refer to what used to be called a 'vulgar fraction', in which the word 'vulgar' simply meant 'common' or 'ordinary'. This sense of the word is archaic, but it was used to distinguish between the kinds of fractions discussed in this chapter, such as $^3/_8$, written with a top number and a bottom number, and decimal fractions, such as 0.375, which are discussed in the next – and which nowadays are usually referred to as 'decimal numbers'. (For those who are interested in the ways in which words shift their meaning, I have an arithmetic book dated 1886 in which the chapter on 'vulgar fractions' concludes with a set of 'promiscuous exercises'!)

The **numerator** and the **denominator** are simply the top number and the bottom number in the fraction notation. So, for example, in the fraction $^3/_8$ the numerator is 3 and the denominator is 8. Most of the time I prefer to call them simply the top number and the bottom number, but by all means use the technical terms if you wish.

The fraction notation for parts of a unit can also be used in a situation such as that shown in Figure 15.6, where there is more than one whole unit to be represented. Altogether here, we have eleven-eighths of a pizza, written $^{11}/_8$. Since eight of these make a whole pizza, this quantity can be written as $1 + ^3/_8$, which is normally abbreviated to $1^3/_8$. This is called a **mixed number**.

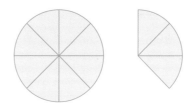

Figure 15.6   A fraction greater than 1

A fraction in which the top number is smaller than the bottom number, such as $^3/_8$, is sometimes called a **proper fraction**, with a fraction such as $^{11}/_8$ being referred to as an **improper fraction**. Proper fractions are therefore those that are less than 1, with improper fractions being those greater than 1. We could refer to improper fractions more informally as 'top-heavy fractions'.

Mixed numbers are also helpfully understood as points on a number line, extending the idea introduced above in Figure 15.2. So, for example, the mixed number $3^5/_8$ (which is $3 + ^5/_8$) is represented by the point indicated in Figure 15.7. To identify this point, we start at 3 and then move on a further $^5/_8$ of a unit.

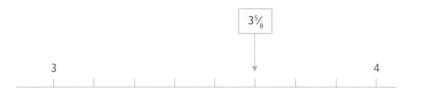

Figure 15.7   The mixed number $3^5/_8$ identified as a point on a number line

WHAT ARE EQUIVALENT FRACTIONS?

The concept of equivalence – which we saw in Chapter 3 to be one of the fundamental processes for understanding mathematics – is one of the key ideas for children to grasp when working with fractions. Using the first idea of a fraction above, that it represents a part of a unit, it is immediately apparent from Figure 15.8, for example, that three-quarters, six-eighths and nine-twelfths all represent the same amount of chocolate bar. This kind of 'fraction chart' is an important teaching aid for explaining the idea of equivalence.

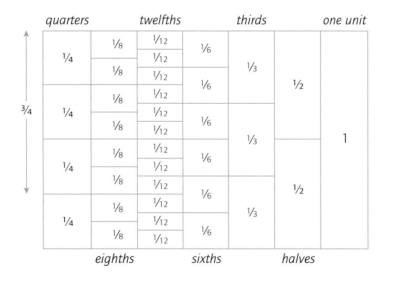

Figure 15.8   A fraction chart showing some equivalent fractions

Sequences of **equivalent fractions** follow a very straightforward pattern. For example, all these fractions are equivalent: $^3/_5$, $^6/_{10}$, $^9/_{15}$, $^{12}/_{20}$, $^{15}/_{25}$, $^{18}/_{30}$, $^{21}/_{35}$, $^{24}/_{40}$ and so on.

The numbers on the top and bottom are simply the 3-times and 5-times tables, respectively. This means that, given a particular fraction, you can always generate an equivalent fraction by multiplying the top and the bottom by the same number; or, vice versa, by dividing by the same number. So, for example:

$^4/_7$ is equivalent to $^{36}/_{63}$ (multiplying top and bottom by 9).

$^{40}/_{70}$ is equivalent to $^4/_7$ (dividing top and bottom by 10).

> **LEARNING AND TEACHING POINT**
>
> Equivalence of fractions is one of the most important ideas to get across to children in primary school. Get them to make a variety of fraction charts (like the one in Figure 15.8) and then to find various examples of equivalent fractions.

HOW DO YOU SIMPLIFY FRACTIONS?

If we remember that the fraction notation can also be interpreted as meaning division of the top number by the bottom number, the principle above is another version of stating the constant ratio principle explained in Chapter 11: that you do not change the answer to a division calculation if you multiply or divide both numbers by the same thing. This is an important method for simplifying fractions: by dividing the top and bottom numbers by any common factors, we can reduce the fraction to its simplest form. This process is often called 'cancelling'.

> **LEARNING AND TEACHING POINT**
>
> Help children to see the pattern in sequences of equivalent fractions and use this to establish the idea that you can change one fraction into an equivalent fraction by multiplying (or dividing) the top and bottom numbers by the same thing.

For example, $^6/_8$ can be simplified to the equivalent fraction $^3/_4$ by dividing top and bottom numbers by their highest common factor, 2 (cancelling 2). Similarly, $^{12}/_{18}$ can be simplified to the equivalent fraction $^2/_3$ by cancelling 6, which is the highest common factor (see Chapter 13) of 12 and 18.

HOW DOES THIS WORK WITH RATIOS?

The principle used for simplifying fractions applies to ratios, of course, because fractions can be interpreted as ratios. If you multiply or divide two numbers by the same thing, the ratio stays the same.

For example, if I am comparing the price of two articles costing £28 and £32 by looking at the ratio of the prices, then the ratio 28:32 can be simplified to the **equivalent ratio** of 7:8 (dividing both numbers by 4). This means that one price is $^7/_8$ (seven-eighths) of the other.

Another example: if I am comparing a journey of 2.8 miles with one of 7 miles, then I could simplify the ratio, 2.8:7, by first multiplying both numbers by 10 (to get 28:70) and then dividing both numbers by 14 (to get 2:5), drawing the conclusion that one journey is $^2/_5$ (two-fifths) of the other.

Often, it is particularly useful to express a ratio as an equivalent ratio in which the first number is 1. For example, the ratio 2:5 used to compare the two journeys in the previous paragraph can be written as the equivalent ratio 1:2.5 (dividing both numbers by 2). This can then be interpreted as 'for every mile in the first journey you have to travel 2.5 miles in the second' or 'the second journey is 2.5 times longer than the first'.

The commonest application of this kind of ratio is to scale drawings and map scales. For example, if a scale drawing of the classroom represents a length of 2 metres by a length of 5 cm then the scale is the ratio of 5 cm to 2 metres, or, writing both lengths in centimetres, 5 cm to 200 cm. The ratio 5:200 can then be simplified to the equivalent ratio of 1:40. This would be the conventional way of expressing the scale, indicating that each length in the original is 40 times the corresponding length in the scale drawing, or that each length in the scale drawing is $^1/_{40}$ of the length of the original. Scale factors for maps are usually much larger than this, of course. For example, the Ordnance Survey Landranger maps of Great Britain use a scale of 1:50,000. This means that a distance of 1 cm on the map represents a distance of 50,000 cm in reality. Since 50,000 cm = 500 m = 0.5 km, then we conclude that 1 cm on the map represents 0.5 km (see Chapter 21).

HOW DO YOU COMPARE ONE FRACTION WITH ANOTHER?

The first point to notice here is that when you increase the bottom number of a fraction you actually make the fraction smaller, and vice versa. Consider, for example, what are called **unit fractions**. These are fractions with numerator '1': a half ($^1/_2$), a third ($^1/_3$), a quarter ($^1/_4$), a fifth ($^1/_5$), a sixth ($^1/_6$) and so on. Important in developing mastery in fractions is to understand that $^1/_2$ is greater than $^1/_3$, which is greater than $^1/_4$, which is greater than $^1/_5$ and so on. This is very obvious if the symbols are interpreted in concrete terms, as bits of pizza or a chocolate bar, for example. Interpreting these unit fractions as points on a number line, as shown in Figure 15.9, also makes it clear that they get smaller as the bottom number gets larger, because the points representing these unit fractions are getting closer to zero. It is very easy for a child to get this wrong, of course, if they simply look at the numbers involved in the fraction notation without thinking about what they mean.

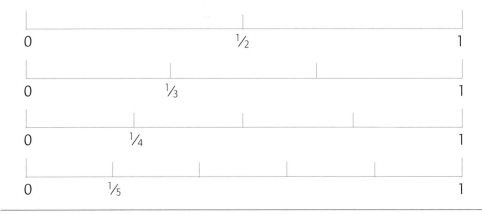

Figure 15.9   Unit fractions shown as points on a number line

Then, second, there is no difficulty in comparing two fractions with the same bottom number. Clearly, five-eighths of a pizza ($^5/_8$) is a greater portion than three-eighths ($^3/_8$), for example.

Generally, to compare two fractions with *different* bottom numbers we may need to convert them to equivalent fractions with the *same* bottom number. This will have to be a common multiple of the two numbers. It might be (but does not have to be) the lowest common multiple (see Chapter 13). For example, which is greater, seven-

tenths ($^7/_{10}$) of a chocolate bar or five-eighths ($^5/_8$)? The lowest common multiple of 10 and 8 is 40, so convert both fractions to fortieths:

$^7/_{10}$ is equivalent to $^{28}/_{40}$ (multiplying top and bottom by 4); and

$^5/_8$ is equivalent to $^{25}/_{40}$ (multiplying top and bottom by 5).

We can then see instantly that, provided you like chocolate, the seven-tenths is the better choice.

WATCH THE PROBLEM SOLVED! VIDEO AT: HTTPS://STUDY.SAGEPUB.COM/HAYLOCK7E

HOW IS A REMAINDER IN A DIVISION CALCULATION INTERPRETED AS A FRACTION?

Take an example: $30 \div 7 = 4$ remainder 2. Depending on the context in which this division calculation arose, it may be possible to deal with the remainder here by dividing that by 7 as well. Using the idea that a fraction can represent a division, we know that $2 \div 7$ is equal to $^2/_7$. So we might give the result of the division as a mixed number: $4^2/_7$. This would be a possible answer if the original (rather artificial) problem had been to find out how much chocolate each person gets if 30 bars are shared equally between 7 people: answer, $4^2/_7$ bars each! It would *not* be an appropriate answer, how-

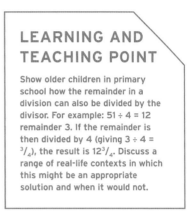

LEARNING AND TEACHING POINT

Show older children in primary school how the remainder in a division can also be divided by the divisor. For example: $51 \div 4 = 12$ remainder 3. If the remainder is then divided by 4 (giving $3 \div 4 = ^3/_4$), the result is $12^3/_4$. Discuss a range of real-life contexts in which this might be an appropriate solution and when it would not.

ever, if the question had been, 'how many vans do you need to transport 30 children if each van holds 7 children?', because you cannot have $^2/_7$ of a van.

HOW DO YOU ADD AND SUBTRACT FRACTIONS?

To be honest, I have to say that there are not many practical situations that genuinely require the addition or subtraction of fractions. In practice, most calculations, such as those arising from measurements, are done with decimals. Questions such as '$^1/_6$ of a class are 7 years of age and $^1/_2$ of the class are 8 years of age – what fraction of the class are 7 or 8 years of age?' do sound very contrived. However, because you may well find yourself in the situation where someone expects you or the children you are teaching to be able to do this kind of thing, here's how it's done.

First, to add two fractions with the same bottom number (denominator) is very simple. Just visualize the fractions as parts of a whole unit. So, for example, if you have one-eighth ($^1/_8$) of a chocolate bar and you add it to three-eighths ($^3/_8$) of a chocolate bar, you have altogether four-eighths ($^4/_8$) of a chocolate bar. So, clearly, $^1/_8 + ^3/_8 = ^4/_8$. This answer can then be simplified to $^1/_2$ by cancelling 4.

Subtraction is equally straightforward when the two fractions have the same denominator. For example, if you have seven-eighths of a pizza and eat five-eighths then you are left with two-eighths. Recording this in fraction notation, $^7/_8 - ^5/_8 = ^2/_8$. This result could, of course, be simplified to $^1/_4$ by cancelling 2.

Sometimes, when adding two or more proper fractions, the result may be an improper fraction. For example, $^3/_8 + {}^5/_8 + {}^7/_8 = {}^{15}/_8$. This result could then be expressed as a mixed number ($1^7/_8$). But please do not ask me to come up with a real-life situation where anyone might need to do a calculation like this, other than when taking an unenlightened mathematics test.

To add or subtract two fractions with different denominators is a bit trickier. Before attempting to combine the fractions, you have to change one or both of them to equivalent fractions so that they finish up with the same bottom number – it's best to use the lowest common multiple for this.

So, for example, to add $^1/_6$ and $^1/_2$, we would spot that the $^1/_2$ is equivalent to $^3/_6$, so both fractions can be expressed as sixths. We go for sixths because 6 is the lowest number that is a multiple of both 2 and 6. In this context, the lowest common multiple of the two denominators is often called the **lowest common denominator**. The calculation then looks like this: $^1/_6 + {}^1/_2 = {}^1/_6 + {}^3/_6 = {}^4/_6 (= {}^2/_3)$.

Here is an example with subtraction: how much more is $^2/_3$ of a litre than $^1/_4$ of a litre? This requires the calculation $^2/_3 - {}^1/_4$. To do this, we change both fractions to twelfths, because 12 is the lowest common multiple of 3 and 4. The $^2/_3$ of a litre is equivalent to $^8/_{12}$ of a litre; and the $^1/_4$ of a litre is equivalent to $^3/_{12}$ of a litre. The difference between $^8/_{12}$ of a litre and $^3/_{12}$ of a litre is clearly $^5/_{12}$ of a litre. Written down, the calculation looks like this: $^2/_3 - {}^1/_4 = {}^8/_{12} - {}^3/_{12} = {}^5/_{12}$.

> **LEARNING AND TEACHING POINT**
>
> A common error in adding fractions is to add the top numbers and add the bottom numbers; for example: $^1/_5 + {}^3/_5 = {}^4/_{10}$. This error only occurs when the learner just responds to the symbols mindlessly without any attempt to connect them to a visual image that makes sense of the fractions. In this case, the correct addition is: $^1/_5 + {}^3/_5 = {}^4/_5$ (one-fifth of a pizza plus three-fifths of a pizza is equal to four-fifths of a pizza).

> **LEARNING AND TEACHING POINT**
>
> A prerequisite for being able to add or subtract fractions with different denominators is the ability to identify the lowest common multiple of two (or more) numbers. Aim to develop mastery of this skill before children move on to identifying the lowest common denominators in the process of the addition and subtraction of fractions.

WHAT CALCULATIONS WITH FRACTIONS OCCUR MOST OFTEN IN EVERYDAY LIFE?

The commonest everyday situations involving calculations with fractions are those where we have to calculate a simple fraction of a set or a quantity. For example, we might say, 'three-fifths of my class of 30 children are boys'. If we see an article priced at £45 offered with one-third off, then the reduced price must be 'two-thirds of £45'. Or we might encounter fractions in measurements such as 'three-quarters of a litre' or 'two-fifths of a metre' and want to change these to millilitres and centimetres respectively.

The process of doing these calculations is straightforward. For example, to find $^3/_5$ of 30, first divide by the 5 to find one-fifth of 30, then multiply by the 3 to obtain three-fifths. Here are some examples of the process:

$^1/_5$ of 30 is 6, so $^3/_5$ of 30 is 18.

$^1/_3$ of £45 is £15, so $^2/_3$ of £45 is £30.

$^1/_4$ of 1000 ml is 250 ml, so $^3/_4$ of 1000 ml (a litre) is 750 ml.

$^1/_5$ of 100 cm is 20 cm, so $^2/_5$ of 100 cm (a metre) is 40 cm.

> **LEARNING AND TEACHING POINT**
>
> Make sure that children have grasped the equivalence of, for example, finding a quarter of a number and dividing by 4; or finding a fifth of a number and dividing by 5; or finding a twelfth of a number and dividing by 12; and so on.

> **LEARNING AND TEACHING POINT**
>
> Explain to children the procedure for finding a fraction of a quantity, by dividing by the bottom number and then multiplying by the top number, applying this procedure to a range of everyday practical contexts, and using a calculator where necessary.

HOW DO YOU MULTIPLY TWO SIMPLE FRACTIONS?

Some primary school children may be required to multiply together two simple fractions. The need to know how to do this only becomes clear much later on in secondary school mathematics when they will be manipulating algebraic expressions and simplifying formulas.

The process for multiplying two fractions can be understood in visual terms by applying the fractions to the area of a square, as shown in Figure 15.10(a). The square has been divided into thirds by the horizontal lines in Figure 15.10(b) and then divided into quarters by the vertical lines drawn in Figure 15.10(c). The square has now been divided into twelfths. Using the idea that the area of a rectangle is given by the product of the two sides, it is now clear that the area of the shaded rectangle in Figure 15.10(d) is equal to $^2/_3$ multiplied by $^3/_4$.

> **LEARNING AND TEACHING POINT**
>
> The rule for multiplying two fractions is very straightforward: multiply the two denominators and multiply the two numerators. But if you teach the rule without helping the children to make connections to visual imagery and appropriate language, then you can expect them to get in a muddle and apply this rule (wrongly) when they are adding fractions.

Since this area is six-twelfths, we have shown that $^2/_3 \times {}^3/_4 = {}^6/_{12}$. (This can then be cancelled down to $^1/_2$.)

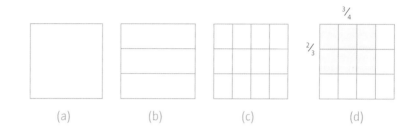

Figure 15.10   The shaded area in (d) represents $^2/_3 \times {}^3/_4 = {}^6/_{12}$

Note that we get twelfths in Figure 15.10(c) because we have 3 sections horizontally and 4 sections vertically. This is effectively multiplying together the denominators of the $^2/_3$ and $^3/_4$. And we get 6 of these twelfths shaded in Figure 15.10(d) because the shaded rectangle arises from 2 of the horizontal sections and 3 of the vertical sections. This is effectively multiplying together the numerators of the $^2/_3$ and $^3/_4$. So, there is a very simple rule for multiplying two fractions: multiply the two denominators and multiply the two numerators. Then give the answer in the simplest form, by cancelling.

Finally, in this example, notice that the shaded rectangle in Figure 15.10(d) can be thought of as two-thirds of three-quarters of the square; or as three-quarters of two-thirds of the square. The word *of* in the language pattern 'a fraction *of* something' is helpfully connected with the symbol of multiplication. So, for example, the fact that $^2/_3$ of 60 is 40 can also be expressed as $^2/_3 \times 60 = 40$. Similarly, $^1/_2 \times {}^1/_4$ can be understood as: what is a half of a quarter?

> **LEARNING AND TEACHING POINT**
>
> A key principle in teaching for mastery of fractions: make explicit and help children to understand the equivalence of, for example, these three statements:
>
> $^1/_5 \times 30 = 6$
> $^1/_5$ of $30 = 6$
> $30 \div 5 = 6$

WHAT OTHER CALCULATIONS WITH FRACTIONS SHOULD I BE ABLE TO DO?

In the final section of this chapter, I am including a few examples of calculations with fractions that might be required of older children in primary school. As with much of this chapter, it is difficult to think of practical applications of most of these skills in a world in which nearly all measurements are done in metric units and expressed as decimal numbers. As mentioned above, there are some manipulations of fractions that become more significant much later on in secondary school algebra, but otherwise it really does seem to me that the main driving force for making calculations with fractions such a prominent part of the primary curriculum is simply nostalgia.

Addition with Mixed Numbers

(a) $3^1/_5 + 2^3/_5$

$= 3 + \,^1/_5 + 2 + \,^3/_5$ (separating the whole number parts and fractional parts)
$= 5 + \,^4/_5$ (adding the whole numbers and the fractions separately)
$= 5^4/_5$ (writing this as a mixed number)

(b) $3^4/_5 + 2^3/_5$

$= 3 + \,^4/_5 + 2 + \,^3/_5$ (separating the whole number parts and fractional parts)
$= 5 + \,^7/_5$ (adding the whole numbers and the fractions separately)
$= 5 + 1 + \,^2/_5$ (changing the improper fraction to a whole number and a fraction)
$= 6 + \,^2/_5$ (adding the 1 to the 5)
$= 6^2/_5$ (writing this as a mixed number)

The additions in (a) and (b) can also be carried out – and understood better – on a number line, as shown in Figure 15.11, in each case by first adding on the 2 and then adding on the $^3/_5$.

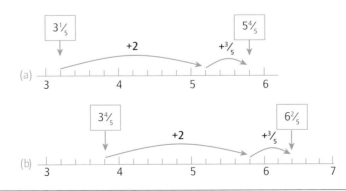

Figure 15.11   Using a number line to find (a) $3^1/_5 + 2^3/_5$ and (b) $3^4/_5 + 2^3/_5$

Subtraction with Mixed Numbers

(a) $5^4/_5 - 3^3/_5$

Think of this as $(5 + {}^4/_5) - (3 + {}^3/_5)$
First deal with the fractional parts: ${}^4/_5 - {}^3/_5 = {}^1/_5$
Then deal with the whole number parts: $5 - 3 = 2$
Combining these as a mixed number, we get $5^4/_5 - 3^3/_5 = 2^1/_5$

(b) $5^2/_5 - 3^3/_5$

Think of this as $(5 + {}^2/_5) - (3 + {}^3/_5)$
This will require the use of a form of decomposition (see subtraction methods in Chapter 9).
First look at the subtraction with the fractional parts: ${}^2/_5 - {}^3/_5$; this would give a negative result.
So, exchange 1 from the 5 for five-fifths: $5 + {}^2/_5 = 4 + 1 + {}^2/_5 = 4 + {}^5/_5 + {}^2/_5 = 4 + {}^7/_5$
We can now complete the subtraction, using the method in (a):
$(4 + {}^2/_5) - (3 + {}^3/_5) = {}^{14}/_5$

Once again, these calculations with mixed numbers may be better understood on a number line; this can be done, for example, by interpreting the subtraction as finding the difference. In Figure 15.12(a), the difference between the two numbers is clearly an interval of 2 added to an interval of ${}^1/_5$, giving the result $2^1/_5$. In Figure 15.12(b), the difference is made up of an interval of 1 plus an interval of ${}^2/_5$ (to get to 5) and a further interval of ${}^2/_5$, giving the result $1^4/_5$.

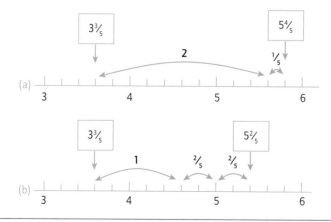

Figure 15.12   Using a number line to find (a) $5^4/_5 - 3^3/_5$ and (b) $5^2/_5 - 3^3/_5$

Divisions with Fractions

(a) Calculate $^4/_5 \div 3$

Remember that dividing by 3 is the same thing as finding a third of something, which is the same as multiplying by $^1/_3$.

So, $^4/_5 \div 3 = ^1/_3 \times ^4/_5$

$= ^4/_{15}$ (using the method for multiplying fractions explained earlier in the chapter).

(b) Calculate $20 \div ^1/_4$

Remember that the result of a division is unchanged if both numbers are multiplied by the same thing (see mental and informal strategies for division in Chapter 11). So, multiply both numbers in $20 \div ^1/_4$ by 4. (Remember that 4 quarters = 1.) Then we get $20 \div ^1/_4 = 80 \div 1$, which is just 80.

The result in (b) makes a lot of sense if you think of the division as the inverse of multiplication. For example, how many quarters of a pizza can you get from 20 pizzas? You can now ask your friends what is twenty divided by a quarter and then enjoy explaining to those who give the answer 5 why the answer is 80.

RESEARCH FOCUS: UNDERSTANDING FRACTION NOTATION

A teaching experiment with children aged 9 years (Boulet, 1998) sheds light on many of the conceptual difficulties that children have with fraction notation. Difficulties were highlighted in what is termed 'equipartition'. In representing simple fractions such as $^1/_3$ by dividing up a given shape such as a rectangle or a circle, often children would use non-equal sections for thirds. The exception was when they could generate the fractions (such as quarters) by halving. Children had difficulty in 'reconstitution'; that is, they did not necessarily recognize that if you put all fractional parts back together again, you would have the 'whole' that you started with. Ordering fractions proved to be a real problem, with the expected error of deducing, for example, that $^1/_4 > ^1/_3$, simply because 4 > 3. Some children responded that which of $^1/_4$ and $^1/_3$ was the greater would depend on the size of the whole. You can see their point. This is an illuminating observation for teachers to note in teaching the ordering of fractions. Some interesting errors of 'quantification' were also noted. For example, to represent the fraction $^1/_7$ with counters, children might make an arrangement of 1 black counter placed above a line of 7 white counters. When given a rectangular strip of 5 squares, the association of the word 'fifth' with ordinal numbers led some children to reject the idea that one of the squares not at the end of the strip could be a 'fifth' of the rectangle. Karika (2020) provides further insights into the errors made by children aged 10–12 years in Hungary. His research shows that children's errors are logically consistent and rule-based and make sense to them, but, of course, the 'rules' that produce the errors are not always mathematically correct. For example, given a square grid of 36 cells, children were asked to colour 3/12 of the grid; many of them coloured 3 rectangles made up of 12 squares. As with Boulet (1998) above, Karika reports that many children would assume that a fraction with a larger denominator would be larger than one with a smaller denominator; for example, 4/6 would be seen as larger than 4/5. Siegler and Pyke (2013) compared the understanding of fractions of children aged 10–11 years with those aged 12–13 years. Their assessment items included positioning fractions on number lines. Siegler and Pike's analysis indicates that the difference between the performances of the lower and higher achievers in tasks such as these increases significantly over these two years; the gap widens. The implication for primary school teachers is the importance of ensuring deep understanding of the concept of fractions for as many children as possible at this age – especially the idea that the size of a fraction depends not on the sizes of the two numbers involved (the numerator and the denominator) but on the relationship between them.

LEARNING RESOURCES

Access activities for your **lesson plans** at: https://study.sagepub.com/haylock7e

Before trying the self-assessment questions below, you should complete the **self-assessment questions** for this chapter at: https://study.sagepub.com/haylock7e

15.1: Give examples where $^4/_5$ represents: (a) a proportion of a whole unit; (b) a proportion of a set; (c) a division using the idea of sharing; and (d) a ratio.

15.2: Find as many different examples of equivalent fractions illustrated in the fraction chart in Figure 15.8 as you can.

15.3: Assuming you like pizza, which would you prefer – three-fifths of a pizza ($^3/_5$) or five-eighths ($^5/_8$)? (Convert both fractions to fortieths.)

15.4: By converting them all to twelfths, put these fractions in order, from the smallest to the largest: $^3/_4$, $^1/_6$, $^1/_3$, $^2/_3$, $^5/_{12}$.

15.5: Make up a problem about prices to which the answer is 'the price of A is three-fifths ($^3/_5$) of the price of B'.

15.6: Walking to work takes me 24 minutes; cycling takes me 9 minutes. Complete this sentence with an appropriate fraction: 'The time it takes to cycle is ... of the time it takes to walk.'

15.7: Find: (a) three-fifths of £100, without using a calculator; and (b) five-eighths of £2500, using a calculator.

15.8: Do the following abstract calculations involving fractions:

    a $^3/_4 + ^5/_8$

    b $1 - ^5/_{12}$

    c $^1/_3 \times ^3/_4$

    d $^{41}/_8 - ^{33}/_8$

    e $^5/_6 \div 2$

FURTHER PRACTICE

Access the website material for

- Knowledge check 10: Finding a fraction of a quantity
- Knowledge check 11: Simplifying ratios

at: https://study.sagepub.com/haylock7e

FROM THE STUDENT WORKBOOK

Questions 15.01–15.20: Checking understanding (fractions and ratios)

Questions 15.21–15.30: Reasoning and problem solving (fractions and ratios)

Questions 15.31–15.42: Learning and teaching (fractions and ratios)

GLOSSARY OF KEY TERMS INTRODUCED IN CHAPTER 11

Fraction A way of (a) representing a part of a whole or unit, (b) representing a point on a line, (c) representing a proportion of a set, (d) modelling a division problem, or (e) expressing a ratio.

Proportion A comparative part of a quantity or set. A proportion (such as 4 out of 10) can be expressed as a fraction ($^2/_5$), as a percentage (40%) or as a decimal (0.4).

Numerator The top number in a fraction.

Denominator The bottom number in a fraction.

Mixed number A way of writing a fraction greater than 1 as a whole number plus a proper fraction. For example, $^{18}/_5$ as a mixed number is $3^3/_5$ (three and three-fifths).

Proper fraction A fraction in which the top number is smaller than the bottom number; a fraction less than 1.

Improper fraction A fraction in which the top number is greater than the bottom number; a fraction greater than 1; informally, a top-heavy fraction.

Equivalent fractions Two or more fractions that represent the same part of a unit or the same ratio. For example, $^2/_3$, $^4/_6$, $^6/_9$, $^8/_{12}$ are all equivalent fractions.

Cancelling The process of dividing the top number and the bottom number in a fraction by a common factor to produce a simpler equivalent fraction.

Equivalent ratios Different ways of expressing the same ratio; for example, the ratio 30:50 can be written as the equivalent ratio 3:5.

Unit fraction A fraction with numerator 1, such as $^1/_2$, $^1/_4$ or $^1/_{10}$.

Lowest common denominator The lowest common multiple of the denominators of two or more fractions. For example, the lowest common denominator of $^1/_3$, $^3/_4$ and $^5/_6$ is 12, because 12 is the smallest number into which all three denominators (3, 4 and 6) will divide exactly.

DECIMAL NUMBERS AND ROUNDING

IN THIS CHAPTER, THERE ARE EXPLANATIONS OF

- the extension of the place-value principle to tenths, hundredths, thousandths;
- how to relate decimal numbers to division of whole numbers by 10, 100 or 1000;
- the decimal point as a separator in the contexts of money and measurement;
- locating decimal numbers on a number line;
- rounding up or down;
- the idea of rounding to the nearest something;
- converting decimal numbers to fractions, and vice versa;
- fractions that are equivalent to recurring decimals;
- division calculations giving exact decimal answers;
- division calculations giving answers that are recurring decimals;
- interpretation of calculator results in real-life contexts;
- how to give answers to so many decimal places or significant digits.

READ THIS CHAPTER'S CURRICULUM LINKS AT: HTTPS://STUDY.SAGEPUB.COM/HAYLOCK7E

HOW DOES THE PLACE-VALUE SYSTEM WORK FOR QUANTITIES LESS THAN ONE?

The place-value system for numbers explained in Chapter 6 works in exactly the same way for numbers less than 1. Once the principle of being able to 'exchange one of these for ten of those' is established, we can continue with it to the right of the units position, with tenths, hundredths, thousandths and so on. These positions are usually referred to as *decimal places* and are separated from the units by the **decimal point**. Numbers containing digits after the decimal point are called **decimal numbers**. Since a tenth and a hundredth are what you get if you divide a unit into ten and a hundred equal parts respectively, it follows that one unit can be exchanged for ten tenths, and one tenth can be exchanged for ten hundredths. In this way, the principle of 'one of these being exchanged for ten of those' continues indefinitely to the right of the decimal point, with the values represented by the places getting progressively smaller by a factor of ten each time. So, for example, the decimal number 3.456 represents 3 units, 4 tenths, 5 hundredths and 6 thousandths.

A useful way to picture decimals is to explore what happens if we decide that the 'flat' piece in the base-ten blocks represents 'one whole unit'.

In this case, the 'longs' represent tenths of this unit and the small cubes represent hundredths. Then the collection of blocks shown in Figure 16.1(a) now represents 3 units, 6 tenths and 6 hundredths. This quantity is represented by the decimal number 3.66. Similarly, the blocks in Figure 16.1(b) would now represent the decimal number 3.07, that is, 3 units, no tenths and 7 hundredths. Look back at Figures 6.7 and 6.9 (in Chapter 6) and notice that these are identical arrangements of blocks for representing 366 and 307.

Arrow cards (see Figure 6.10 in Chapter 6) can also be used to show the values represented by the digits in a decimal number. Figure 16.2 shows how 3.45 is made up of 3 (3 units) + 0.4 (4 tenths) + 0.05 (5 hundredths). Note that it is conventional to put a zero in front of the decimal point (as here in 0.4 and 0.05) if there are no units; this is to ensure that the decimal point does not get overlooked.

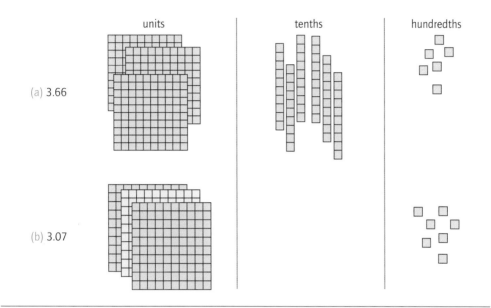

Figure 16.1   Representing (a) 3.66 and (b) 3.07 using base-ten Dienes blocks

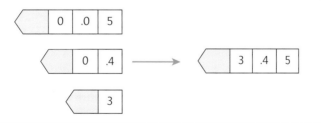

Figure 16.2   The decimal number 3.45 shown with arrow cards

WATCH THE PROBLEM SOLVED! VIDEO AT: HTTPS://STUDY.SAGEPUB.COM/HAYLOCK7E

HOW ARE DECIMAL NUMBERS PRODUCED BY DIVISION BY 10, 100 OR 1000?

Imagine there are 7 pizzas to be shared equally between 10 friends. As we saw in Chapter 15, each friend will get $^7/_{10}$ (seven tenths) of a pizza. Seven tenths is the decimal number 0.7. So, one way of thinking about the decimal number 0.7 is that it is what you get if you divide 7 by 10. Similarly, 0.07 (seven hundredths) is what you get if you divide 7 by 100. And 0.007 is what you get if you divide 7 by 1000. These are important relationships in the understanding and mastery of decimals.

> **LEARNING AND TEACHING POINT**
>
> Help children to make the connections between decimal numbers and the division of whole numbers by 10, 100 and 1000. For example, 0.8 is equal to 8 divided by 10; 0.75 is equal to 75 divided by 100.

This idea extends to understanding the decimal number 0.27. This is 2 tenths add 7 hundredths. But the 2 tenths are equivalent to 20 hundredths, so 0.27 is equal to 27 hundredths. Hence, 0.27 is the decimal number produced by dividing 27 by 100. Similarly, 0.027 (which is 27 thousandths) is the result of dividing 27 by 1000; and 0.327 (which is 327 thousandths) is the result of dividing 327 by 1000.

The patterns and principles involved in multiplying and dividing decimal numbers by 10, 100 and 1000 are discussed more fully in Chapter 17.

HOW DOES DECIMAL NOTATION APPLY IN THE CONTEXTS OF MONEY AND MEASUREMENT?

The collection of base-ten counters shown in Figure 16.3 represents the number 3.66, with three units (ones), six-tenths and six-hundredths. In terms of decimal numbers in general, the function of the decimal point is to indicate the transition from units to tenths. Because of this, a decimal number such as 3.66 is read as 'three point six six', with the first digit after the point indicating the number of tenths and the next the number of hundredths. It would be confusing to read it as 'three point sixty-six', since this might be taken to mean three units and sixty-six tenths.

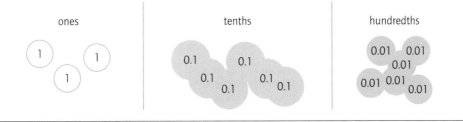

ones   tenths   hundredths

Figure 16.3 Representing 3.66 using base-ten counters representing ones, tenths and hundredths

There is a different convention, however, when using the decimal point in recording amounts of money: the amount £4.85 is read as 'four pounds eighty-five'. In this case, there is no confusion about what the 'eighty-five' refers to: the context makes clear that it is 'eighty-five pence'. In practice, it is possible that it is in money notation like this that children will first encounter the decimal point. In this form, they use the decimal point quite simply as something that separates the pounds from the pennies – so that £4.85 represents simply 4 whole pounds and 85 pence – without any overt awareness necessarily that the 8 represents 8 tenths of a pound and the 5 represents 5 hundredths. The decimal point here is effectively no more than a **separator** of the pounds from the pennies; so we have the convention of always writing two digits after the point when recording amounts of money in pounds. So, for example, we would write £3.20 rather than £3.2, and read it as 'three pounds twenty (meaning twenty pence)'. By contrast, if we were working with pure decimal numbers then we would simply write 3.2, meaning '3 units and 2 tenths'.

> **LEARNING AND TEACHING POINT**
>
> Children will first encounter the decimal point as a separator in the context of money (pounds and pence) and then in the context of length (metres and centimetres), with two digits after the point. They can use the notation in these contexts initially without necessarily having any real awareness of the digits representing tenths and hundredths.

Anticipating the explanation of prefixes in units of measurement in Chapter 21, let us just note here that there are a hundred centimetres (cm) in a metre (m), just like a hundred pence in a pound. This means that the measurement of length in centimetres and metres offers a close parallel to recording money in pounds and pence. So, for example, a length of 366 cm can also be written in metres, as 3.66 m. Once again, the decimal point may be seen simply as something that separates the 3 whole metres from the 66 centimetres. In this context, it is helpful to exploit children's familiarity with money notation and press the parallel quite strongly, following the same convention of writing two digits after the point when expressing lengths in metres, for example writing 3.20 m rather than 3.2 m. We can then interpret this simply as three metres and twenty centimetres. I shall explain in Chapter 17 how this convention is very useful when dealing with additions and subtractions involving decimals.

> **LEARNING AND TEACHING POINT**
>
> The use of the decimal point as a separator can extend to further experience of decimal notation in the contexts of mass (kilograms and grams) and liquid volume and capacity (litres and millilitres), with three digits after the point.

This principle then extends to the measurement of mass (or, colloquially, weight: see Chapter 21) where, because there are a thousand grams (g) in a kilogram (kg), it is best, at least to begin with, to write a mass measured in kilograms with three digits after the point. For example, 3450 g written in kg is 3.450 kg. The decimal point can then simply be seen as something that separates the 3 whole kilograms from the 450 grams. Similarly, in recording liquid volume and capacity, where there

> **LEARNING AND TEACHING POINT**
>
> Once children are confident with using the decimal point as a separator in money and measurement, these contexts can be used to reinforce the explanation of the idea that the digits after the point represent tenths, hundredths and thousandths.

are a thousand millilitres (ml) in a litre, a volume of 2500 ml is also written as 2.500 litres, with the decimal point separating the 2 whole litres from the 500 millilitres.

So, when working with primary school children, it is not necessary initially to explain about tenths and hundredths when using the decimal point in the context of money, length and other measurement contexts. To begin with, we can use it simply as a separator and build up the children's confidence in handling the decimal notation in these familiar and meaningful contexts. Later, of course, money and measurement in general will provide fertile contexts for explaining the ideas of tenths, hundredths and thousandths. For example, a decimal number such as 1.35 can be explained in the context of length by laying out in a line 1 metre stick, 3 decimetre rods (tenths of a metre) and 5 centimetre pieces (hundredths of a metre), as shown in Figure 16.4.

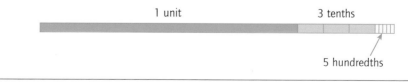

Figure 16.4   The decimal number 1.35 shown as a length

HOW ARE DECIMAL NUMBERS LOCATED ON A NUMBER LINE?

Having connected decimal numbers with lengths, as in Figure 16.4, we can then connect them with the number-line image of numbers, where 1.35 is now represented by a point on a line, as shown in Figure 16.5. Notice how the divisions on the number line are based on the principle of 'one of these is ten of those'. Each unit length from one whole number to the next can be divided into ten equal parts, each of

LEARNING AND TEACHING POINT

Children should learn how to locate decimal numbers on the number line. To do this, they have to understand how the unit lengths can be divided into tenths, and then how each of these tenths can be divided into hundredths (and so on).

which is a tenth. Figure 16.5 shows the interval from 1 to 2 divided in this way, producing the points labelled 1.1, 1.2, 1.3, 1.4, 1.5, 1.6, 1.7, 1.8 and 1.9. The number line shows clearly that, for example, 1.3 is 1 and 3 tenths. These tenths can then be divided into ten smaller divisions, each of which is a hundredth. This has been done in Figure 16.5 for the interval from 1.3 to 1.4.

Figure 16.5 The decimal number 1.35 as a point on a number line

The number 1.35 is now seen as 1 add 3 tenths add 5 hundredths. You get to this number if you start at zero, count along 1 unit, then 3 tenths and then 5 hundredths. This process can then be extended further, by dividing the hundredths into ten equal parts, each of which would be a thousandth. For example, if the interval from 1.35 to 1.36 was divided into ten equal parts, the new divisions on the number line would represent 1.351, 1.352, 1.353, 1.354, 1.355, 1.356, 1.357, 1.358 and 1.359.

Note again how this image of the number line enables us to appreciate the position of the number 1.35 in relation to other numbers, using the ordinal aspect of number: for example, 1.35 lies between 1 and 2; it lies between 1.3 and 1.4; it lies between 1.34 and 1.36. Or, using inequality signs: $1 < 1.35 < 2$; then, more precisely, $1.3 < 1.35 < 1.4$; and, even more precisely, $1.34 < 1.35 < 1.36$.

WATCH THE PROBLEM SOLVED! VIDEO AT: HTTPS://STUDY.SAGEPUB.COM/HAYLOCK7E

WHAT ABOUT ROUNDING?

Numbers obtained from measurements or as the results of calculations in practical situations often require **rounding** in some way for us to be able to make sense of them and to use them. For example, we might round the cost of a jumper given as £24.95 by saying that it costs about £25; here we have rounded the cost upwards. Or we might say that the decimal number 0.612 is approximately 0.6; in doing this, we are rounding it downwards. Sometimes we round up and sometimes we round down. The first consideration in this process must always be the *context* that gives rise to the numbers.

For example, when buying wallpaper most people would find it helpful to round *up* in their calculation of how many rolls to purchase, to be on the safe side and to avoid an extra trip to the DIY store. On the other hand, if we were planning to catch the 8:48 train we might well decide to round this time *down* to 'about a quarter to nine', since rounding this time *up* to 'about ten to nine' might result in our missing the train. So sometimes the context requires us to round numbers *down* and sometimes to round them *up.* This consideration of the context is much more important than any rule we might remember from secondary school about digits being greater or smaller than five.

So, in practice, it is the context that most often determines that we should round up or that we should round down. In some situations, however, there is a convention that we should round an answer *to the nearest something*. In fact, such situations are fairly rare in practice, since there is more often than not a contextual reason for rounding up or down, particularly when dealing with money. One situation where 'rounding to the nearest something' is often employed is in recording measurements.

Think of examples from everyday life where measurements are recorded in some form. It is nearly always the case that the measurement is actually recorded to the nearest something. For example, when the petrol pump says that I have put 15.8 litres in my tank this presumably means that the amount of petrol I have taken is really 15.8 litres to the nearest tenth of a litre. So it could actually be slightly more than that or slightly less. When I stand on the bathroom scales and record my weight as 69.5 kg, I am actually recording my weight to the nearest half a kilogram. When I read that someone has run 100 metres in 9.84 seconds, I would deduce from the two decimal places in the answer that the time is rounded to the nearest one-hundredth of a second. In all these examples, the measurement is recorded to the nearest something. The reasons for this are sometimes the limitations of the measuring device and sometimes simply that no practical purpose would be served by having a more accurate measurement.

It is in the handling of statistical data that the convention of rounding to the nearest something is employed most frequently. For example, you may well need to round your answer in this way when calculating averages (means). If over one cricket season I have six innings and score 50, 34, 0, 12, 0 and 43, I would calculate my average (mean) score by

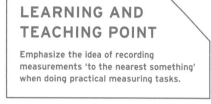

LEARNING AND TEACHING POINT

Emphasize the idea of recording measurements 'to the nearest something' when doing practical measuring tasks.

adding these up and dividing by 6 (see Chapter 27 for a discussion of averages). Doing this on a calculator I get the result 23.166666. Normally, I would not wish to record all these digits, so I might round my average score to the nearest whole number, which would be 23, or possibly to the nearest tenth, which would be 23.2.

HOW DOES THE PROCESS OF ROUNDING TO THE NEAREST SOMETHING WORK?

It is helpful to note that the concept of 'nearest' is essentially a spatial one. It helps in this process, therefore, to imagine the position of the number concerned on a number line, as shown in Figure 16.6. Clearly, the decision to round down or up is determined by whether the number is less or more than halfway along the line between two marks on the scale. The crucial questions in this process are always:

- What number would be halfway between the two marks on the scale?
- Is my number less than or greater than this?

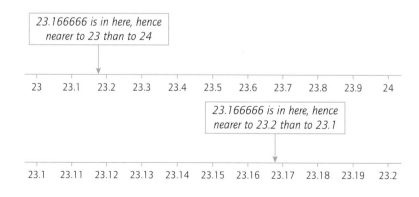

Figure 16.6 Rounding to the nearest something

For example, to express 23.166666 to the nearest whole number, we have to decide between 23 and 24. The number halfway between 23 and 24 is 23.5, so we round *down* to 23, because our number is *less* than 23.5. But to express 23.166666 to the nearest tenth, we have to decide between 23.1 and 23.2. The number halfway between 23.1 and 23.2 is 23.15, so we round *up* to 23.2, because our number is greater than 23.15.

This is just the same principle as when you are rounding to the nearest ten, the nearest hundred, the nearest thousand and so on. For

example, imagine I had worked out the estimated cost for renovating a classroom to be £8357. In discussions with the governors, I might find it appropriate to round this amount to the nearest hundred, or even the nearest thousand. To the nearest hundred, I have to choose between £8300 and £8400. Halfway between these two amounts is £8350. My cost of £8357 is more than this, so I would round *up* and say that we estimate the cost to be about £8400. But if it is sufficient to discuss the cost to the nearest thousand, then I now have to choose between £8000 and £9000. Once again, I ask what is halfway between these two amounts. The answer is clearly £8500 and, since my estimated cost of £8357 is less than this, we would round *down* and say that the cost is going to be about £8000, to the nearest thousand.

At this point, someone always asks what you do if the number you are dealing with is exactly halfway between your two marks on the scale. My answer is normally not to do anything. There really is not much point, for example, in pretending to round an amount like £8500 to the nearest thousand, because there is not a 'nearest' thousand! Just give it as £8500 and leave it at that. If I normally work my batting average out to the nearest whole number and one year it happens to come out exactly as 23.5, well, I think I will just leave it like that: at least I will be able to deduce from my records that I did a bit better than last year when my average to the nearest whole number was 23.

HOW DO YOU CHANGE DECIMALS INTO FRACTIONS?

In Chapter 15, we explained fraction notation and how to handle different kinds of fractions. The part of a decimal number that follows the decimal point is simply another example of a proper fraction (see Chapter 15), where the denominator is 10 or 100 or 1000 and so on. First, recall that the decimal 0.3 means 'three-tenths', so it is clearly equivalent to the fraction $^3/_{10}$. Likewise, 0.07 means 'seven-hundredths' and is equivalent to the fraction $^7/_{100}$. Then to deal with, say, 0.37 (3 tenths and 7 hundredths), we have to recognize that the 3 tenths can be exchanged for 30 hundredths, which, together with the 7 hundredths, makes a total of 37 hundredths, which is the fraction $^{37}/_{100}$. This continues into thousandths and so on. For example, 0.375 is 375 thousandths, which is the fraction $^{375}/_{1000}$. The only slight variation in all this is that sometimes the fraction obtained can be changed to an equivalent, simpler fraction, by cancelling (dividing top and bottom numbers by a common factor). Here are a few examples:

0.6 becomes	$^6/_{10}$	which is equivalent to	$^3/_5$	(dividing top and bottom by 2)
0.04 becomes	$^4/_{100}$	which is equivalent to	$^1/_{25}$	(dividing top and bottom by 4)
0.45 becomes	$^{45}/_{100}$	which is equivalent to	$^9/_{20}$	(dividing top and bottom by 5)
0.44 becomes	$^{44}/_{100}$	which is equivalent to	$^{11}/_{25}$	(dividing top and bottom by 4)
0.375 becomes	$^{375}/_{1000}$	which is equivalent to	$^3/_8$	(dividing top and bottom by 125).

HOW DO YOU CHANGE FRACTIONS INTO DECIMALS?

Fractions such as tenths, hundredths and thousandths, where the denominator (the bottom number) is a power of ten, can be written directly as decimals. Here are some examples to show how this works:

$^3/_{10}$ = 0.3 (0.3 means 3 tenths)

$^{23}/_{10}$ = 2.3 (23 tenths make 2 whole units and 3 tenths)

$^3/_{100}$ = 0.03 (0.03 means 3 hundredths)

$^{23}/_{100}$ = 0.23 (23 hundredths make 2 tenths and 3 hundredths)

$^{123}/_{100}$ = 1.23 (123 hundredths make 1 whole unit and 23 hundredths)

$^3/_{1000}$ = 0.003 (0.003 means 3 thousandths)

$^{23}/_{1000}$ = 0.023 (23 thousandths make 2 hundredths and 3 thousandths).

Then there are those fractions that we can readily change into an equivalent fraction with a denominator of 10 or 100. For example, $^1/_5$ is equivalent to $^2/_{10}$ (multiplying top and bottom by 2), which, of course, is written as a decimal fraction as 0.2. Similarly, $^3/_{25}$ is equivalent to $^{12}/_{100}$ (multiplying top and bottom by 4), which then becomes 0.12. Here are some further examples, many of which should be memorized:

$^1/_2$ is equivalent to $^5/_{10}$ which as a decimal is 0.5
$^4/_5$ is equivalent to $^8/_{10}$ which as a decimal is 0.8
$^1/_4$ is equivalent to $^{25}/_{100}$ which as a decimal is 0.25
$^3/_4$ is equivalent to $^{75}/_{100}$ which as a decimal is 0.75
$^1/_{20}$ is equivalent to $^5/_{100}$ which as a decimal is 0.05
$^7/_{20}$ is equivalent to $^{35}/_{100}$ which as a decimal is 0.35
$^1/_{50}$ is equivalent to $^2/_{100}$ which as a decimal is 0.02
$^3/_{50}$ is equivalent to $^6/_{100}$ which as a decimal is 0.06
$^1/_{25}$ is equivalent to $^4/_{100}$ which as a decimal is 0.04.

In a similar way, some fractions readily convert to equivalent fractions with 1000 as the denominator. The most familiar examples are the eighths. We need to change these to thousandths because 8 does not divide exactly into 10 or into 100, but it does divide exactly into 1000 (1000 ÷ 8 = 125). So, in each case, multiplying both the numerator and the denominator by 125, we get the following equivalences:

$^1/_8$ is equivalent to $^{125}/_{1000}$ which as a decimal is 0.125
$^3/_8$ is equivalent to $^{375}/_{1000}$ which as a decimal is 0.375
$^5/_8$ is equivalent to $^{625}/_{1000}$ which as a decimal is 0.625
$^7/_8$ is equivalent to $^{875}/_{1000}$ which as a decimal is 0.875.

Figure 16.7 shows how the number line can provide an image of the equivalences (a) between quarters and hundredths, and (b) between eighths and thousandths. The relationships here could also be shown using *bar-modelling* (see Figure 7.9), with a bar representing one unit divided into (a) 4 equal sections and (b) 8 equal sections. The equivalences shown in Figure 6.7 turn up in

many measuring contexts. For example, when you see a wine bottle that contains 0.75 litres of wine you should recognize this as three-quarters of a litre. When you see a pack of 0.125 kg of butter you should immediately spot that this is one-eighth of a kilogram.

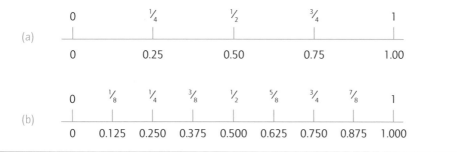

Figure 16.7   Some equivalences between fractions and decimals shown on number lines

IS THERE A MORE GENERAL WAY OF CONVERTING FRACTIONS TO DECIMALS?

A general approach to changing fractions to decimals is to make use of the idea that one of the meanings of the fraction notation is division (see Chapter 15). So, any fraction can be changed to a decimal just by dividing the top number by the bottom number – when appropriate this can be done on a calculator. Sometimes the result obtained will be an *exact decimal*, but often it will be a **recurring decimal**. For example, using a calculator I get $^{17}/_{25}$ = 17 ÷ 25 = 0.68. This is an exact answer. But for the fraction $^{2}/_{9}$, using a calculator I get $^{2}/_{9}$ = 2 ÷ 9 = 0.22222222. This is not an exact answer, because 2 ÷ 9 is equal to the recurring decimal 0.222… with the digit 2 repeated over and over again, endlessly, forever! We refer to it as 'zero point two recurring'. This calculator cannot display more than 8 digits after the decimal point, so it ditches all the digits after these. This does not make any practical difference, because the bits being discarded are extremely small in magnitude.

To express the fraction $^{7}/_{9}$ as a decimal, if I enter '7 ÷ 9 =' on one calculator I get the result 0.77777777, but on another calculator I get 0.77777778. This is the best that these calculators can do with 'zero point seven recurring'. In both cases, the calculator can give the result to only 8 digits after the decimal point, but in the second case the calculator automatically rounds the answer: see the explanation of rounding to so many decimal places below.

If the division is being done by a written method then short or long division can be applied, extending the process beyond the decimal point. An example is given in Figure 16.8(a) where the fraction $^{21}/_{40}$ is converted to a decimal number (0.525) by dividing 21 by 40. Note that the number being divided, 21, is written 21.000, with as many zeros as necessary to complete the calculation. The actual division begins with 21.0 being divided by 40 – but we can think of 21.0 as 210 tenths so we are then effectively dividing 210 by 40. The 5 that results is 5 tenths and so this is written in the tenths place in the answer above the line. Note again that it is helpful to write a zero in front of the decimal point in the answer, so that the decimal point does not get overlooked.

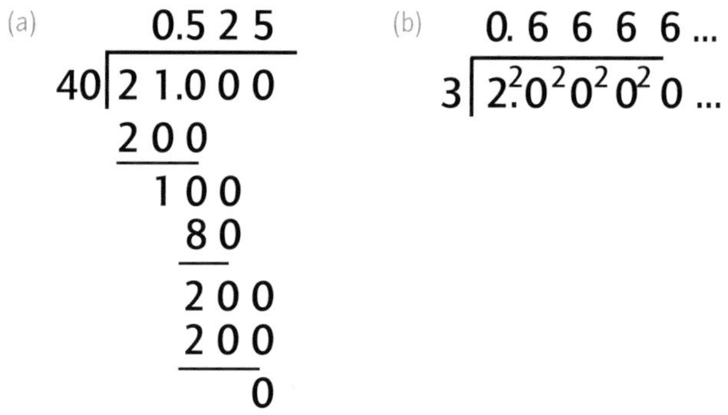

(a)
$$0.5\ 2\ 5$$
$$40\overline{)2\ 1.0\ 0\ 0}$$
$$\underline{2\ 0\ 0}$$
$$1\ 0\ 0$$
$$\underline{8\ 0}$$
$$2\ 0\ 0$$
$$\underline{2\ 0\ 0}$$
$$0$$

(b)
$$0.\ 6\ 6\ 6\ 6\ ...$$
$$3\overline{)2.^2 0^2 0^2 0^2\ 0\ ...}$$

Figure 16.8 Using formal division to change fractions to decimals

In this example, 21 divided by 40 works out exactly as a decimal. Contrast this with the calculation of $^2/_3$ as a decimal, shown in Figure 16.8(b). This requires the division of 2 by 3. After a few steps, you begin to realize that you are going to do $20 \div 3 = 6$, with 2 remaining, over and over again, forever. So, written as a decimal, $^2/_3$ is 0.6666... with the digit 6 being repeated endlessly. Check this on a calculator, by entering '2 ÷ 3 =' and see what you get. This is another example of a recurring decimal. In words, we read it as 'zero point six recurring'. Sometimes recurring decimals have two or more digits repeated endlessly. For example, by dividing 3 by 11 we find that $^3/_{11}$ is equal to the decimal number 0.272727..., with the '27' repeating over and over again. This is 'zero point two seven recurring'. Again, you should check this on a calculator by entering '3 ÷ 11 =' and see what result is obtained.

WHEN IS A FRACTION EQUIVALENT TO A RECURRING DECIMAL, AND VICE VERSA?

If the denominator of a fraction is a factor of 10, 100, 1000, or any power of 10, then it will work out exactly as a non-recurring decimal. For example, because 250 is a factor of 1000 we can be sure that any fraction with 250 as the denominator would be equivalent to a non-recurring decimal. In fact, because there are four 250s in 1000, you can convert 250ths into thousandths just by multiplying by 4. For example, $^{121}/_{250}$ = $^{484}/_{1000}$ = 0.484.

If the fraction in its simplest form (after cancelling) has a denominator that does not divide exactly into 10 or a power of 10, then it is equivalent to a recurring decimal. This would include all fractions in their simplest form that have denominators 3, 6, 7, 9, 11, 12, 13, 14, 15, 17, 18, 19 and so on. At a glance, then, we can tell that all these fractions will give recurring decimals if we divide the bottom number into the top number: $^{1}/_{3}$, $^{2}/_{3}$, $^{1}/_{6}$, $^{5}/_{6}$, $^{1}/_{7}$, $^{2}/_{7}$, $^{3}/_{7}$, $^{4}/_{7}$, $^{5}/_{7}$, $^{6}/_{7}$, $^{1}/_{9}$, $^{2}/_{9}$, $^{4}/_{9}$ and so on.

Now here is a remarkable fact: all recurring decimals are actually rational numbers. In other words, they are equivalent to the ratio of two integers, a fraction. For example: 0.333333..., with the 3 recurring forever, is $^{1}/_{3}$; 0.279279279..., with the '279' recurring forever, is 31/111.

WHAT IS THERE TO KNOW ABOUT THE INTERPRETATION OF A CALCULATOR RESULT?

The interpretation of a calculator result involving decimals is far from being an insignificant aspect of the process of modelling a real-life problem (see Chapter 5). It will often require a decision about what to do with all the digits after the decimal point. To summarize what we have seen so far in this chapter, when we carry out a calculation on a calculator to solve a practical problem, particularly those modelled by division, we can get three kinds of answer:

- an exact, appropriate answer;
- an exact but inappropriate answer;
- an answer that is a **truncation**.

In the last two cases, we normally have to round the answer in some way to make it appropriate to the real-world situation.

For example, consider these three problems, all with the same mathematical structure:

Problem 1: How many lengths of 15 cm can I cut from 90 cm of ribbon?

Problem 2: How many lengths of 24 cm can I cut from 90 cm of ribbon?

Problem 3: How many lengths of 21 cm can I cut from 90 cm of ribbon?

For Problem 1, the mathematical model is $90 \div 15$, so we might enter '90 ÷ 15 =' on a calculator and obtain the result 6. In this case, there is no difficulty in interpreting this result back in the real world. The answer to the problem is indeed exactly 6 of the required lengths of ribbon. The calculator result is both exact and appropriate.

For Problem 2, we might enter '90 ÷ 24 =' and obtain the result 3.75. This is the exact answer to the calculation that was entered on the calculator. In other words, it is the solution to the mathematical problem that we used to model the real-world problem. It is correctly interpreted as 3.75 lengths of ribbon. But, since we want only lengths of 24 cm of ribbon, then clearly the answer is not appropriate. In the final step of the modelling process – checking the solution against the constraints of reality – the 3.75 lengths must be rounded down to a whole number. We would conclude that the solution to Problem 2 is that we can get only 3 of the required lengths, with a smaller bit of ribbon left over.

For Problem 3, we might enter '90 ÷ 21 =' into the calculator and find that we get the result 4.2857142. This is not an exact answer. Dividing 90 by 21 produces a recurring decimal, namely 4.285714285714 …, with the 285714 repeating over and over again without ever coming to an end. Since a simple calculator can display only eight digits, it *truncates* the result, by throwing away all the extra digits. Of course, in this case, the bits that are thrown away are relatively tiny and the error involved in this truncation process is fairly insignificant. We can interpret the result displayed on the calculator as 'about 4.2857142 lengths of ribbon'. But again, when we compare this with the real-world situation, we must recognize the constraints of the actual problem and round the answer down to give a whole number of lengths of ribbon. The obvious conclusion is that we can cut only 4 of the required lengths.

LEARNING AND TEACHING POINT

Discuss with children real-life problems, particularly in the context of money, that produce calculator answers that require different kinds of interpretation, including those with: (a) an exact, appropriate answer; (b) an exact but inappropriate answer; and (c) an answer that has been truncated.

LEARNING AND TEACHING POINT

When considering practical division problems that do not work out exactly when done on a calculator, discuss from the context whether to round up or to round down.

In Problem 3, the calculator result was rounded down to 4 lengths of ribbon. Now, here is another problem that is modelled by the same calculation, namely $90 \div 21 = 4.2857142$, but in this case the context determines that we have to round the answer up.

Problem 4: On a school trip, you can fit 21 children into a dormitory. How many dormitories do you need if you have 90 children to be accommodated?

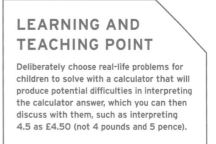

In Problem 4, the calculator result is interpreted as 'about 4.2857142 dormitories'. But you cannot in practice book 4.2857142 dormitories. Since four of these rooms would not be enough for every child to get a bed, you conclude that five are needed. In this case, the context determines that you round the 4.2857142 *up* to the next whole number, which is 5. In explaining what is going on to children, we could talk about the '.2857142' in the answer as representing 'a bit of a room'. We need 4 rooms and a bit of a room: so we must book 5 rooms, otherwise some people will have to sleep in the corridor.

Problems 3 and 4 are in contexts where the answers have to be rounded either up or down. Now consider Problem 5, which is again modelled by $90 \div 21 = 4.2857142$, this time in the context of equal sharing.

Problem 5: If 21 reading books are sold for £90, what is the cost per book?

The £90 has to be shared equally between the 21 books. This problem is in a context where the answer would normally be rounded to the nearest something. In this case, it would be normal to round to the nearest penny, which requires two digits after the decimal point, 4.29. This is interpreted as '£4.29 per book'. This is an example of rounding to two decimal places, which is explained in the next section.

WHAT ABOUT ROUNDING TO SO MANY DECIMAL PLACES AND SIGNIFICANT DIGITS?

When we are rounding a number like 23.166666, it is unnecessarily complex to talk about rounding it to the nearest tenth, hundredth, thousandth, ten-thousandth and so on. It is much easier to talk about rounding *to so many decimal places*. Rounding to one decimal place is the same as rounding to the nearest tenth. Rounding to two decimal places is the same as rounding to the nearest hundredth. And so on.

For example, 23.166666 is:

23, when rounded to the nearest whole number
23.2, when rounded to one decimal place
23.17, when rounded to two decimal places
23.167, when rounded to three decimal places
23.1667, when rounded to four decimal places.

A bit trickier, but often more appropriate, is the idea of rounding to so many **significant digits**. A full consideration of this is properly beyond the scope of this book, but for the sake of completeness I will try to clarify the main idea. The point is that in many practical situations we are really only interested in the approximate size of an answer to a calculation, and just the first two or three digits are normally enough.

If you are applying for a new job, you are not going to get terribly concerned about whether the salary offered is £28,437 or £28,431. In either case, the salary is about £28,400. When you are talking about thousands of pounds, the choice between an additional £37 or £31 is not likely to be particularly significant. If, however, I am buying a pair of trousers then the difference between £37 and £31 becomes quite significant. It is this kind of reasoning that is the basis of the practice of rounding to a certain number of significant digits.

Giving the two salaries above as 'about £28,400' is the result of rounding them *to three significant digits*. The first significant digit – indeed the most significant digit – is the 2, since this represents the largest part of the salary, namely, twenty thousand pounds. Then the 78 and the 4 are the second and third significant digits respectively. This might be all we want to know, since we would reckon that the digits in the tens and units columns are relatively insignificant. In most practical situations like these, when approximate answers are required it is rarely useful to use more than three significant digits. Below are some simple examples of rounding to significant digits.

On my monthly trip to the supermarket, I spend £270.95. When I get home, my wife asks me how big the bill was.

- Exact statement: The bill came to £270.95.
- Using just one significant digit: The bill came to about £300. (This is to the nearest hundred pounds.)
- Using two significant digits: The bill came to about £270. (This is to the nearest ten pounds.)
- Using three significant digits: The bill came to about £271. (This is to the nearest pound.)

A father left £270,550 to be shared equally between four children:

- Mathematical model: 270,550 ÷ 4 = 67,637.5.
- Exact solution: They each inherit £67,637.50.
- To two significant digits: They each inherit about £68,000. (This is giving the amount of money to the nearest thousand pounds.)
- To three significant digits: They each inherit about £67,600. (This is giving the amount of money to the nearest hundred pounds.)
- To four significant digits: They each inherit about £67,640. (This is giving the amount of money to the nearest ten pounds.)

RESEARCH FOCUS: GENERATING DIVISION PROBLEMS WITH REMAINDERS

A reasonable expectation is that a primary school teacher should be able, spontaneously, to generate a range of division problems with remainders in real-life contexts that shows an awareness of the different structures of division and the various ways of interpreting the remainder, depending on the contexts that are discussed in this chapter. An interesting and relevant piece of research conducted with elementary school trainee teachers in the USA by Silver and Burkett (1994) focused on their ability to relate division problems with remainders to real-life situations. The trainees were asked to generate a range of problems in real-life contexts that corresponded to the following division: 540 ÷ 40 = 13, remainder 20. The responses were categorized into those having a 'partitive division structure' (what I have called the equal-sharing structure) and those with a 'quotitive division structure' (what I have called the inverse-of-multiplication structure). Within these, they identified those problems in which the correct approach is:

a   'to increment the quotient if a remainder occurs' (that is, to round up); for example, '40 ants can fit on a leaf – if there are 540 ants trying to cross the river, how many leaves must the ants gather?'

b   'to ignore the remainder' (that is, effectively, to round down); for example, 'There are 540 balls – how many bags can you fill with 40 balls per bag?'

c   'to give the remainder as the solution'; for example, 'If you have 540 cans of food and each of 40 needy families must get the same number of cans, how many cans are left over?'

Only two-thirds of the trainees were able to generate a problem that showed a proper understanding of the relationship between the division and at least one structure. A common error in posing problems in the equal-sharing structure was the failure to specify sharing equally. Additionally, about 25% of the problems posed had some unreasonable aspect to them, such as a highly implausible condition (like each student in a school having 13 lockers) or solutions that required breaking up objects like eggs, children or dogs into fractional parts. The authors suggest that this is indicative of a tendency to dissociate formal school mathematics from the real world and express their concern that elementary school teachers without a better understanding of the relationship between mathematical computation and real-world contexts might perpetuate this dissociation in their own children's learning of the subject. See also the Research Focus for Chapter 10 and Boaler (2020), chapter 5, 'Learning without Reality'.

LEARNING RESOURCES

Access activities for your **lesson plans** at: https://study.sagepub.com/haylock7e

Before trying the self-assessment questions below, you should complete the **self-assessment questions** for this chapter at: https://study.sagepub.com/haylock7e

16.1: Interpret these decimal numbers as collections of base-ten blocks, using a 'flat' to represent a unit, a 'long' to represent a tenth and so on: 3.2, 3.05, 3.15, 3.10. Then arrange them in order from the smallest to the largest.

16.2: Fill in the boxes with single digits: on a number line, 3.608 lies between □ and □; it lies between □.□ and □.□; it lies between □.□□ and □.□□; it lies between □.□□□ and □.□□□. Give answers that are as close as possible to 3.608.

16.3: A teacher wants to order 124 copies of a mathematics book costing £5.95. She must report the total cost of the order to the year-group meeting. What is the mathematical model of this problem? Obtain the mathematical solution, using a calculator. Interpret the mathematical solution as an exact statement back in the real world. What would the teacher say if she reports the total cost to the nearest ten pounds? What would the teacher say if she reports the total cost to three significant digits?

16.4: Would you round your answers up or down in the following situations requiring a division?

    a   There are 327 children on a school trip and a coach can hold 40. How many coaches are needed? ($327 \div 40 = ?$)

    b   How many cakes costing 65p each can I buy with £5? ($500 \div 65 = ?$)

16.5: Here is a division calculation: $320 \div 50 = 6$, remainder 20. Make up two realistic problems that correspond to this division, one of which (a) has the answer 7, and the other of which (b) has the answer 6.

16.6: What does each person pay if three people share equally a restaurant bill for £27.90? Use a calculator to answer this question. Is the calculator result: (a) an exact, appropriate answer; (b) an exact but inappropriate answer; or (c) an answer that has been truncated? Identify the steps in the process of mathematical modelling in what you have done.

16.7: Repeat question 16.6 with a bill for £39.70.

16.8: What is the average (mean) height, to *the nearest centimetre*, of 9 children with heights, 114 cm, 121 cm, 122 cm, 130 cm, 131 cm, 136 cm, 139 cm, 146 cm, 148 cm? (Add the heights; divide by 9; and round to the nearest centimetre.)

16.9: One season, there were 1459 goals in 462 matches in a football league. How many goals is this per match, on average? ($1459 \div 462 = ?$) Do this on a calculator and give the answer: (a) rounded to the nearest whole number; (b) rounded to one decimal place; and (c) rounded to two decimal places.

FURTHER PRACTICE

Access the website material for

- Knowledge check 12: Rounding answers
- Knowledge check 13: Very large and very small numbers

at: https://study.sagepub.com/haylock7e

FROM THE STUDENT WORKBOOK

Questions 16.01–16.24: Checking understanding (decimal numbers and rounding)

Questions 16.25–16.42: Reasoning and problem solving (decimal numbers and rounding)

Questions 16.43–16.50: Learning and teaching (decimal numbers and rounding)

GLOSSARY OF KEY TERMS INTRODUCED IN CHAPTER 16

Decimal point   A punctuation mark (.) required when the numeration system is extended to include tenths, hundredths and so on; it is placed between the digits representing units and tenths.

Decimal number   A number written with a decimal point and one or more digits after it.

Separator   The function of the decimal point in the contexts of money and other units of measurements, where it serves to separate, for example, pounds from pence, or metres from centimetres.

Rounding   The process of approximating a measurement or an answer to a calculation to an appropriate degree of accuracy; this can be done by rounding up or rounding down or rounding to the nearest something. For example, £25.37 rounded *up* to the next ten pence is £25.40 and rounded *down* to the next ten pence is £25.30; rounded to the *nearest* ten pence it is £25.40, because it is nearer to this than to £25.30.

Recurring decimal   A decimal, which might be the result of a division calculation, where one or more digits after the decimal point repeat over and over again, forever. For example, 48 ÷ 11 is equal to 'four point three six recurring' (4.36363636 … with the '36' being repeated over and over again, forever).

Truncation   This is what a calculator does when it has to cut short an answer to a calculation by throwing away some of the digits after the decimal point, because it does not have room to display them all. For example, a calculator with space for only 8 digits in the display might truncate the result 987.654321 to 987.65432.

Significant digits   The digits in a number as you read from left to right; for example, in 25.37 the first significant digit is the 2, then the 5, then the 3 and then the 7. This number rounded to two significant digits is 25 and rounded to three significant digits is 25.4.

CALCULATIONS WITH DECIMALS

IN THIS CHAPTER, THERE ARE EXPLANATIONS OF

- the procedures for addition and subtraction with decimal numbers;
- the contexts that might give rise to the need for calculations with decimals;
- checking the reasonableness of answers by making estimates, using approximations;
- multiplication and division of a decimal number by an integer in real-life contexts;
- the results of repeatedly multiplying or dividing decimal numbers by 10;
- multiplying an integer by a decimal;
- how to obtain decimal answers in division calculations;
- how the remainder in a written division relates to the decimal answer in various contexts;
- how to deal with the multiplication of two decimals;
- some simple examples of division involving decimals.

READ THIS CHAPTER'S CURRICULUM LINKS AT: HTTPS://STUDY.SAGEPUB.COM/HAYLOCK7E

WHAT ARE THE PROCEDURES FOR ADDITION AND SUBTRACTION WITH DECIMALS?

The procedures for addition and subtraction with decimals are effectively the same as for whole numbers. Difficulties would arise only if you were to forget about the principles of place value outlined in Chapters 6 and 16. Provided you remember which digits are units, tens and hundreds, or tenths, hundredths and so on, then the algorithms (and adhocorithms) employed for addition and subtraction of whole numbers (see Chapters 8 and 9) work in an identical fashion for decimals – with the principle that 'one of these can be exchanged for ten of those' guiding the whole process.

LEARNING AND TEACHING POINT

When adding or subtracting with decimals, show children how the principle of 'one of these can be exchanged for ten of those' works in the same way as when working with integers.

A useful tip with decimals is to ensure that the two numbers in an addition or a subtraction have the same number of digits after the decimal point. If one has fewer digits than the other then fill up the empty places with zeros, acting as 'place holders' (see Chapter 6). So, for example: 1.45 + 1.8 would be written as 1.45 + 1.80; 1.5 – 1.28 would be written as 1.50 – 1.28; and 15 – 4.25 would be written as 15.00 – 4.25. This makes the standard algorithms for addition and subtraction look just the same as when working with whole numbers, but with the decimal points in the two numbers lined up, one above the other, as shown in Figure 17.1. The reader should note that if the decimal points are removed, these calculations now look exactly like 286 + 404, 145 + 180, 150 – 128 and 1500 – 425. There really is nothing new to learn here, apart from the need to line up the decimal points carefully and to fill up empty spaces with zeros.

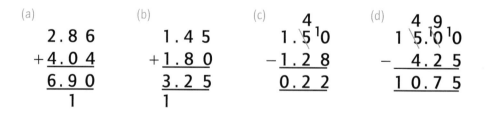

Figure 17.1   Additions and subtractions with decimals

In practice, additions and subtractions like these with decimals would usually be employed to model real-life situations, particularly those related to money or measurement. In Chapter 16, I discussed the convention of putting two digits after the decimal point when recording money in pounds. We saw also that when dealing with measurements of length in centimetres and metres, with a hundred centimetres in a metre, it is often a good idea to adopt the same convention, for example writing 180 cm as 1.80 m, rather than 1.8 m. Similarly, when handling liquid volume and capacity, where we have 1000 millilitres in a litre, or mass, where we have 1000 grams in a kilogram, the convention would often be to write measurements in litres or kilograms with three digits after the point. This means that, if this convention is followed, the decimal numbers will arrive for the calculation already written in the required form, that is with each of the two numbers in the addition or subtraction having the same number of digits after the point, with zeros used to fill up empty places.

For example, the four calculations shown in Figure 17.1 might correspond to the following real-life situations:

a   Find the total length of wall space taken up by a cupboard that is 2.86 m wide and bookshelving that is 4.04 m wide.

b   If a child saves £1.45 one month and £1.80 the next month, how much have they saved altogether?

1   What is the difference in height between a girl who is 1.50 m tall and a boy who is 1.28 m tall?

2   What is the change from £15.00 if you spend £4.25?

Note that example (d) here is most easily done by adding on from £4.25 to get to £15: first £0.75 to get to £5, then a further £10 to get to £15, giving the result £10.75. This illustrates again that often an informal calculation strategy will be easier to handle than formal vertical layout (as in Figure 17.1(d)), particularly when the calculation is in a real-life context.

LEARNING AND TEACHING POINT

Locate calculations with decimals in realistic contexts where the decimals represent money or measurements. Emphasize the usefulness of having the same number of digits after the decimal point when adding or subtracting money or measurements written in decimal notation. Explain to children about putting in extra zeros as place holders where necessary.

HOW CAN YOU CHECK THAT THE ANSWER TO AN ADDITION OR A SUBTRACTION IS REASONABLE?

Always check the reasonableness of your answer to a calculation involving decimals by making a mental estimate based on simple approximations. In example (a) above, the measurements could be approximated to 3 metres and 4 metres to the nearest metre, so we would expect an answer around 7 metres. In example (b), the amounts of money are about £1 and £2, so we would expect an answer around £3. In example (c), we might approximate the heights to 150 cm and 130 cm, so an answer around 20 cm would be expected. And in example (d), approximating this to £15 – £4, we would expect the change to be around £11. Notice how I have tended to round the numbers to the nearest something in these examples (see Chapter 16).

For the benefit of the reader, here's a word of caution, going beyond what might be expected of primary school children: when you are adding two numbers, remember that you will be adding the rounding errors. If both numbers have been rounded up (or both down), this could lead to quite a significant error in your estimate. A safer procedure is first to round both up and then to round both down, thus determining limits within which the sum must lie.

> **LEARNING AND TEACHING POINT**
>
> Get children into the habit of checking the reasonableness of their answers to calculations by making estimates based on approximations of the numbers involved. Encourage them to round the numbers involved to the nearest whole number, or the nearest ten, the nearest tenth, the nearest hundred and so on.

An addition example: 16.47 + 7.39
Round both up: 17 + 8 = 25
Round both down: 6 + 7 = 23
So, the answer lies between 23 and 25.

A similar comment applies when subtracting two numbers when one has been rounded up and the other rounded down. For subtraction, we obtain limits within which the answer must lie by rounding the first number up and the second one down (to get an answer which is clearly too large) and then rounding the first number down and the second one up (to get an answer which is clearly too small).

A subtraction example: 16.47 – 7.39
Round first number up, second one down: 17 – 7 = 10

Round second number up, first one down: 16 − 8 = 8
So, the answer lies between 8 and 10.

WHAT ABOUT MULTIPLICATIONS AND DIVISIONS INVOLVING DECIMALS IN A REAL-LIFE CONTEXT?

Consider the multiplication of a decimal number by a whole number. Once again, the key point is to think about the practical contexts that would give rise to the need to do multiplications of this kind. In the context of money, we might need to find the cost of a number of articles at a given price. For example, find the cost of 12 rolls of sticky tape at £1.35 per roll. The mathematical model of this real-world problem is the multiplication, 1.35 × 12. On a calculator, we enter 1.35, ×, 12, =, read off the mathematical solution (16.2) and then interpret this as a total cost of £16.20.

There is some potential to get in a muddle with the decimal point when doing this kind of calculation by written methods. So, a useful tip is, if you can, avoid multiplying the decimal numbers altogether. In the example above, this is easily achieved, simply by rephrasing the situation as 12 rolls at 135p per roll, hence writing the cost in pence rather than in pounds. We then multiply 135 by 12, by whatever methods we prefer (see Chapters 11 and 12), to get the answer 1620, interpret this as 1620p and, finally, write the answer as £16.20.

Almost all the multiplications involving decimals we have to do in practice can be tackled like this. Here's another example: find the length of wall space required to display eight posters, each 1.19 m wide. Rather than tackle 1.19 × 8, rewrite the length as 119 cm, calculate 119 × 8 and convert the result (952 cm) back to metres (9.52 m). So, 1.19 × 8 = 9.52.

> ## LEARNING AND TEACHING POINT
>
> Realistic multiplication and division problems with decimals involving money or measurements can often be recast into calculations with whole numbers by changing the units (for example, pounds to pence, metres to centimetres). Teach children how to do this.

The same tip applies when dividing a decimal number by a whole number. The context from which the calculation has arisen will suggest a way of handling it without the use of decimals. For example, a calculation such as 4.35 ÷ 3 could have arisen from a problem about sharing £4.35 between 3 people. We can simply rewrite this as the problem of sharing 435p between 3 people and deal with it by whatever division process is appropriate (see Chapters 11 and 12), concluding that each person gets 145p each. The final step is to put this back into pounds notation, as £1.45. So 4.35 ÷ 3 = 1.45.

HOW DO YOU EXPLAIN THE BUSINESS OF MOVING THE DECIMAL POINT WHEN YOU MULTIPLY AND DIVIDE BY 10 OR 100 AND SO ON?

On a basic calculator, enter the following key sequence and watch carefully the display: 10, ×, 1.2345678, =, =, =, =, =, =, =. This procedure is making use of the constant facility, which is built into most basic calculators, to multiply repeatedly by 10. (If this does not work on your calculator, try 1.2345678, ×, 10, =, =, =, =, =, =, =. The calculator on my desk requires the first option, the one on my computer the second.) The results are shown in Figure 17.2(a). It certainly looks on the calculator as though the decimal point is gradually moving along, one place at a time, to the right.

(a)	(b)
1.2345678	1.2345678
12.345678	12.345678
123.45678	123.45678
1234.5678	1234.5678
12345.678	12345.678
123456.78	123456.78
1234567.8	1234567.8
12345678.	12345678

Figure 17.2 The results of repeatedly multiplying by 10: (a) as displayed on a calculator, (b) arranged on the basis of place value

Now, without clearing your calculator, enter: ÷, 10, =, =, =, =, =, =, =. This procedure is repeatedly dividing by 10, thus undoing the effect of multiplying by 10 and sending the decimal point back to where it started, one place at a time. Because of this phenomenon, we tend to think of the effect of multiplying a decimal number by 10 to be to move the decimal point one place to the right, and the effect of dividing by 10 to be to move the decimal point one place to the left. Since multiplying (dividing) by 100 is equivalent to multiplying (dividing) by 10 and by 10 again, this results in the point moving two places. Similarly, multiplying or dividing by 1000 will shift it three places and so on for other powers of 10.

However, to understand this phenomenon, rather than just observing it, it is more helpful to suggest that it is not the decimal point that is moving, but the digits. This is shown by the results displayed on the basis of place value, as shown in Figure 17.2(b). Each time we multiply by 10, the digits all move one place to the left. The decimal point stays where it is. To understand why this happens, trace the progress of one of the digits, for example the 3. In the original number, it represents 3 hundredths. When we multiply the number by 10, each hundredth becomes a tenth, because ten hundredths can be exchanged for a tenth. This is, once again, the principle that 'ten of these can be exchanged for one of those', as we move right to left. So, the 3 hundredths become 3 tenths and the digit 3 moves

> **LEARNING AND TEACHING POINT**
>
> In calculations with natural numbers children will have picked up the idea that to multiply by 10 you just put a zero on the end of the number (for obvious reasons avoid saying 'add a zero'). So 27 × 10 = 270, for example. Make sure children are explicitly aware that this does not work with decimal numbers. So, 27.5 × 10 does NOT equal 27.50. Discuss with them why this is the case.

> **LEARNING AND TEACHING POINT**
>
> Base your explanation of multiplication and division of decimal numbers by 10 (and 100 and 1000) on the principle of place value. Talk about the digits moving, rather than the decimal point. Allow children to explore repeated multiplications and divisions by 10 with a calculator, making use of the constant facility.

from the hundredths position to the tenths position. Next time we multiply by 10, these 3 tenths become three whole units, and the 3 shifts to the units position. Next time we multiply by 10, these 3 units become 3 tens, and so on. Because the principle that 'ten of these can be exchanged for one of those', as you move from right to left, applies to any position, each digit moves one place to the left every time we multiply by 10.

Since dividing by 10 is the inverse of multiplying by 10 (in other words, one operation undoes the effect of the other), clearly the effect of dividing by 10 is to move each digit one place to the right.

HOW DO YOU USE THIS WHEN MULTIPLYING BY A DECIMAL NUMBER?

Older children in primary school should be able to handle multiplications such as (a) 7 × 0.4, (b) 6 × 0.03, (c) 24 × 0.6 and (d) 75 × 0.02. These calculations can, of course, be done by reinterpreting them in the context of money, as explained above (7 items at

40p each, 6 items at 3p each and so on). But, in preparation for more complex multi-plications involving decimals that they will have to learn in secondary school, it is help-ful even at this level of difficulty to learn how to manage these as abstract calculations.

The key principle is that a multiplication by a decimal number can always be done as a multiplication by a whole number and then division by a power of 10. I will show how this principle works with the four examples here:

(a)   Because $0.4 = 4 \div 10$
      to multiply by 0.4, first multiply by 4 and then divide by 10
      $7 \times 4 = 28$, so $7 \times 0.4 = 2.8$

(b)   Because $0.03 = 3 \div 100$
      to multiply by 0.03, first multiply by 3 and then divide by 100
      $6 \times 3 = 18$, so $6 \times 0.03 = 0.18$

(c)   Because $0.6 = 6 \div 10$
      to multiply by 0.6, first multiply by 6 and then divide by 10
      $24 \times 6 = 144$, so $24 \times 0.6 = 14.4$

(d)   Because $0.02 = 2 \div 100$
      to multiply by 0.02, first multiply by 2 and then divide by 100
      $75 \times 2 = 150$, so $75 \times 0.02 = 1.50$

In (d) here, because there is no context for the calculation, the 0 at the end of 1.50 tells us simply that there are no hundredths. But nor are there any thousandths, or ten-thou-sandths and so on. This 'trailing zero' can therefore be dropped and the answer given just as 1.5. In terms of abstract numbers, 1.50, 1.500, 1.5000 and so on are all equal to 1.5. In some contexts, these trailing zeros are retained by convention (as in £1.50) or as an indication that a quantity has been rounded (for example, 1.498 rounded to two decimal places is 1.50).

> **LEARNING AND TEACHING POINT**
>
> A prerequisite for understanding how to multiply and divide with decimal numbers is to be confident in multiplying and dividing by 10 and 100 (and other powers of 10).

HOW CAN THE REMAINDER IN A WRITTEN DIVISION CALCULATION BE EXPRESSED AS A DECIMAL?

In Chapter 16, we saw how dividing the numerator of a fraction by the denominator gave a decimal number equivalent to the fraction. For example, we divided 21 by 40 and found that $^{21}/_{40}$ was equal to 0.525. This means that in any division calculation that does not work out exactly as a whole number answer we can express the remainder as a decimal. All that is involved is to continue the division process beyond the decimal point. Figure 17.3 provides two examples. In (a) 647 is divided by 5 using the method of short division. Instead of stopping at the point where the result would be 129 remainder 2, note that this 2 has been changed into 20 tenths. This is facilitated by writing in an additional zero in the tenths position. Then this 20 divided by 5 gives 4 in the tenths position in the answer. Hence, we get 647 ÷ 5 = 129.4. Again, we should remember that all genuine calculations arise from some context and the context will determine whether it is appropriate to give the result in the form of a decimal. For example, if the problem had been to divide a length of 647 m of rope into 5 equal parts, then to say that the length of each part would be 129.4 m is a very appropriate result. But if the problem had been to find out how many lengths of 5 m could be cut from a length of 647 m, then the appropriate solution is 129 lengths with 2 m of rope unused (the remainder).

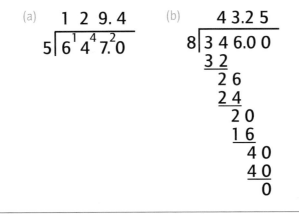

Figure 17.3 Examples of division continuing beyond the decimal point

WATCH THE PROBLEM SOLVED! VIDEO AT: HTTPS://STUDY.SAGEPUB.COM/HAYLOCK7E

Note that the 0.4 in the result for 647 ÷ 5 is the decimal equivalent of the fraction $^{2}/_{5}$. In Chapter 15 we saw how the remainder of a division problem could be interpreted as a fraction. So we now have three possible exact results for this division, which would be used as appropriate to the context for the calculation:

647 ÷ 5 = 129 remainder 2
647 ÷ 5 = 129$^{2}/_{5}$
647 ÷ 5 = 129.4

Figure 17.3(b) shows the written calculation of 346 ÷ 8, using the long division format, again extending the process beyond the decimal point. This time we need to write in two additional zeros in order to be able to complete the division and to get the answer 43.25. Once again, it is helpful to note the three different ways of expressing the exact outcome here:

$346 \div 8 = 43$ remainder 2

$346 \div 8 = 43^2/_8$ (which can be simplified to $43^1/_4$)

$346 \div 8 = 43.25$

HOW DO THE DIGITS AFTER THE DECIMAL POINT IN THE RESULT OF A DIVISION CALCULATION RELATE TO THE REMAINDER?

Of the three outcomes for the division 346 ÷ 8 obtained above, note that the first of these would be the appropriate solution if calculating how many teams of 8 could be made from 346 children (43 teams, 2 children remaining); the second, if calculating how many cakes each person gets if 346 cakes are shared equally between 8 people (43 cakes and a quarter of a cake each); the

> **LEARNING AND TEACHING POINT**
>
> Explicitly discuss with children the difference in meaning between the remainder in a division problem and the digits after the decimal point in the calculator answer.

third, if sharing £346 equally between 8 people (£43.25 each). The different solutions are appropriate for different kinds of problems. We explore these ideas further by considering three problems (A–C below), all involving division, but with different structure.

A   How many coaches seating 60 children each are needed to transport 250 children?

Using the process of mathematical modelling (see Chapter 5), we could represent this problem with the division, 250 ÷ 60. If we divide 250 by 60, using whatever mental or written procedure we are confident with, we could get to the result '4 remainder 10'. Because there is a remainder, we should see immediately that four coaches will not be sufficient and we will have to have 5. Note that the 'remainder 10' does not stand for 10 coaches; it represents the 10 children who would have no seats if you ordered only

4 coaches. When we do the calculation, either by short division extended beyond the decimal point or on a calculator, we get the recurring decimal answer '4.1666…' or, say, 4.167 rounded to three decimal places (see Chapter 16). The '.167' represents a bit of a coach, not a bit of a child. So, in this example, the digits after the decimal point represent a bit of a coach, whereas the remainder represents a number of children. They are certainly not the same thing. This is a very significant observation, requiring careful explanation to children, through discussion of a variety of examples.

B   If I share 150 bananas equally between 18 people, how many bananas do they each get?

The model for this problem is the division, 150 ÷ 18. The result of this division is either '8 remainder 6' or the recurring decimal answer '8.3333333', or, say, 8.333 to three decimal places. In interpreting the answer '8 remainder 6', the remainder represents the 6 bananas left over after you have done the sharing, giving 8 bananas to each person. But the digits after the decimal point in the calculator answer (the '.333' part of 8.333) represent the portion of a banana that each person would get if you *were* able to share out the remainder equally. This would, of course, involve cutting up the bananas into bits. So, in this example, both the remainder and the digits after the decimal point refer to bananas: the first refers to bananas left over, and the second to portions of banana received if the left-overs are shared out. (Whether or not you can actually share out the remainder in practice will, of course, depend on what you are dealing with. For example, people are not usually cut up into smaller bits, but lengths, areas, weights and volumes can usually be subdivided further into smaller units.)

C   How many boxes that hold 18 pencils each do you need to store 150 pencils?

The model for this problem is the same division as in B: 150 ÷ 18. The remainder 6 now represents the surplus of pencils after you have filled up 8 boxes of 18. The digits after the decimal point in the calculator answer (the '.333' part of 8.333) represent what fraction this surplus is of a full set of 18. In our example, the '.333' means the fraction of a box that would be taken up by the 6 remaining pencils. So, the *remainder* in this division problem represents surplus pencils, but

LEARNING AND TEACHING POINT

Older children can explore a range of examples of real-life division problems that do not work out exactly, to be done both by a mental or written method producing a remainder and on a calculator. Each time, discuss what the remainder means and what the digits after the decimal point mean.

the *digits after the decimal point* represent 'a bit of a box'. In simple language, we might say either that 'we need eight boxes and there will be six pencils left over', referring to the answer with the remainder, or 'we need eight boxes and a bit of a box', referring to the answer with digits after the decimal point. Of course, in practice we need 9 boxes, since you cannot purchase a bit of a box.

HOW CAN YOU GET FROM THE DECIMAL ANSWER TO THE REMAINDER, AND VICE VERSA?

Health warning: this section and those that follow are beyond primary school level mathematics, but are included to deepen the understanding and skills of the reader.

First, we will compare the calculator answer for a division calculation ($100 \div 7$) with the answer with a remainder obtained by some written or mental method:

- Calculator: $100 \div 7 = 14.285714$.
- Written or mental method: $100 \div 7 = 14$, remainder 2.

Note that the calculator answer is truncated (see Chapter 5). If the calculator were able to display more than eight digits, the result would actually prove to be a recurring decimal, 14.285714285714285714…, with the sequence of digits '285714' recurring.

Now, the 14 is common to both answers. The question is, how does the .285714 in the first answer relate to the remainder 2 in the second? The easiest way to see this is to imagine that the division is modelling a problem with an equal-sharing structure in a measuring context; for example, sharing 100 units of wine equally between 7 glasses. The 14 represents 14 whole units of wine in each portion, and the remainder 2 represents 2 units of wine left over. If these 2 units are also shared out between the 7 glasses, then each glass will get a further portion of wine corresponding to the answer to the division, $2 \div 7$. Doing this on the calculator gives the result 0.2857142. Yes, it is the same as the bit after the decimal point in the calculator answer to the original question, allowing for the calculator's truncation of the result. So, the digits after the decimal point are the result of *dividing the remainder by the divisor*.

This means, of course, that, since multiplication and division are inverse processes, the remainder should be the result of multiplying the bit after the decimal point by the divisor. Checking this with the above example, we would multiply 0.2857142 by 7 and expect to get the remainder, 2. What I actually get on my calculator is 1.9999994, which is *nearly* equal to 2, but not quite. The discrepancy is due to the tiny bit of the answer to $2 \div 7$ that the calculator discarded when it truncated the result.

Another way of getting from the calculator answer (14.285714) to the remainder (2) is simply to multiply the whole number part of the calculator answer (14) by the divisor (7) and subtract the result (98) from the dividend (100). The reasoning here is that the 14 represents the 14 whole units of wine in each glass. Since there are 7 glasses, this amounts to 98 units (since $14 \times 7 = 98$). Subtracting these 98 units from the original 100 units leaves the remainder of 2 units.

So, in summary, allowing for small errors resulting from truncation:

- The digits after the decimal point are the result of dividing the remainder by the divisor.
- The remainder is the result of multiplying the digits after the decimal point by the divisor.
- The remainder is also obtained by multiplying the whole number part of the calculator answer by the divisor and subtracting the result from the dividend.

HOW DO YOU MULTIPLY TOGETHER TWO DECIMAL NUMBERS?

Imagine that we want to find the area in square metres of a rectangular lawn, 3.45 m wide and 4.50 m long. The calculation required is 3.45×4.50 (see Chapter 22). For multiplications, there is no particular value in retaining trailing zeros, so we can rewrite the 4.50 as 4.5, giving us this calculation to complete: 3.45×4.5. This is a pretty difficult calculation. In practice, most people would sensibly do this on a calculator and read off the answer as 15.525 square metres. But it will be instructive to look at how to tackle it without a calculator. There are three steps involved:

1   Get rid of the decimals by multiplying each number by 10 as many times as necessary.
2   Multiply together the two integers.
3   Divide the result by 10 as many times in total as you multiplied by 10 in step 1.

So, the first step is to get rid of the decimals altogether, using our knowledge of multiplying decimals repeatedly by 10, as follows:

$3.45 \times 10 \times 10 = 345$
$4.5 \times 10 = 45$

Hence, by doing '× 10' *three* times in total we have changed the multiplication into 345 × 45, a fairly straightforward calculation with integers. The second step is to work this out, using whatever method is preferred (see Chapters 11 and 12), to get the result 15,525. Finally, we simply undo the effect of *multiplying* by 10 three times by *dividing* by 10 three times, shifting the digits three places to the right and producing the required result: 15.525.

Figure 17.4 shows how a whole collection of results can be deduced from one multiplication result with integers, using as examples: (a) 4 × 36 = 144; and (b) 5 × 44 = 220, to show that the procedure works just the same when there is a zero in the result.

(a)

×	36	3.6	0.36	0.036
4	144	14.4	1.44	0.144
0.4	14.4	1.44	0.144	0.0144
0.04	1.44	0.144	0.0144	0.00144
0.004	0.144	0.0144	0.00144	0.000144

(b)

×	44	4.4	0.44	0.044
5	220	22	2.2	0.22
0.5	22	2.2	0.22	0.022
0.05	2.2	0.22	0.022	0.0022
0.005	0.22	0.022	0.0022	0.00022

Figure 17.4   Multiplication tables for decimal numbers derived from (a) 4 × 36 = 144; and (b) 5 × 44 = 220

WATCH THE PROBLEM SOLVED! VIDEO AT: HTTPS://STUDY.SAGEPUB.COM/HAYLOCK7E

HOW DO YOU CHECK THE REASONABLENESS OF THE RESULT OF A MULTIPLICATION?

As with addition and subtraction, we should always remember to check the reasonableness of our answers to multiplication calculations by using approximations. For example, 3.45 × 4.5 should give us an answer fairly close to 3 × 5 = 15. So the answer of 15.525 looks reasonable, whereas an answer of 1.5525 or 155.25 would suggest we made an error with the decimal point.

We need to be particularly alert to the problems of rounding errors when estimating answers to multiplications. If both numbers are rounded up (or both down), we can generate much more significant errors in our estimate than was the case with addition. Often, it makes most sense to round one up and one down. So, an estimate for 1.67 × 6.39 (which equals 10.6713) might be 2 × 6 (12). A more general procedure is to round both up (to get an answer clearly too large) and then round both down (to get an answer clearly too small), to determine limits within which the answer must lie:

A multiplication example: 1.67× 6.39
Round both up: 2 × 7 = 14
Round both down: 1× 6 = 6
So, the answer lies between 6 and 14.

AND WHAT ABOUT DIVIDING A DECIMAL BY A DECIMAL?

As with the previous three sections, this is an area where the teacher's own level of skill should perhaps be markedly higher than the level they have to teach to their children. I have four suggestions here for handling divisions by decimals. I am assuming, of course, that we are looking to get a decimal answer to the divisions involved, not a whole number with a remainder.

> WATCH THE PROBLEM SOLVED! VIDEO AT: HTTPS://STUDY.SAGEPUB.COM/HAYLOCK7E

Suggestion 1: In measuring contexts, you can often work with whole numbers by changing the units appropriately.

Suggestion 2: You can often transform a division question involving decimals into a much easier equivalent calculation by multiplying both numbers by 10 or 100 or 1000.

Suggestion 3: You can always start from a simpler example involving the same digits and work your way gradually to the required result by multiplying and dividing by 10s.

Suggestion 4: Remember to check whether the answers are reasonable by using approximation.

HOW DOES SUGGESTION 1 HELP WITH DIVIDING DECIMALS?

The need to divide a decimal number by a decimal number might occur in a real-life situation with the inverse-of-multiplication division structure (see Chapter 10). This could be, for example, in the context of money or measurement. My first suggestion then is that in these cases we can usually recast the problem in units that dispense with the need for decimals altogether. For example, to find how many payments of

£3.25 we need to make to reach a target of £52 (£52.00), we might at first be inclined to model the problem with the division, 52.00 ÷ 3.25. This would be straightforward if using a calculator. However, without a calculator we might get in a muddle with the decimal points. So, what we could do is to rewrite the problem in pence, which gets rid of the decimal points altogether: how many payments of 325p do we need to reach 5200p? The calculation is now 5200 ÷ 325, which can then be done by whatever method is appropriate.

Similarly, to find out how many portions of 0.125 litres we can pour from a 2.5-litre container, we could model the problem with the division, 2.500 ÷ 0.125. But it's much less daunting if we change the measurement to millilitres, so that the calculation becomes 2500 ÷ 125, with no decimals involved at all (see Chapter 21 for units of measurement and prefixes).

AND WHAT ABOUT SUGGESTION 2?

In fact, what we are doing above, in changing, for example, 52.00 ÷ 3.25 into 5200 ÷ 325, is multiplying both numbers by 100. This is using the principle established in Chapter 11, that you do not change the result of a division calculation if you multiply both numbers by the same thing. This is my second suggestion: that we can often use this principle when we have to divide by a decimal number. Here are two examples:

1   To find 4 ÷ 0.8, simply multiply both numbers by 10, to get the equivalent calculation 40 ÷ 8: so the answer is 5.
2   To find 2.4 ÷ 0.08, multiply both numbers by 100, to get the equivalent calculation 240 ÷ 8: so the answer is 30.

AND WHEN DOES SUGGESTION 3 HELP?

I find that an innocent-looking calculation like 0.46 ÷ 20 can cause considerable confusion amongst students. Using the method suggested above, multiplying both numbers by 100 changes the question into the equivalent calculation 46 ÷ 2000. But how do you deal with this? This is where my third suggestion comes to the rescue. I will first make one observation about division: *the smaller the divisor, the larger the answer (the quotient)*.

Note the pattern in the results obtained when, for example, 10 is divided by 2, 0.2, 0.02, 0.002 and so on:

$10 \div 2 = 5$
$10 \div 0.2 = 50$
$10 \div 0.02 = 500$
$10 \div 0.002 = 5000$
$10 \div 0.0002 = 50,000$ and so on.

Each time the divisor gets 10 times smaller, the quotient gets 10 times bigger. This is such a significant property of division – which, for some reason, often surprises people – that it is worth drawing specific attention to it from time to time. It's easy enough to make sense of this property if you think of $a \div b$ as meaning 'how many bs make a?' Is it not obvious that the smaller the number b, the greater the number of bs that you can get from a?

And, of course, the reverse principle is true as well: *the larger the divisor, the smaller the answer.* So, we could construct a similar pattern for dividing 10 successively by 2, 20, 200, 2000 and so on:

$10 \div 2 = 5$
$10 \div 20 = 0.5$
$10 \div 200 = 0.05$
$10 \div 2000 = 0.005$ and so on.

Each time the divisor gets 10 times larger, the answer gets 10 times smaller; in other words, it is divided by 10. So my third suggestion is that we can use these two principles to handle a division like the $0.46 \div 20$ above. Start with what you can do ... $46 \div 2 = 23$. Then work step by step to the required calculation. Here are two examples:

1   To calculate $0.46 \div 20$:

$0.46 \div 20 = 46 \div 2000$ (multiplying both numbers by 100)
$46 \div 2 = 23$
$46 \div 20 = 2.3$
$46 \div 200 = 0.23$
$46 \div 2000 = 0.023$, so $0.46 \div 20 = 0.023$

2   To calculate $0.05 \div 25$:

$0.05 \div 25 = 5 \div 2500$ (multiplying both numbers by 100)

$5 \div 2.5 = 2$

$5 \div 25 = 0.2$

$5 \div 250 = 0.02$

$5 \div 2500 = 0.002$, so $0.05 \div 25 = 0.002$

AND WHAT ABOUT SUGGESTION 4: CHECKING THE REASONABLENESS OF THE ANSWER?

My final suggestion for being successful at division with decimals, as with all calculations, is to remember to check the reasonableness of the answers by making estimates using approximations. For example, in working out how many payments of £3.25 are needed to reach £52, we should expect the answer to be somewhere between 10 and 20, since £3 × 10 = £30 and £3 × 20 = £60. So, if we get the answer to be 160 or 1.6 rather than 16 we have obviously made a mistake with the decimal point. Similarly, for the calculation $4 \div 0.8$, we would expect the answer to be fairly close to $4 \div 1$, which is 4. (In fact, it should be greater than this, because the divisor is less than 1.) So, we are not surprised to get the answer 5. However, an answer of 0.5 or 50 would suggest again that we had made an error with the decimal point.

As with multiplication, we need to be particularly alert to the problems of compounding rounding errors when estimating answers to divisions. If one number in a division is rounded up and the other rounded down, we can generate significant errors in our estimate. Often, it makes most sense to round both numbers up or to round both down. So, an estimate for $20.67 \div 3.39$ (which equals approximately 6.097) might be $20 \div 3$ (which is 6.7 to one decimal

> ## LEARNING AND TEACHING POINT
>
> Although much of the material in some of the later parts of this chapter may be beyond what is taught in primary school to most children, if you find that your personal confidence is boosted by explanations based on understanding, rather than on recipes learnt by rote, then adopt this principle in your own teaching of mathematics.

place). A more sophisticated procedure is to find limits within which the answer must lie by rounding the first number up and the second one down (to get an answer which is clearly too large) and then rounding the first number down and the second one up (to get an answer which is clearly too small).

A division example: $20.67 \div 3.39$

Round first number up, second one down: $21 \div 3 = 7$

Round first number down, second one up: $20 \div 4 = 5$

So, the answer lies between 5 and 7.

RESEARCH FOCUS: CONTEXTUALIZING DECIMALS

Working with children aged 11 and 12 from a lower economic area in New Zealand, Irwin (2001) investigated the role of children's everyday experience of decimals in supporting the development of their knowledge of decimals in school. The children worked in pairs (one member of each pair a more able student and one a less able student) to solve problems related to common misconceptions about decimal fractions. Half the pairs worked on problems presented in familiar everyday contexts and half worked on problems presented without context. The children who worked on contextual problems made significantly more progress in their knowledge of decimals than did those who worked on non-contextual problems. Irwin also analysed the conversations between the pairs of students during problem solving. The pairs working on the contextualized problems worked much better together, because the less able students were able to contribute from their everyday knowledge and experience of decimals in the world outside the classroom.

LEARNING RESOURCES

Access activities for your **lesson plans** at: https://study.sagepub.com/haylock7e

Before trying the self-assessment questions below, you should complete the **self-assessment questions** for this chapter at: https://study.sagepub.com/haylock7e

17.1: Without using a calculator, find the values of (a) 0.08 + 1.22 + 0.015; and (b) 10.5 – 1.05.

17.2: Pose a problem in the context of money that is modelled by 3.99 × 4. Solve your problem without using a calculator.

17.3: Pose a problem in the context of length that might be modelled by 4.40 ÷ 8. Solve your problem without using a calculator.

17.4: Without using a calculator, find: (a) 145 ÷ 8, giving the answer as an exact decimal number; (b) 145 ÷ 7, giving the answer as a decimal number rounded to 2 decimal places.

17.5: A batch of 3500 books is to be shared equally between 17 shops. Use a calculator to find 3500 ÷ 17. How many books are there for each shop? What is the remainder?

17.6: Use approximations to spot the errors that have been made in placing the decimal points in the answers to the following calculations: (a) 2.8 × 0.95 = 0.266; (b) 12.05 × 0.08 = 9.64; and (c) 27.9 ÷ 0.9 = 3.1.

FURTHER PRACTICE

Access the website material for

* Knowledge check 14: Fractions to decimals and vice versa

* Knowledge check 15: Adding and subtracting decimals

* Knowledge check 16: Multiplication with decimals

* Knowledge check 17: Division with decimals

at: https://study.sagepub.com/haylock7e

FROM THE STUDENT WORKBOOK

Questions 17.01–17.19: Checking understanding (calculations with decimals)

Questions 17.20–17.25: Reasoning and problem solving (calculations with decimals)

Questions 17.26–17.36: Learning and teaching (calculations with decimals)

PROPORTIONALITY AND PERCENTAGES

HOW DO YOU SOLVE DIRECT PROPORTION PROBLEMS?

In Chapter 15 we saw how the word 'proportion' was used to refer to a fraction, in the sense of a share or part of a unit or set. We now encounter the other use of this word in mathematics to describe particular kinds of relationships.

Consider the following five problems, all of which have exactly the same mathematical structure:

Recipe problem 1: A recipe for 6 people requires 12 eggs. Adapt it for 8 people.
Recipe problem 2: A recipe for 6 people requires 4 eggs. Adapt it for 9 people.
Recipe problem 3: A recipe for 6 people requires 120 g of flour. Adapt it for 7 people.
Recipe problem 4: A recipe for 8 people requires 500 g of flour. Adapt it for 6 people.
Recipe problem 5: A recipe for 6 people requires 140 g of flour. Adapt it for 14 people.

These have the classic structure of a problem of **direct proportion**. Such a problem involves four numbers, three of which are known and one of which is to be found. The structure can be represented by the four-cell diagram shown in Figure 18.1, in which it is assumed that the three numbers w, x and y are known and the fourth number z is to be found.

Variable A Variable B

w	x
y	$z?$

Figure 18.1   The structure of a problem of direct proportion

▶ WATCH THE PROBLEM SOLVED! VIDEO AT: HTTPS://STUDY.SAGEPUB.COM/HAYLOCK7E

The problems always involve two **variables** (see glossary at the end of this chapter), which I have called variable A and variable B in Figure 18.1. For example, in problem (1)

above, variable A would be the number of people and variable B would be the number of eggs. The numbers w and y are values of variable A, and the numbers x and z are values of variable B. So, for example, in recipe problem (1) above, $w = 6$ and $y = 8$, $x = 12$ and z is to be found. (See Figure 18.2(a).) In situations like the recipe problems above, we say that the two variables are 'in direct proportion' (or, often, just 'in proportion' or 'proportional to each other'), meaning that the ratio of variable A to variable B (for example, the ratio of people to eggs) is always the same. This means that the ratio $w{:}x$ must be equal to the ratio $y{:}z$. It is also true that the ratio $w{:}y$ is equal to the ratio $x{:}z$. As a consequence, when it comes to solving problems of this kind you can work with either the left-to-right ratios or the top-to-bottom ratios, depending on the numbers involved. It is important to note, therefore, that the most efficient way of solving one of these problems will be determined by the numbers involved, as illustrated in Figure 18.2(a), (b), (c) and (d).

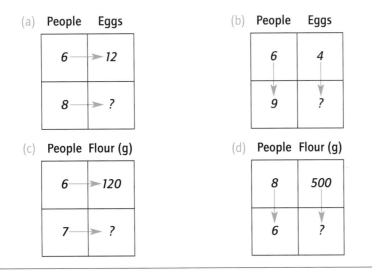

Figure 18.2 Solving the recipe problems

Recipe problem (1) (Figure 18.2a) – here I am attracted immediately by the simple relationship between 6 and 12. Double 6 gives me 12. So, I work from left to right, doubling the 8, to get 16. Answer: 16 eggs.

Recipe problem (2) (Figure 18.2b) – this time it's the relationship between the 6 and the 9 that attracts me. Halving 6 and multiplying by 3 gives 9. So, I work from top to bottom, and do the same thing to the 4, halving it and multiplying by 3, to get 6. Answer: 6 eggs.

Recipe problem (3) (Figure 18.2c) – the left-to-right relationship is the easier to work with here: multiplying 6 by 20 gives 120. So, do the same to the 7, to get 140. Answer: 140 g of flour.

Recipe problem (4) (Figure 18.2d) – I think it's easier to use the ratio of 8 to 6 than 8 to 500. So I'll work from top to bottom, going from 8 people to 4 to 2, and then to 6. Now, 8 people need 500 g, so 4 people need 250 g, so 2 people need 125 g. Adding the results for 4 people and 2 people, 6 people need 375 g.

These informal, ad hoc approaches are the ways in which most people solve the problems of ratio and direct proportion that they encounter in everyday life, including problems involving percentages such as those discussed below. We should encourage their use by children. The four-cell diagrams used in Figures 18.1 and 18.2 provide a useful starting point for organizing the data in a structured way that makes the relationships between the numbers more transparent. However, sometimes there is no easy or obvious relationship between the numbers in the problem. In such a case, we may call on a calculator to help us in using the method shown below.

> **LEARNING AND TEACHING POINT**
>
> Use the four-cell diagram to make clear the structure of direct proportion problems and encourage children to use the most obvious relationships between the three given numbers to find the fourth number.

Recipe problem (5):

6 people require 140 g

So, 1 person requires $140 \div 6 = 23.333$ g (using a calculator)

So, 14 people require $23.333 \times 14 = 326.662$ g (using a calculator)

Answer: approximately 327 g.

WHAT DOES 'PER CENT' MEAN?

Per cent means 'in each hundred' or 'for each hundred'. The Latin root *cent*, meaning 'a hundred', is used in many English words, such as 'century', 'centurion', 'centigrade' and 'centipede'. We can use the concept of 'per cent' to describe a proportion of a quantity or of a set, just as we did with fractional notation in Chapter 15. So, for example, if there are 300 children in a school and on a particular day 180 of them arrive at school by car, we might describe the proportion of children arriving by car as 'sixty

> **LEARNING AND TEACHING POINT**
>
> Repeatedly emphasize the meaning of *per cent* as 'for each hundred' and show how percentages are used to describe a proportion of a quantity or of a set.

per cent' (written as 60%) of the school population. This means simply that for each 100 children there are 60 arriving by car. If on a car journey of 200 miles a total of 140 miles is single carriageway, we could say that 70% (seventy per cent) of the journey is single carriageway, meaning 70 miles for each 100 miles. In effect, we have here the structure shown in Figure 18.1 for a direct proportion problem, but where one of the numbers must be 100, as shown in Figure 18.3.

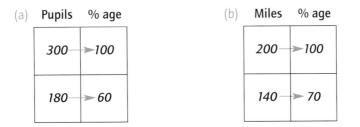

Figure 18.3   Percentage seen as direct proportion

HOW DO YOU USE AD HOC METHODS TO EXPRESS A PROPORTION AS A PERCENTAGE?

What we have been doing in the preceding examples is just what we did with fractions in Chapter 15, when we used fraction notation to represent a proportion of a unit or a part of a set. The concept of a percentage is simply a special case of a fraction, with 100 as the bottom number. So, 60% is an abbreviation for $^{60}/_{100}$ and 70% for $^{70}/_{100}$. In examples like these, it is fairly obvious how to express the proportions involved as 'so many per hundred'. This is not always the case. The following examples demonstrate a number of approaches to expressing a proportion as a percentage, using ad hoc methods, when the numbers can be related easily to 100. The trick is to find an equivalent proportion for a population of 100, by multiplying or dividing by appropriate numbers.

1   In an infant school with 50 children, there are 30 girls. What percentage are girls? What percentage are not girls?

   30 girls out of 50 children is the same proportion as 60 out of 100.

   So, 60% of the population are girls.

   This means that 40% are not girls (since the total population must be 100%).

2   In a primary school population of 250, there are 130 boys. What percentage are boys? What percentage are not boys?

130 boys out of 250 children is the same proportion as 260 out of 500.
260 boys out of 500 children is the same as 52 per 100 (dividing 260 by 5).
So, 52% of the population are boys.
This means that 48% are not boys (52% + 48% = 100%).

3   In an infant school population of 75, there are 30 girls. What percentage are girls? What percentage are not girls?

30 girls out of 75 children is the same proportion as 60 out of 150.
60 girls out of 150 children is the same proportion as 120 out of 300.
120 girls out of 300 children is the same as 40 per 100 (dividing 120 by 3).
So, 40% of the population are girls, and therefore 60% are not girls.

4   In a junior school population of 140, there are 77 girls. What percentage are girls? What percentage are not girls?

77 girls out of 140 children is the same proportion as 11 out of 20 (dividing by 7).

11 girls out of 20 children is the same proportion as 55 per 100 (multiplying by 5).
So, 55% of the population are girls, and therefore 45% are not girls.

> **LEARNING AND TEACHING POINT**
>
> Encourage the use of ad hoc methods for expressing a proportion as a percentage, using numbers that relate easily to 100.

HOW DO YOU USE A CALCULATOR TO EXPRESS A PROPORTION AS A PERCENTAGE?

▶ **WATCH THE PROBLEM SOLVED! VIDEO AT: HTTPS://STUDY.SAGEPUB.COM/HAYLOCK7E**

When the numbers do not relate so easily to 100 as in the examples above, the procedure is more complicated and is best done with the aid of a calculator, as in the following example:

1   In a school population of 140, there are 73 girls. What percentage are girls? What percentage are boys?

73 girls out of 140 children means that $^{73}/_{140}$ are girls.
The equivalent proportion for a population of 100 children is $^{73}/_{140}$ of 100.

Work this out on a calculator. (Key sequence: 73, ÷, 140, ×, 100, =.)

Interpret the display (52.142857): just over 52% of the population are girls.

This means that just under 48% are boys.

Most calculators have a percentage key (labelled %) which enables this last example to be done without any complicated reasoning at all. On my calculator, for example, I could use the following key sequence: 73, ÷, 140, %. This makes finding percentages with a calculator very easy indeed, although the precise key sequences involved may vary.

WHY ARE PERCENTAGES USED SO MUCH?

Percentages are used extensively, in newspapers, in advertising and so on. We are all familiar with claims such as '90% of cats prefer Kittymeat' and '20% of seven-year-olds cannot do subtraction'. There is certainly no shortage of material in the media for us to use with children to make this topic relevant to everyday life. Teachers will also find that they require considerable facility with percentages to make sense of such areas of their professional lives as assessment data, inspection reports and salary claims – which is why in this chapter I go some way beyond what would be taught to primary school children.

The convention of always relating everything to 100 enables us to make comparisons in a very straightforward manner. It is much easier, for example, to compare 44% with 40%, than to compare $^4/_9$ with $^2/_5$. This is why percentages are used so much: they provide us with a standard way of comparing various proportions. Consider this example:

School A spends £190,300 of its annual budget of £247,780 on teaching-staff salaries.

School B is larger and spends £341,340 out of an annual budget of £450,700.

It is very difficult to take in figures like these. The standard way of comparing these would be to express the proportions of the budget spent on teaching-staff salaries as percentages:

School A spends about 76.80% of its annual budget on salaries.

School B spends about 75.74% of its annual budget on salaries.

Now we can make a direct comparison: school A spends about £76.80 in every £100; school B spends about £75.74 in every £100. There are other occasions, however, when the tendency always to put proportions into percentage terms seems a bit daft, particularly when the numbers involved are very small. For example, I come from a family of 3 boys and 1 girl. I could say that 75% of the children in my family were boys. I suppose there's no harm in this, but remember that what this is saying is effectively: '75 out of every 100 children in my family were boys'!

HOW DO PERCENTAGES RELATE TO DECIMALS?

In Chapter 16 we saw how a decimal such as 0.37 means 37 hundredths. Since 37 hundredths also means 37 per cent, we can see a direct relationship between decimals with two digits after the point and percentages. So, 0.37 and 37% are two ways of expressing the same thing. Here are some other examples: 0.50 is equivalent to 50%, 0.05 is equivalent to 5% and 0.42 is equivalent to 42%. It really is as easy as that: you just move the digits two places to the left, because effectively what we are doing is multiplying the decimal number by 100. It's just like changing pence into pounds and vice versa. This works even if there are more than two digits, for example 0.125 = 12.5% and 1.01 = 101%.

This means that we have effectively three ways of expressing proportions of a quantity or of a set: using a fraction, using a decimal or using a percentage. It is useful to learn by heart some of the most common equivalences, such as the following:

Fraction	Decimal	Percentage
1/2	0.5 (0.50)	50%
1/4	0.25	25%
3/4	0.75	75%
1/5	0.2 (0.20)	20%
2/5	0.4 (0.40)	40%
3/5	0.6 (0.60)	60%
4/5	0.8 (0.80)	80%
1/10	0.1 (0.10)	10%
3/10	0.3 (0.30)	30%
7/10	0.7 (0.70)	70%
9/10	0.9 (0.90)	90%
1/20	0.05	5%
1/3	0.33 (approximately)	33% (approximately)

Other equivalences can be quickly deduced from these results. For example, we can use the fact that an eighth is half of a quarter (think of a quarter as 0.250) to work out that $^1/_8$ written as a decimal is 0.125 and therefore equivalent to 12.5% as a percentage. We can then multiply by 3 to deduce $^3/_8$ is equal to 0.375, which is equivalent to 37.5%. (See also self-assessment question 18.4 at the end of this chapter.)

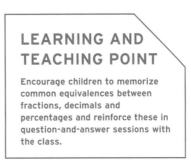

LEARNING AND TEACHING POINT

Encourage children to memorize common equivalences between fractions, decimals and percentages and reinforce these in question-and-answer sessions with the class.

Knowledge of these equivalences is useful for estimating percentages. For instance, if 130 out of 250 children in a school are girls, then, because this is just over half the population, I would expect the percentage of girls to be just a bit more than 50% (it is 52%). If the proportion of girls is 145 out of 450 children, then because this is a bit less than a third, I would expect the percentage of girls to be around 33% (to two decimal places it is 32.22%).

Also, because it is so easy to change a decimal into a percentage, and since we can convert a fraction to a decimal (see Chapter 16) just by dividing the top number by the bottom number, this gives us another direct way of expressing a fraction or a proportion as a percentage. For example, 23 out of 37 corresponds to the fraction $^{23}/_{37}$. On a calculator, enter: 23, ÷, 37, =. This gives the approximate decimal equivalent, 0.6216216. Since the first two decimal places correspond to the percentage, we can just read this straight off as 'about 62%'. If we wish to be more precise, we could include a couple more digits, giving the result as about 62.16%. (See Chapter 16 for a discussion of rounding.)

So, in summary, we now have these four ways of expressing a proportion of '*A* out of *B*' as a percentage:

1   Use ad hoc multiplication and division to change the proportion to an equivalent number out of 100.
2   Work out A/B × 100, using a calculator, if necessary.
3   Enter on a calculator: *A*, ÷, *B*, % (but note that calculators may vary in the precise key sequence to be used).
4   Use a calculator to find $A \div B$ and read off the decimal answer as a percentage, by shifting all the digits two places to the left.

HOW CAN YOU HAVE A HUNDRED AND ONE PER CENT?

Since 100% represents the whole quantity being considered, or the whole population, it does seem a bit odd at first to talk about percentages greater than 100. Football managers are well known for abusing mathematics in this way; for example, by talking about their team having to give a hundred and one per cent – meaning, presumably, everything they have plus a little more.

However, there are perfectly correct uses of percentages greater than 100, not for expressing a proportion of a whole unit but for comparing two quantities. Just as we use fractions to represent the ratio of two quantities, we can also use percentages in this way. So, for example, if in January a window cleaner earns £2000 and in February he earns only £1600, one way of comparing the two months' earnings would be to say that February's were 80% of January's. The 80% here is simply an equivalent way of saying four-fifths ($^4/_5$). If then in March he was to earn £2400, we could quite appropriately record that March's earnings were 120% of January's. This is equivalent to saying 'one and a fifth' of January's earnings.

HOW DO YOU USE AD HOC METHODS TO CALCULATE A PERCENTAGE OF A QUANTITY?

The most common calculation we have to do with percentages is to find a percentage of a given quantity, particularly in the context of money. In cases where the percentage can be converted to a simple equivalent fraction, there are often very obvious ad hoc methods of doing this. For example, to find 25% of £48, simply change this to $\frac{1}{4}$ of £48, which is £12.

Then, when the quantity in question is a nice multiple of 100, we can often find easy ways to work out percentages. First, let us note that, for example, 37% of 100 is 37. So, if I had to work out 37% of £600, we could reason that, since 37% of £100 is £37, we simply need to multiply £37 by 6 to get the answer required, namely £222.

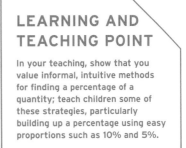

LEARNING AND TEACHING POINT

In your teaching, show that you value informal, intuitive methods for finding a percentage of a quantity; teach children some of these strategies, particularly building up a percentage using easy proportions such as 10% and 5%.

It is also possible to develop ad hoc methods for building up a percentage, using easy components. One of the easiest percentages to find is 10% and most people intuitively start with this. (Note, however, that the fact that 10% is the same as a tenth makes it a very special case: 5% is not a fifth, 7% is not a seventh and so on.) So, to find, say, 35% of £80, I could build up the answer like this:

    10% of £80 is £8
so 20% of £80 is £16   (doubling the 10%)
and 5% of £80 is £4   (halving the 10%)
Adding the 10%, 20% and 5% gives me the 35% required: £28.

It is often the case that 'intuitive' approaches to finding percentages, such as this one, are neglected in school, in favour of more formal procedures. This is a pity, since success with this kind of manipulation contributes to greater confidence with numbers generally.

LEARNING AND TEACHING POINT

Make a point of explaining to children that 10% being equivalent to 1/10 is a special case and warn them not to fall into the trap of thinking that, for example, 20% is equal to 1/20.

WHAT ABOUT USING A CALCULATOR TO FIND A PERCENTAGE OF A QUANTITY?

When the numbers are too difficult for intuitive methods such as those described above, then we should turn to a calculator or a calculator app. For example:

To find 37% of £946:
We need to find $\frac{37}{100}$ of 946.

On a calculator enter: 37, ÷, 100, ×, 946, =.

This gives the result, 350.02, so we conclude that 37% of £946 is £350.02.

Personally, I use a more direct method on the calculator. Since 37% is equivalent to 0.37, I just enter the key sequence: 0.37, ×, 946, =. There is another way, of course, using the percentage key. On my calculator, the appropriate key sequence is: 946, ×, 37, %.

WHAT ABOUT PERCENTAGE INCREASES AND DECREASES?

We are now definitely in the area of material provided for the professional development of the teacher, not for teaching to primary schoolchildren. One of the most common uses of percentages is to describe the size of a *change* in a given quantity, by expressing it as a proportion of the starting value in the form of a **percentage increase** or **percentage decrease**. We are familiar with percentage increases in salaries, for example. So, if your monthly salary of £1500 is increased by 5%, to find your new salary you would have to find 5% of £1500 (£75) and add this to the existing salary.

There is a more direct way of doing this: since your new salary is the existing salary (100%) plus 5%, it must be 105% of the existing salary. So, you could get your new salary by finding 105% of the existing salary, that is, by multiplying by 1.05 (remember that 105% = 1.05 as a decimal).

Similarly, if an article costing £200 is reduced by 15%, then to find the new price we have to find 15% of £200 (£30) and deduct this from the existing price, giving the new price as £170. More directly, we could reason that the new price is the existing price (100%) less 15%, so it must be 85% of the existing salary. Hence, we could just find 85% of £200, for example, by multiplying 200 by 0.85.

The trickiest problem is when you are told the price *after* a percentage increase or decrease and you have to work backwards to get the original price. For example, if the price of an article has been reduced by 20% and now costs £44, what was its original price? This problem is represented in Figure 18.4. The £44 must be 80% of the original price. We have to find what is 100% of that original price. It's fairly easy now to get from 80% (£44) to 20% (£11) and then to 100% (£55).

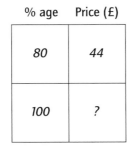

% age	Price (£)
80	44
100	?

Figure 18.4 If 80% is £44 find 100%

There is an interesting phenomenon, related to percentage increases and decreases, that often puzzles people. If you apply a given percentage increase and then apply the same percentage decrease, you do not get back to where you started! For example, the price of an article is £200. The price is increased one month by 10%. The next month the price is decreased by 10%. What is the final price? Well, after the 10% increase, the price has gone up to £220. Now we apply the 10% decrease to this. This is a decrease of £22, not £20, because the percentage change always applies to the existing value. So the article finishes up costing £198.

RESEARCH FOCUS: PROPORTIONAL REASONING

Cramer and Post (1993) provide a helpful summary of the different kinds of problems and contexts requiring proportional reasoning, and the various strategies that might be employed. Their conclusion is that research findings indicate that teachers in the primary school age range should focus more on intuitive strategies. These would include using the unit rate (per) and identifying factors in the relationships between the numbers involved. The power of intuitive approaches was highlighted in an important research study by Carraher, T. et al. (1985). They looked at the extraordinary ability of Brazilian street children, with little or no formal education, to perform complicated calculations, which they had learnt from necessity in the meaningful context of their work as street vendors. In particular, some of these children handled proportional reasoning with a startling facility. For example, one nine-year-old child (who apparently did not know that 35 × 10 was 350) nevertheless worked out the cost of 10 coconuts from knowing that 3 cost 105 cruzerios, reasoning as follows: 'Three will be a hundred and five. With three more that will be two hundred and ten … I need four more. That is … three hundred and fifteen … and … I think it is three hundred and fifty' (Carraher, T. et al., 1985: 23). This child shows a grasp of the principles of proportion, simultaneously co-coordinating the pro rata increases in the two variables involved, the number of coconuts and the cost, using a combination of adding and doubling. This research is one of the most significant pieces of evidence for the effectiveness of children learning mathematics through purposeful activity in meaningful contexts.

LEARNING RESOURCES

Access activities for your **lesson plans** at: https://study.sagepub.com/haylock7e

Before trying the self-assessment questions below, you should complete the **self-assessment questions** for this chapter at: https://study.sagepub.com/haylock7e

18.1: A department store is advertising '25% off' for some items and 'one-third off' for others. Which is the greater reduction?

18.2: Use ad hoc methods to find what percentage of children in a year group did not achieve a particular standard in a mathematics test, if there are:

(a) 50 children in the year group and 13 did not achieve the standard;

(b) 300 in the year group and 57 did not achieve the standard;

(c) 80 in the year group and 24 did not achieve the standard;

(d) 130 in the year group and 26 did not achieve the standard.

In each case, state what percentage of children achieved the standard.

18.3: On a page of an English textbook there are 1249 letters, of which 527 are vowels. In an Italian text, it is found that there are 277 vowels in a page of 565 letters.

Approximately what percentage of letters are vowels in each case? Use a calculator if you wish.

18.4: Given that $^1/_8$ equals 0.125 and is equivalent to 12.5%, find decimal and percentage equivalences for $^5/_8$ and $^7/_8$.

18.5: The price of a television costing £275 is increased by 12% one month and decreased by 12% the next. What is the final price? Use a calculator, if necessary.

FURTHER PRACTICE

Access the website material for

- Knowledge check 18: Mental calculations, changing proportions to percentages

- Knowledge check 19: Mental calculations, changing more proportions to percentages

- Knowledge check 20: Decimals and percentages

- Knowledge check 21: Expressing a percentage in fraction notation

- Knowledge check 22: Using a calculator to express a proportion as a percentage

- Knowledge check 23: Mental calculations, finding a percentage of a quantity

- Knowledge check 24: Finding a percentage of a quantity using a calculator

- Knowledge check 25: Sharing a quantity in a given ratio

- Knowledge check 26: Increasing or decreasing by a percentage

- Knowledge check 27: Expressing an increase or decrease as a percentage

- Knowledge check 28: Finding the original value after a percentage increase or decrease

at: https://study.sagepub.com/haylock7e

FURTHER PRACTICE

FROM THE STUDENT WORKBOOK

Questions 18.01–18.15: Checking understanding (proportionality and percentages)

Questions 18.16–18.23: Reasoning and problem solving (proportionality and percentages)

Questions 18.24–18.33: Learning and teaching (proportionality and percentages)

GLOSSARY OF KEY TERMS INTRODUCED IN CHAPTER 18

Direct proportion   The relationship between two variables where the ratio of one to the other is constant. For example, the number of cows' legs in a field and the number of cows would normally be in direct proportion.

Variable   A quantity the value or size of which can vary; for example, the number of children in a school is a variable, whereas the number of letters in the word 'school' is not.

Per cent (%)   In (or 'for') each hundred; for example, 87% means 87 in each hundred.

Percentage increase or percentage decrease   An increase or decrease expressed as a percentage of the original value.

SUGGESTIONS FOR FURTHER READING FOR SECTION D

1   Take a break from mathematics and read a novel! Doxiadis (2001) features Goldbach's Conjecture (the conjecture that every even number after 2 is the sum of two primes). It is the delightful story of Uncle Petros, an ageing recluse who was once a brilliant mathematician and who staked everything on being able to solve a problem that had defied all attempts at proof for over three centuries.

2   Develop your enthusiasm for the wonderful world of numbers by reading *Alex's Adventures in Numberland* (Bellos, 2020), an engaging and popular book about mathematics.

3   Some straightforward material for brushing up your understanding of square numbers, multiples, factors and other aspects of number, is presented in a clear and accessible way in Chapter 3 of Section 1 of Haighton et al. (2020). Further

practice of rounding to so many decimal places and to so many significant digits can be found in Chapter 4, and further opportunity to strengthen your understanding of decimal notation and calculations in Chapter 9.

4   Chapter 7 of Anghileri (2008) is a thoughtful chapter on decimals, fractions and percentages.

5   Hansen (2020) contains some interesting material on children's errors in handling fractions.

6   Singer, Kohn and Resnick provide a fascinating study of the intuitive bases for ratio and proportion that are available to young children in a chapter entitled 'Knowing about proportions in different contexts', in Nunes and Bryant (1996). The study addresses the question of how children can be helped to build on this informal grasp of these ideas in the context of formalizing mathematical concepts and procedures.

7   Boaler (2022) in her book, *Mathematical Mindsets,* 2nd edition, talks about the importance of hearing from children about how they visualize patterns developing. She explains in detail in Chapter 5 how creative their different approaches were and their importance in supporting the later development of understanding of algebra.

8   Chapter 6 of English, R. (2013) provides plenty of guidance on teaching fractions, ratios, decimals and percentages in primary school.

9   There are two chapters in Haylock and Warburton (2013) where we explain the concepts of proportionality and percentages in the context of healthcare practice. I recommend these to primary teachers, to help them to realize the importance of the material in this chapter and how developing understanding of these concepts is vital.

10 Chapter 1 of Du Sautoy (2011) is an enjoyable exploration of the mysteries of prime numbers.

SECTION E

ALGEBRA

19   Algebraic Reasoning   333
20   Coordinates and Linear Relationships   353

WATCH THE SECTION OPENER VIDEO AT: HTTPS://STUDY.SAGEPUB.COM/HAYLOCK7E

ALGEBRAIC REASONING

READ THIS CHAPTER'S CURRICULUM LINKS AT: HTTPS://STUDY.SAGEPUB.COM/HAYLOCK7E

IN THIS CHAPTER, THERE ARE EXPLANATIONS OF

- the nature of algebraic reasoning and its presence in mathematics across the primary-school age range;
- the idea of a letter representing a variable used to express generalizations;
- simple equations with one variable and their solutions;
- solving equations by trial and improvement;
- simple equations with two variables and pairs of solutions;
- the use of the equals sign in algebra;
- precedence of operators;
- developing the idea of a letter as a variable using tabulation;
- sequential and global generalizations;
- independent and dependent variables;
- formulas and substituting values of the independent variable(s) into a formula;
- tabulation and generalization in mathematical investigations;
- the meaning of the word 'mapping' in an algebraic context;
- some differences between arithmetic thinking and algebraic thinking;
- the distinction between solving a problem and representing it.

ALGEBRA? ISN'T THAT SECONDARY SCHOOL MATHEMATICS?

My main aim in writing this chapter is to disabuse the reader of this notion. By the time you get to the end, I hope you will have a better understanding of the true nature of algebraic thinking. It's not just about simplifying expressions like $2x + 3x$ to get $5x$, rearranging formulas and solving quadratic equations. This view of algebraic thinking might be picked up from the mathematics programmes of study for the National Curriculum in England (DfE, 2013), for example, where 'algebra' is not mentioned specifically until Year 6 and is introduced as being about using letters to represent variables and unknown numbers.

Most mathematics educators, by contrast, would argue that genuine algebraic thinking starts in the exploration of patterns in both shape and number in the early years of primary school. For example, the enlightened Australian mathematics curriculum for the foundation stage (ages 4–5 years) includes a section headed 'Algebra', with the expectation that children will learn to 'recognize, copy and continue repeating patterns represented in different ways' (ACARA, 2022). Below are some examples of things that children up to the age of 7 years might do that are the beginnings of thinking algebraically:

- make a repeated pattern using coloured beads on a string; such as blue, blue, red; blue, blue, red; blue, blue, red … and so on;
- explore, discuss and continue patterns in odd and even numbers;
- understand a statement like 'all numbers ending in 5 or 0 can be grouped into fives' and check whether this is true in particular cases, using counters;
- on a grid 3 squares wide, colour in the numbers 1, 4, 7, 10, 13, 16 … and describe the pattern that emerges (see Figure 19.1);
- recognize the relationship between doubling and halving;
- for each of the numbers from 1 to 10 apply a 'double and add 1' rule and discuss the pattern of numbers that emerges.

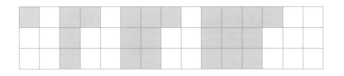

Figure 19.1   The pattern for the sequence 1, 4, 7, 10 …

Examples such as these, none of which uses letters for numbers, are about making *generalizations* (see Chapter 4). A generalization is an assertion that something is always the case. We have already encountered numerous examples of generalizations in this book. To say (as we did in Chapter 13) that the digital sum for any multiple of 9 is always a multiple of 9 is to make a generalization. To say that all square numbers have an odd number of factors (as we did in Chapter 13) is to make a generalization. This, as we shall see, is the essence of algebraic thinking. We shall see in this chapter that the letters that are used in algebra are introduced to enable us to articulate generalizations.

SO WHAT DO THE LETTERS USED IN ALGEBRA ACTUALLY MEAN?

To answer this important question, I will pose the reader a problem, using letters as symbols. Readers are invited to write down their answer to the following before reading on:

In Denmark you can exchange 7 kroner for 1 euro.

You have e euros.

You exchange this money for k kroner.

What is the relationship between e and k?

Most people I give this problem to write down $7k = e$ (or $7k = 1e$ or $e = 7k$ or $1e = 7k$, which are all different ways of saying the same thing). If you have done that, then please forgive me for deliberately leading you astray! The correct answer is actually $k = 7e$. Let me explain. Figure 19.2 shows various values for e and k. For example, if I have 1 euro ($e = 1$) I can exchange this for 7 kroner ($k = 7$); if I have 2 euros ($e = 2$) I can exchange them for 14 kroner ($k = 14$); similarly, when $e = 3$, $k = 21$ and so on. The table in Figure 19.2 makes it clear that whatever number is chosen for e (1, 2, 3, …) the value of k is 7 times this number (7, 14, 21, …). This is precisely what is meant by the algebraic statement, $k = 7e$.

No. of euros e	No. of kroner k
1	7
2	14
3	21
4	28
5	35

Figure 19.2 Tabulating values for e and k

Many people who are quite well qualified in mathematics get this answer the wrong way round when this problem is given to them, so you need not feel too bad if you did as well. It is instructive to analyse the thinking that leads to this misunderstanding. When we wrongly write down $7k = e$ what we are thinking, of course, is that we are writing a statement that is saying '7 kroner make a euro'. In our minds, the k and the e are being used as abbreviations for 'kroner' and 'euro'. This is, of course, how we use letters in

arithmetic, when they are actually abbreviations for fixed quantities or measurements. When we write 10p for ten pence or 5 m for five metres, the p stands for 'a penny' and the m stands for 'a metre'. But this is precisely what the letters do *not* mean in **algebra**. They are not abbreviations for measurements. They do not represent 'a thing' or an object. They usually represent *variables* (see the glossary at the end of Chapter 18).

The letter e in the currency problem stands for 'whatever number of euros you choose'. It does not stand for a euro, but for the *number* of euros. The value of e can be any number; and whatever number is chosen, the value of k is 7 times this. So, the relationship is $k = 7e$. This means 'the number of kroner is 7 times the number of euro' or, referring to the **tabulation** of values in Figure 19.2, 'the number in column k is 7 times whatever the corresponding number is in column e'.

It is understandable that so many people get the algebraic statements in these problems the wrong way round. First, the choice of e and k as letters to represent the variables in the problems is actually unhelpful (deliberately, I have to admit: sorry!). Using the first letters of the words 'euro' and 'kroner' does rather suggest that they are abbreviations for these things. (Fewer people get these relationships the wrong way round if other letters are used for the variables, such as n and m.) Then, so many of us have been subjected to explanations in 'algebra lessons' that reinforce this misconception that the letters stand for things. For example, it does not help to explain $2a + 3a = 5a$ by saying '2 apples plus 3 apples makes 5 apples'. This again makes us think of a as an abbreviation for apples. What this statement means is: whatever number you choose for a, then 'a multiplied by 2' plus 'a multiplied by 3' is the same as 'a multiplied by 5'.

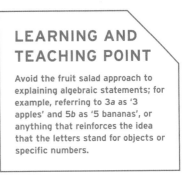

LEARNING AND TEACHING POINT

Avoid the fruit salad approach to explaining algebraic statements; for example, referring to 3a as '3 apples' and 5b as '5 bananas', or anything that reinforces the idea that the letters stand for objects or specific numbers.

Some of the generalizations we have already discussed in this book have been expressed not just in words, but also with letters representing variables; for example, the axioms that underpin operations with numbers, the various commutative, associative and distributive laws (see Chapters 7, 8, 10 and 11). For instance, we had the associative law for addition: $a + (b + c) = (a + b) + c$. This law tells us that we can choose *any* values for the variables a, b and c, and the value of $a + (b + c)$ will be equal to that of $(a + b) + c$.

SO WHAT IS AN EQUATION?

The algebraic generalizations given in the previous section are statements that are true for *all* values of the variables. An **equation** is an algebraic statement that is true for one or some specific values of the variable(s). For example, $x + 7 = 11 - x$ is a statement involving just one variable, indicated by the letter x. This variable can theoretically take any value. But the statement is only *true* when x takes the value 2 (because $2 + 7 = 11 - 2$). This value is called the **solution** of the equation. To solve an equation then is to find the value(s) of the variables that make the statement true. This is sometimes referred to as finding values that 'satisfy' the equation.

Primary school children meet the basic idea of an equation when they are faced with finding the number that should go in the box in statements like $3 + \square = 5$ (which is equivalent to the equation $3 + x = 5$). They can be later introduced to equations by using 'what is my number?' puzzles, expressed first in words and then in algebraic form. For example, 'I am thinking of a number. If you add 7 to my number and subtract my number from 11 you get the same answer. What is my number?' This is, of course, the equation discussed above put into words: $x + 7 = 11 - x$. Primary school children might solve this initially simply by trying various values of the variable, as below:

> Try $x = 0$. Left-hand side = $0 + 7 = 7$; right-hand side = $11 - 0 = 11$.
> Try $x = 1$. Left-hand side = $1 + 7 = 8$; right-hand side = $11 - 1 = 10$.
> Try $x = 2$. Left-hand side = $2 + 7 = 9$; right-hand side = $11 - 2 = 9$.
> So the solution is $x = 2$.

HOW ARE EQUATIONS SOLVED BY TRIAL AND IMPROVEMENT?

We would not usually introduce formal algebraic processes for solving equations in primary school – especially since the techniques involved can so easily reinforce the idea that the letters stand for 'things', or even specific numbers, rather than variables. What is appropriate, however, is to give children the opportunity to develop their algebraic reasoning by solving problems through a **trial and improvement** approach. These can be purely numerical problems that cannot be solved by a simple arithmetic procedure, or they can be equations that arise from 'What is my number?' puzzles, or they can be equations that arise from practical problems. Here is an example:

> If you subtract 19 from my number and multiply by my number, you get 666. What is my number?

Expressed algebraically, this problem asks us to solve the equation $(n - 19) \times n = 666$. Figure 19.3 shows how this might be solved by trial and improvement, using a calculator to do any tricky multiplications, thus allowing the child to focus on the process and mathematical reasoning involved. The first column shows the value of the variable n being tried; the second shows the corresponding value of

$n - 19$; the third shows the product of these. We are aiming to get 666 in the third column! In this case, the first value tried is $n = 50$. The result (1550) is far too large. So, next the value $n = 30$ is tried, giving a result that is too small (330). So, we try something in between: $n = 40$, which gives 840. That's too large. And so on, with $n = 35$, then 36 and finally 37, which gives the answer required. So, we have solved the equation: $n = 37$.

n	$n - 19$	$(n - 19) \times n$
50	31	1550
30	11	330
40	21	840
35	16	560
36	17	612
37	18	666

Figure 19.3   Solving (n - 19) × n = 666 using trial and improvement

WATCH THE PROBLEM SOLVED! VIDEO AT: HTTPS://STUDY.SAGEPUB.COM/HAYLOCK7E

WHAT ABOUT EQUATIONS WITH TWO VARIABLES?

Equations can, of course, involve two (or more) variables. For example, 'I am thinking of two numbers that add up to 12. What might they be?' Using x and y for the two variables here, the equation is $x + y = 12$. This equation is satisfied by lots of pairs of values for x and y. For example, when $x = 0$, $y = 12$; when $x = 1$, $y = 11$; when $x = 2$, $y = 10$; when $x = 3$, $y = 9$; when $x = 4$, $y = 8$; when $x = 5$, $y = 7$; when $x = 6$, $y = 6$; when $x = 7$, $y = 5$ and so on. If we include negative and rational numbers, there are limitless possible solutions for this equation. For example, when $x = 3.7$, $y = 8.3$; and when $x = 32$, $y = -20$. We shall see in the next chapter that each pair of values for the two variables in an equation like this can be interpreted as the coordinates of a point in a graphical representation of the relationship between the variables.

HOW IS THE EQUALS SIGN USED IN ALGEBRA?

The distinction between the meaning of letters used in algebra (as variables) and in arithmetic (as abbreviations) is undoubtedly one of the most crucial differences between these two branches of mathematics. Another key difference is the way in which the equals sign is used.

What the equals sign means strictly in mathematical terms is not the same thing necessarily as the way it is interpreted in practice. When doing arithmetic, that is manipulating numbers, most children (and especially younger children) think of the equals sign as an instruction to do something with some numbers, to perform an operation. They see '3 + 5 =' and respond by doing something: adding the 3 and the 5 to get 8. So, given the question, 3 + □ = 5, many young children (incorrectly) put 8 in the box; this is because they see the equals sign as an instruction to perform an operation on the numbers in the question, and naturally respond to the '+' sign by adding them up.

Children also use the equals sign simply as a device for connecting the calculation they have performed with the result of the calculation. It means simply, 'This is what I did and this is what I got …'. Given the problem, 'You have £28, earn £5 and spend £8, how much do you have now?' children will quite happily write something like: 28 + 5 = 33 − 8 = 25. This way of recording the calculation is mathematically incorrect, because 28

> **LEARNING AND TEACHING POINT**
>
> Reinforce through your own language the idea that the equals sign means 'is the same as' (is equivalent to), even in the early stages of recording the results of calculations. Avoid using the word 'makes' as a substitute for 'equals'.

+ 5 does not equal 33 − 8. But this is not what the child means, of course. What is written down here represents the child's thinking about the problem, or the buttons he or she has pressed on a calculator to solve it. It simply means something like, 'I added 28 and 5, and got the answer 33, and then I subtracted 8 and this came to 25.'

In algebra, however, the equals sign must be seen as representing *equivalence* – one of the fundamental concepts in mathematics (see Chapter 5). It means that what is written on one side 'is the same as' what is written on the other side. Of course, it has this meaning in arithmetic as well: 3 + 5 = 8 does mean that 3 + 5 is the same as 8. But children rarely use it to mean this; their experience reinforces the perception of the equals sign as an instruction to perform an operation with some numbers. But in algebraic statements it is the idea of equivalence that is strongest in the way the equals sign is used. For example, when we write $p + q = r$, this is not actually an instruction to add p and q. In fact, we may not have to do anything at all with the statement. It is simply a statement of equivalence between one variable and the sum of two others. This apparent lack of closure is a cause of consternation to some learners. If the answer to an algebra question is $p + q$, they will have the feeling that there is still something to be done, because they are so wedded to the idea that the addition sign is an instruction to do something to the p and the q.

WHAT IS MEANT BY 'PRECEDENCE OF OPERATORS'?

An expression like $3 + 5 \times 2$ is potentially ambiguous. If you do the addition first, the answer is 16. But if you do the multiplication first, the answer is 13. Which is correct?

Well, if you enter this calculation as it stands on the kind of basic calculator that might be used in primary school (using the key sequence: 3, +, 5, ×, 2, =), you get 16. The calculator does the operations in the order they are entered. However, if you use a more advanced, scientific calculator, with the same key sequence, you will probably get the answer 13. These calculators use what is called an **algebraic operating system** (as do many computer applications, such as spreadsheet programs). This means that they adopt the convention of giving precedence to the operations of multiplication and division. So, when you enter 3, +, 5, the calculator waits to determine whether there is a multiplication or a division following the 5; if there is, this is done first. If you actually mean to do the addition first, you would have to use brackets to indicate this: for example, $(3 + 5) \times 2$.

This convention of **precedence of operators** (sometimes called 'the hierarchy of operations') is always applied strictly in algebra and is essential for avoiding ambiguity, particularly because of the way symbols are used in algebra to represent problems, not just to solve them. So, for example, $x + y \times z$ definitely stands for 'x added to the product of y and z'. To represent 'x added to y, then multiply by z', we would write $(x + y) \times z$.

But in number work in primary school we do not often need this convention. The calculations we have to do should normally arise from a practical context and this will naturally determine the order in which the various operations have to be performed, so there is not usually any ambiguity. Ambiguity arises only in context-free calculations. But my feeling is that since the basic calculators we use in primary school deal with operations in the order they arrive, there is little point in giving children calculations like $32 + 8 \times 5$ and expecting them to remember the convention that this means you do the multiplication first. If that is what we want, then write either $8 \times 5 + 32$ or $32 + (8 \times 5)$. But as soon as we get into using algebra to express generalizations, we definitely need this convention for precedence of operators. At this stage, children will have to learn to recognize it and to use brackets as necessary to override it.

This is an opportune moment to mention the convention of dropping the multiplication sign in algebraic expressions: so instead of writing $a \times b$, we can write just ab. We can do this sometimes in arithmetic calculations where brackets remove any ambiguity. For example, the calculation $(3 + 5) \times 2$, using the commutative law, can be written as $2 \times (3 + 5)$; and then, using the convention, this

LEARNING AND TEACHING POINT

When doing number work, the context that gives rise to the calculation will determine the appropriate sequence of operations. But the need to make this principle explicit arises when algebraic notation is introduced. Explain the different systems used by various calculators, and show how brackets can be used to make clear which operation has to be done first.

could also be written as $2(3 + 5)$. The multiplication sign is omitted but understood. Similarly, $(x + y) \times z$ would normally be written as $z(x + y)$ and $x + y \times z$ would normally be written as $x + yz$.

HOW CAN THE IDEA OF A LETTER BEING A VARIABLE BE DEVELOPED?

The central principle in algebra is the use of letters to represent *variables*, which enable us to express *generalizations*. This idea of, say, x being a variable is much more sophisticated and powerful than the idea that 'x stands for an unknown number'. I remember being told this at school and spending a whole year doing things like $2x + 3x = 5x$, all the time believing that the teacher actually knew what this unknown number was and that one day he would tell us. Children should therefore have plenty of opportunities to use letters as algebraic symbols representing variables. The most effective way of doing this is through the tabulation of number patterns in columns, with the problem being to express the pattern in the numbers, first in words and later in symbols. A useful game in this context is *What's my rule?*

Figure 19.4 shows some examples of this game. In each case, the children are challenged to say what is the rule that is being used to find the numbers in column B and then to use this rule to find the number in column B when the number in column A is 100. In example (a), children usually observe first that the rule is 'adding 2'. Here they have spotted what I refer to when talking to children as the 'up-and-down rule'. When talking to teachers, I call it the **sequential generalization**. This is the pattern that determines how to continue the sequence.

(a)

A	B
1	3
2	5
3	7
4	9
5	11
6	13
7	15
8	17
9	19
10	21
100	?

(b)

A	B
1	3
2	7
3	11
4	15
5	19
6	23
7	27
8	31
9	35
10	39
100	?

(c)

A	B
1	3
2	8
3	13
4	18
5	23
6	28
7	33
8	38
9	43
10	48
100	?

(d)

A	B
1	99
2	98
3	97
4	96
5	95
6	94
7	93
8	92
9	91
10	90
100	?

Figure 19.4 What's my rule?

Asking what answer you get when the number in A is 100, or some other large number, makes us realize the inadequacy of the sequential generalization. We need a 'left-to-right rule': a rule that tells us what to do to the numbers in A to get the numbers in B. This is what I shall refer to as the **global generalization**. Many children towards the top end of the primary range can usually determine that when the number in A is 100, the number in B is 201, and this helps them to recognize that the rule is 'double and add 1'.

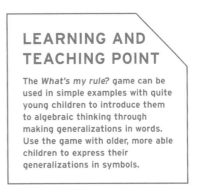

LEARNING AND TEACHING POINT

The *What's my rule?* game can be used in simple examples with quite young children to introduce them to algebraic thinking through making generalizations in words. Use the game with older, more able children to express their generalizations in symbols.

Later, this can be expressed algebraically. If we use x to stand for 'any number in column A' and y to stand for the corresponding number in column B, then the generalization is $y = x \times 2 + 1$, or $y = 2x + 1$. This clearly uses the idea of letters as variables, expressing generalizations. The statement means essentially, 'The number in column B is whatever number is in column A multiplied by 2, add 1.'

Similarly, in Figure 19.4(b), the sequential generalization, 'add 4', is easily spotted. More difficult is the global generalization, 'multiply by 4 and subtract 1', although, again, working out what is in B when 100 is in A helps to make this rule explicit. This leads to the algebraic statement, $y = (x \times 4) - 1$, or $y = 4x - 1$.

The global generalizations for Figure 19.4(c) and (d) are left for the reader to find in self-assessment question 19.5 at the end of this chapter.

In these kinds of examples, where x is chosen and a rule is used to determine y, x is called the **independent variable** and y is called the **dependent variable**.

WHAT IS A FORMULA?

Any algebraic rule that enables you to determine the value of a dependent variable from given values of one or more independent variables is a **formula**. So, earlier in this chapter, the rule $k = 7e$ for converting any number of euros into the corresponding number of kroner could be called a *formula* for finding the value of k. The independent variable here is e and the dependent variable is k.

Here is a simple example of a formula with two independent variables: the number, m, halfway between two numbers a and b on a number line is given by the formula $m = \frac{1}{2}(a + b)$ (see Figure 19.5). The numbers a and b here are the independent variables and these can take any values. The value of m, the dependent variable, is then determined by using the formula. So, for example, if $a = 15$ and $b = 73$, then $m = \frac{1}{2}(15 + 73) = \frac{1}{2}$ of $88 = 44$. So, the formula tells us that the point halfway between 15 and 73 is 44. This process is called 'substituting' values into the formula.

$$\begin{array}{ccc} a & m & b \end{array}$$

Figure 19.5   m = ½(a + b)

Formulas are used in Excel spreadsheets to relate cells together, with the contents of various cell locations being the variables. For example, if you put the formula C2 × 10 + D2 into cell E2 then the number in cell E2 will always equal the number in C2 multiplied by 10, plus the number in D2. Change the numbers in C2 and D2 and the number in cell E2 changes accordingly. A well-known formula used in healthcare is the one for calculating the body-mass index (BMI) for an individual: BMI $= m/b^2$. The independent variables are m, a person's mass (weight) measured in kilograms, and b, their

LEARNING AND TEACHING POINT

Here are some very simple examples that can be used to introduce formulas:
If x is a number on a hundred square and y is the number immediately below it, then $y = x + 10$.
If n is the number of bicycles in the rack and m is the number of wheels, then $m = 2n$.
If m is the number of matchsticks available and n is the number of separate triangles you can make with them, then $n = m/3$. (The value obtained will need to be rounded down for some values of m.)
The number of squares, n, in an array of p squares by q squares is given by the formula $n = pq$.
The number of points scored in the Premier Football League is $3w + d$, where w is the number of games won and d is the number of games drawn.

height measured in metres. The symbol '/' is used to represent division, reflecting the interpretation of fraction as a division (see Chapter 15). So, for one colleague, $m = 65$ and $h = 1.81$. The formula gives BMI $= 65 \div 1.81^2$. Using a calculator, I get the result, BMI $= 65 \div 3.2761 = 19.84$ (rounded to one decimal place). Try this for yourself: a BMI value between 18.5 and 25 is in the 'healthy' range.

WHERE ELSE ARE TABULATION AND ALGEBRAIC GENERALIZATION USED?

The experience of tabulation and finding generalizations to describe the patterns that emerge occurs very often in *mathematical investigations*, particularly those involving a sequence of geometric shapes. An example is the investigation of square picture frames in Chapter 4 (see Figure 4.1 and the associated text), where the global generalization was given as $f = 4n - 4$.

Other examples would be the patterns of shapes discussed in Chapter 13. For example, children could try to find a general rule for working out the number of counters in the sequence of shapes given in Figure 13.9(a) in that chapter. If the shapes are called shape 1, shape 2, shape 3, shape 4 and so on, then we want a rule for the number of counters in shape n. This will be a formula for the nth odd number. Finding this rule is left for the reader in self-assessment question 19.9 at the end of this chapter.

An investigation into the pattern in the triangle numbers shown in Figure 13.10 in Chapter 13 could lead to the generalization that the nth triangle number, which is the sum of $1 + 2 + 3 + 4 + 5 \ldots + n$, is equal to $\frac{1}{2} n (n + 1)$. The reader is invited to confirm this formula for triangle numbers in self-assessment question 19.8.

Figure 19.6 provides another example: the problem is to determine how many children can sit around various numbers of tables, arranged side

> **LEARNING AND TEACHING POINT**
>
> Encourage children to tabulate results from investigations, to enable them to find and articulate patterns in the sequence of numbers obtained.

> **LEARNING AND TEACHING POINT**
>
> Take children through this procedure; expect those of differing abilities to reach different stages; tabulate results in an orderly fashion; articulate the up-and-down rule; check this with a few more results; predict the result for a big number, such as 100; articulate the left-to-right rule in words; check this against some results you know; and, for the most able children, express the left-to-right rule in symbols.

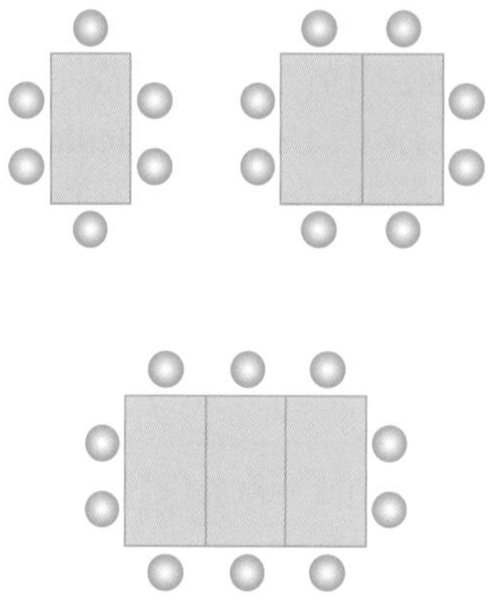

number of tables x	number of children y
1	6
2	8
3	10
4	
5	
6	
7	
8	
9	
10	
100	?

Figure 19.6 An investigation leading to a generalization

by side, if six children can sit around one table. The number of tables here is the independent variable and the number of children the dependent variable.

With 2 tables we can seat 8 children; with 3 tables we can seat 10. These results are already tabulated. The tabulation can then be completed for other numbers of tables, the sequential generalization can be articulated, the answer for 100 tables can be predicted, and, finally, the global generalization can be formulated. This will be first in words ('the number y is equal to the number x multiplied by …') and then in symbols ($y = \ldots$), with x being the independent variable (the number of tables) and y the dependent variable (the corresponding number of children). This is left as an exercise for the reader, in self-assessment question 19.6.

LEARNING AND TEACHING POINT

Introduce older primary school children to the idea of an imaginary function machine, into which one or more numbers are fed and a rule applied to produce an output. A simple example would be that the rule is double the input plus 1. Put 7 into this machine and turn the handle (or press the button); what number comes out? Then ask questions like, 23 is the output, what was the input? Then consider function machines that have a rule for processing two input numbers (for example, double the first number and add the second). Eventually, you may be able to pose questions like this: 'If I input 2 and 3 the output is 11; if I input 3 and 3 the output is 12; and if I input 3 and 4 the output is 15; what rule does the machine use?'

WHAT IS A MAPPING?

In the examples of tabulation used above, there have always been the following three components: a set of input numbers (the values of the independent variable), a rule for doing something to these numbers and a set of output numbers (the values of the dependent variable). These three components put together – input set, rule, output set – constitute what is sometimes called a **mapping**. It is also sometimes called a *functional relationship* and the dependent variable is said to be a **function** of the independent variable.

This idea of a mapping, illustrated in Figure 19.7, is an all-pervading idea in algebra. In fact, most of what we have to learn to do in algebra fits into this simple structure of input, rule and output. Sometimes we are given the input and the rule and we have to find the output: this is substituting into a formula. Then sometimes we are given the input and the output and our task is to find the rule: this is the process of generalizing (as in the examples of tabulation above). Then, finally, we can be given the output and the rule and be required to find the input: this is the process of solving an equation. That just about summarizes the whole of algebra!

Figure 19.7   A mapping

WHAT IS DIFFERENT ABOUT ARITHMETIC AND ALGEBRA IN RELATION TO RECOGNIZING THE MATHEMATICAL STRUCTURE OF A PROBLEM?

In the last two sections of this chapter, I aim to encourage the reader to reflect further on the distinctive nature of algebraic reasoning. In arithmetic, children often succeed through adopting informal, intuitive, context-bound approaches to solving problems. Often, they do this without having to be aware explicitly of the underlying mathematical structure. For example, many children will be able to solve, 'How much for 10 grams of chocolate if you can get 2 grams per penny?', without recognizing the formal structure of the problem as that of division. So, even with a calculator, they may then be unable to solve the same problem with more difficult numbers: 'How much for 75 grams of chocolate if you can get 1.35 grams per penny?'

The corresponding algebraic problem is a generalization of all problems in this context with this same structure: 'How much for p grams of chocolate if you can get q grams per penny?' It is often no use trying to explain this by just putting in some simple numbers for p and q and asking what you do to these numbers to answer the question – because the children have to recognize the existence of a division structure here. They have to connect the language pattern in this statement with the division of p by q. Primitive, intuitive thinking about the arithmetic problem with simple numbers does not make the mathematical structure explicit in a way that supports the algebraic generalization, $p \div q$. But you cannot construct the algebraic generalization without recognizing the formal structure of the problem. It is partly because of this that I have put so

much stress on the *structures* of addition, subtraction, multiplication and division in Chapters 7 and 10, and on the need to make connections between situations involving these structures with corresponding language and symbols.

WHAT IS THE DISTINCTION BETWEEN SOLVING A PROBLEM AND REPRESENTING IT?

The discussion above leads to a further significant difference between arithmetic and algebra. Given a problem to solve in arithmetic involving more than one operation, the question we ask ourselves is, 'What sequence of operations is needed to *solve* this problem?' In algebra, the question is, 'What sequence of operations is needed to *represent* this problem?' For example, consider this problem:

A plumber's call-out charge is £15; then you pay £12 an hour.

How many hours' work would cost £75?

The arithmetic thinking might be: $75 - 15 = 60$, then $60 \div 12 = 5$. This is the sequence of operations required to *solve* the problem: subtract 15, then divide by 12. But the algebraic approach would be to let n stand for the number of hours (which is therefore a variable and can take any value) and then to write down: $12n + 15 = 75$. This is the sequence of operations that *represents* the problem: multiply by 12, then add 15. Then to solve the problem we have to find the value of n that makes the algebraic representation true.

It seems, therefore, that the two approaches result in the use of inverse operations, as illustrated here. To solve the problem, we think: subtraction, then division; but to represent the problem algebraically, we think: multiplication, then addition. It is this difference in the thinking involved which makes it so difficult for many learners to make generalized statements using words or algebraic symbols, even of the simplest kind.

RESEARCH FOCUS: ALGEBRAIC REASONING IN YOUNG CHILDREN

At the beginning of this chapter, I argued that algebraic reasoning is an appropriate area of mathematics for children to experience across the years of primary education and that 'algebra' is not just something that enters children's mathematical experience when they reach the age of 10 or 11 years. Mason (2008), basing his argument on an extensive review of research and examples of children's responses to mathematical tasks, outlines how teachers can make use of what he calls 'the natural powers' of young children to produce and develop algebraic thinking. Mason explains how children make sense of their experiences by using natural powers 'to collect, classify, assimilate, accommodate, and even reject sensations, whether physical or imagined, remembered or constructed, literal or metaphoric' (p. 59). Amongst the natural powers that are relevant to mathematics and particularly to algebraic reasoning, he identifies, for example, imagining and expressing, focusing and de-focusing, specializing and generalizing, conjecturing and convincing, and classifying and characterizing. Mason suggests that teachers should be explicitly aware of the significance of these natural powers and he shows how teachers might provide opportunities for young children to develop and use them in mathematical tasks. See also the Research Focus for Chapter 20.

LEARNING RESOURCES

Access activities for your **lesson plans** at: https://study.sagepub.com/haylock7e

Before trying the self-assessment questions below, you should complete the **self-assessment questions** for this chapter at: https://study.sagepub.com/haylock7e

19.1: Using old-fashioned units, there are three feet in one yard. The length of a garden is f feet. Measured in yards, it is y yards long. What is the relationship between f and y? What criticism could you make of this question?

19.2: If I buy a apples at 10p each and b bananas at 12p each, what is the meaning of: (a) $a + b$; (b) $10a$; (c) $12b$; and (d) $10a + 12b$? What criticism could you make of this question?

19.3: The first 5 rides in a fair are free. The charge for all the other rides is £2 each. (a) Jenny has £12 to spend, so how many rides can she have? Identify the arithmetic steps you used to answer this. (b) How would you represent algebraically the cost of n rides? (c) Use your answer to (b) to write down an equation for which the solution is the number of rides that Jenny can have.

19.4: What answer would you get if you entered $25 - 5 \times 3$ on: (a) a basic calculator; and (b) a scientific calculator using an algebraic operating system?

19.5: For each of Figures 19.4(c) and 19.4(d), using x for any number in column A and y for the corresponding number in column B, write down:

a the sequential generalization;

b the value of y when x is 100;

c the global generalization in words; and

d the global generalization in symbols ($y = \ldots$).

19.6: Complete the tabulation of results in Figure 19.6. Then write down:

a the sequential generalization;

b the number of children if there are 100 tables;

c the global generalization in words; and

d the global generalization in symbols ($y = \ldots$).

19.7: I choose a number, double it, add 3 and multiply the answer by my number. The result is 3654. What is my number? Use a calculator and the trial and improvement method to answer this. What equation have you solved?

19.8: List the first 10 triangle numbers: 1, 3, 6, 10 and so on (see Figure 13.10 in Chapter 13). Now double these to get 2, 6, 12, 20 and so on. Express each of these numbers as products of two factors, using a pattern starting with 1×2, 2×3, 3×4. Hence, obtain a generalization for the nth triangle number (the sum of the first n natural numbers). What is the one-hundredth triangle number (that is, the sum of all the natural numbers from 1 to 100)?

19.9: Use the patterns in Figure 13.9(a) in Chapter 13 to help you to state a rule for the nth odd number.

FURTHER PRACTICE

Access the website material for

- Knowledge check 29: Using a four-function calculator, precedence of operators
- Knowledge check 30: Substituting into formulas

at: https://study.sagepub.com/haylock7e

FROM THE STUDENT WORKBOOK

Questions 19.01–19.27: Checking understanding (algebraic reasoning)

Questions 19.28–19.38: Reasoning and problem solving (algebraic reasoning)

Questions 19.39–19.51: Learning and teaching (algebraic reasoning)

GLOSSARY OF KEY TERMS INTRODUCED IN CHAPTER 19

Algebra   A branch of mathematics in which letters are used to represent variables in order to express generalizations.

Tabulation   The process of putting corresponding values of two or more variables in a table, to show the relationships between the variables and to enable generalizations to be identified.

Equation   A statement of equivalence involving one or more variables, which may or may not be true for any particular value of the variable(s).

Solution   A value of a variable in an equation that makes the equivalence true. For example, $2x + 1 = 16 - x$ is an equation with the solution $x = 5$. Finding all the solutions is called solving the equation.

Trial and improvement   A procedure for finding the solution to a mathematical problem by means of successive approximations (trials) which gradually close in on the required solution; it can be used to find solutions to simple equations.

Algebraic operating system   A system used by scientific calculators and spreadsheet software that follows the algebraic conventions of precedence of operators.

Precedence of operators   A convention that, unless otherwise indicated by brackets, the operations of multiplication and division should have precedence over addition and subtraction. This convention is always used in algebraic expressions.

Sequential generalization   When the input and output sets of a mapping are tabulated, a rule for getting the next value of the dependent variable from the previous one(s); the up-and-down rule.

Global generalization   When the input and output sets of a mapping are tabulated, a rule for getting the value of the dependent variable from any value of the independent variable; the left-to-right rule.

Independent variable   In a relationship between two variables, the variable whose values may be chosen freely from the given input set, and are then put into the rule to generate the values of the dependent variable in the output set.

Dependent variable   In a relationship between two variables, the one whose values are determined by the value of the independent variable and the rule.

Formula   An algebraic rule involving one or more independent variables, used to determine the value of a dependent variable.

Mapping   A system consisting of an input set, a rule and an output set.

Function   In a mapping, the relationship between the dependent variable and the independent variable. For example, if $y = 2x + 1$, then y is a function of x.

COORDINATES AND LINEAR RELATIONSHIPS

IN THIS CHAPTER, THERE ARE EXPLANATIONS OF

- how a rectangular grid represents combinations of values of two variables;
- how the coordinate system enables us to specify location in a plane;
- axis, x-coordinate and y-coordinate, origin;
- the meaning of 'quadrant' in the context of coordinates;
- how to plot an algebraic relationship as a graph;
- linear relationships, including those where one variable is directly proportional to another;
- simple distance-time graphs.

READ THIS CHAPTER'S CURRICULUM LINKS AT: HTTPS://STUDY.SAGEPUB.COM/HAYLOCK7E

HOW DOES A GRID REPRESENT THE VALUES OF TWO VARIABLES?

In this chapter, we introduce the coordinate system. The key idea here is that a point in two-dimensional space can be specified by the values of two variables. This simple but powerful principle is first met when we use a similar system to label the spaces in a rectangular grid. For example, in a number of paper-and-pencil games children will use a grid like the one shown in Figure 20.1(a). The shaded square in this grid, for example, might be labelled D5. The D here indicates which column the square lies in and the 5 indicates which row. This kind of labelling of spaces in a grid is used extensively in board games, street maps and theatre seats. It is also the way in which cells are labelled in a computer-based spreadsheet. The location of any of the small squares in Figure 20.1(a) is determined by the value of two variables: the choice of column and the choice of row. In this example, there are 6 possible values for the choice of column (A–F) and 6 possible values for the choice of row (1–6), giving a total of 36 squares, each one representing a different combination of the values of the two variables.

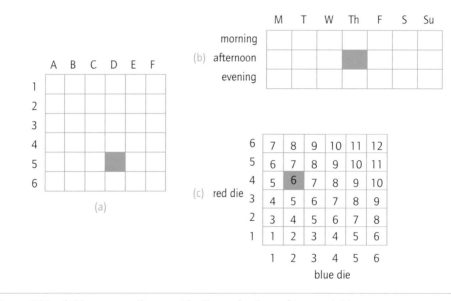

Figure 20.1   Grids representing combinations of values of two variables

Using this idea, a rectangular grid can be used to enumerate possible combinations of the values of two variables. For example, Figure 20.1(b) shows how a weekly diary for a person's commitments might be set out, using two variables: the day of the week (7 values, of course) and the time of day (using just 3 values: morning, afternoon and evening). Structured like this there are 21 slots in the week. The shaded cell shows an appointment on Thursday afternoon. Similarly, Figure 20.1(c) shows all the possible outcomes when two dice are thrown, one red and one blue. The two variables are the score on the red die and the score on the blue die, with each square in the grid representing a unique combination of red score and blue score. There are therefore 36 possible outcomes. The shaded square represents a score of 4 on the red die and 2 on the blue die.

Interpreting any kind of two-way reference chart is an important skill, fundamental to understanding a coordinate system and to interpreting various kinds of charts. (See, for example, Figure 26.4 in Chapter 26.) Various kinds of scatter diagrams (also called scatter plots, scattergrams and scattergraphs) use two-way reference to show the relationship between two variables for the same members of a population. Figure 20.2 shows an elementary form of scatter diagram. A number of children in a school are sorted according to two variables: how they travelled to school that day and their year group. Each of these variables takes four values. The data is presented in a 4 by 4 grid, with each child represented by a small cross, placed inside one of the cells of the grid.

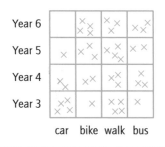

Figure 20.2   A simple scatter diagram

HOW DOES THE COORDINATE SYSTEM WORK AND WHAT ARE QUADRANTS?

The coordinate system is a wonderfully simple but elegant device for specifying location in two dimensions. An important feature of this system is that it is the points in the plane that are labelled by the coordinates, not the squares in a grid. Moving on from labelling squares in a grid to labelling points in two-dimensional space is like the young child moving from using a number strip – where numbers label a chain of squares – to a number line – where numbers label points on the line.

Two number lines are drawn at right angles to each other, as shown in Figure 20.3. These are called **axes** (pronounced ax-eez, the plural of 'axis'). Of course, the lines can continue as far as we wish at either end. The point where the two lines meet (called the **origin**) is taken as the zero for both number lines. The points on the axes correspond to the values of the two variables involved. Here, I have used the conventional x and y to represent the two variables, but of course any letters could be used. It is also conventional for the horizontal axis to represent the values of the independent variable x and the vertical line the value of the dependent variable y (see glossary for Chapter 19). The horizontal line is then called the x-axis, and the vertical line the y-axis. Any point in the plane can then be specified by two numbers, called its **coordinates**. The x-coordinate of a point is the distance moved along the x-axis, and the y-coordinate is the distance moved vertically, in order to get from the origin to the point in question. For example, to reach the point P shown in Figure 20.3, we would move 3 units along the x-axis and then 4 units vertically, so the x-coordinate of P is 3 and the y-coordinate is 4. We then state that the coordinates of P are (3, 4). The convention is always to give the x-coordinate first and the y-coordinate second.

The axes divide the plane into four sections, called **quadrants**. The **first quadrant** consists of all the points that have a positive number for each of their two coordinates. The points P and Q in Figure 20.3 are in the first quadrant. The point R (−2, 1) is in the second quadrant, S (−3, −3) in the third quadrant and T (2, −2) in the fourth quadrant.

> **LEARNING AND TEACHING POINT**
>
> The use of coordinates to specify the location of points in a plane, rather than spaces, is a significant point to be explained to children carefully.

> **LEARNING AND TEACHING POINT**
>
> To begin with, primary school children experience coordinates only in the first quadrant. The principles are the same when they move onto using coordinates in the other quadrants; these provide some useful experience of interpreting and applying negative numbers.

> **LEARNING AND TEACHING POINT**
>
> Give children the chance to play simple games where they use the coordinate system to describe movements from one point to another.

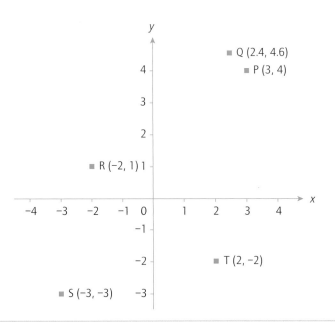

Figure 20.3   The coordinate system

The beauty of this system is that we can now refer specifically to any point in the plane. And, of course, we are not limited to integers, as is shown by the point Q, with coordinates (2.4, 4.6). We can also use the coordinate system to describe the movement from one point to another. For example, from R to P is a movement of 5 units in the *x*-direction and 3 units in the *y*-direction; from T to S is a movement of −5 in the *x*-direction and −1 in the *y*-direction. In Chapter 24, we shall see that one kind of transformation of a shape (sliding without turning) can be described in this way.

WHAT ARE LINEAR RELATIONSHIPS?

The system described above is sometimes called the Cartesian coordinate system. It takes its name from René Descartes (1596–1650), a prodigious French mathematician, who first made use of the system to connect geometry and algebra. He discovered that by interpreting the inputs and outputs from an algebraic mapping (see Chapter 19) as coordinates, and then plotting these as points, you could generate a geometric picture of the relationship. Then by the reverse process, starting with a geometric picture drawn on a coordinate system, you can generate an algebraic representation of the geometric properties. At primary school level, we can only just touch on these massive mathematical ideas, so a couple of simple examples will suffice here.

First, we can take any simple algebraic relationship of the kind considered in Chapter 19 and explore the corresponding geometric picture. For example, the table shown in Figure 19.4(a) in Chapter 19 is generated by the algebraic rule, $y = 2x + 1$. In this case, x is the independent variable and is represented by the x-axis, and y, the dependent variable, is represented by the y-axis. The pairs of values in the table can be written as coordinates, as follows: (1, 3), (2, 5), (3, 7) and so on. When these are plotted, as shown in Figure 20.4(a), it is clear that they lie on a straight line. These points can then be joined up and the line continued indefinitely, as shown in Figure 20.4(b). This straight line is a powerful geometric image of the way in which the two variables are related. An algebraic rule like $y = 2x + 1$, which produces a straight-line graph, is called a **linear relationship**.

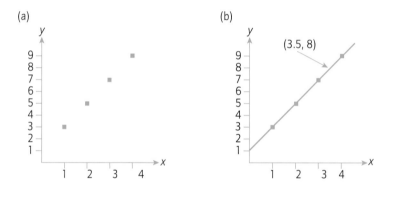

Figure 20.4   The rule y = 2x + 1 represented by coordinates

We can use the straight-line graph to read off related values of x and y other than those plotted; for example, the arrow in Figure 20.4(b) shows that when $y = 8$, $x = 3.5$. What we have done here is to find the value of the variable x for which $2x + 1 = 8$; in other words, we have *solved the equation* $2x + 1 = 8$. This can be an introduction to the important mathematical method of solving equations by drawing graphs and reading off values.

Exploring the graphs of different algebraic rules leads us to recognize a linear relationship as one in which the rule is simply a combination of multiplying or dividing by a particular number and addition or subtraction. For example, all these rules are linear relationships:

divide by 6 and add 4	$(y = x/6 + 4)$
multiply by 7 and subtract 5	$(y = 7x - 5)$
multiply by 3 and subtract from 100	$(y = 100 - 3x)$
add 1 and multiply by 2	$(y = 2(x + 1))$

There are, of course, relationships that are non-linear. These have other kinds of rules; for example, those involving squares of the independent variable (multiply the input by itself) and other powers (cubes and so on), which produce sets of coordinates that do not lie on straight lines. Typically, these kinds of relationships produce curved graphs. Non-linear relationships are beyond the scope of this book.

> ### LEARNING AND TEACHING POINT
>
> Science experiments give children opportunities to investigate whether or not a particular relationship approximates to a straight-line graph when plotted as coordinates. Use simple data-handling software to record the data and to plot the points on a scattergraph.

WHAT HAPPENS WHEN Y IS DIRECTLY PROPORTIONAL TO X?

The simplest kind of linear relationship is where y is directly proportional to x (see Chapter 18). This means that the ratio of y to x is constant, or, to put it another way, y is obtained by multiplying x by a constant factor. Examples of this kind of relationship abound in everyday life.

If a bottle of beer costs £3, the independent variable x is the number of bottles I buy and the dependent variable y is the total cost in pounds, then the rule for finding y is 'multiply x by 3' or $y = 3x$. This rule generates the coordinates: (1, 3), (2, 6), (3, 9) and so on, and, including the possibility

> ### LEARNING AND TEACHING POINT
>
> Give children examples of how a relationship, where one variable is directly proportional to another, can be shown as a straight-line graph passing through the origin.

that I do not buy any bottles, (0, 0). As shown in Figure 20.5, this rule produces a straight-line graph, which passes through the origin (0, 0).

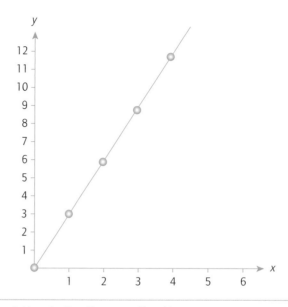

Figure 20.5   The variable y is directly proportional to x

Here are two simple tips for spotting that two variables are directly proportional:

1   If one variable takes the value 0, so does the other one.
2   If you double one variable, you double the other one.

Any rule of this kind, in which y is directly proportional to x, such as $y = 7x$, $y = 0.5x$, $y = 2.75x$, will produce a straight-line graph passing through the origin. This means that zero of one kind of unit of measurement corresponds to zero of the other kind. So, an exception, for example, would be converting temperatures between °F and °C (because 0°F does not equal 0°C: see Chapter 21); these two variables are *not* directly proportional.

An interesting point for discussion in the example above about bottles of beer relates to the fact that the number of bottles bought can only be a whole number: you cannot buy 3.6 bottles, for example. (In Chapter 26 we will refer to such a variable as a 'discrete variable'.) This means that the points on the line between the whole number values do not actually have any meaning. By contrast, if x had been the number of litres of petrol being bought at £3 per litre (price used for demonstration purposes only), then the rule would have been the same, $y = 3x$, but this time x would be a continuous variable and all the points on the straight-line graph in the first quadrant would have meaning. For example, when x is 3.6, y is 10.8, corresponding to a charge of £10.80 for 3.6 litres of petrol.

This provides us with a practical method for solving direct proportion problems. For example, most conversions from one unit of measurement to another provide examples of two variables that are directly proportional, and will therefore generate a straight-line graph passing through the origin.

Consider exchanging British pounds for US dollars, for example, assuming the tourist exchange rate is given as $1.45 dollars to the pound. A quick bit of mental arithmetic tells us that $29 is equivalent to £20. Plotting this as the point with coordinates (29, 20) and drawing the straight line through this point and the origin, produce a standard **conversion graph**, as shown in Figure 20.6. This can then be used to do other conversions. The arrows, for example, show (a) how you would convert $36 to just under £25, and (b) £35 to just over $50. All problems of direct proportion, such as those tackled by arithmetic methods in Chapter 18, can also be solved by this graphical method (see, for example, self-assessment question 20.5 below).

> ## LEARNING AND TEACHING POINT
>
> When a real-life relationship produces a straight-line graph, children should discuss whether or not the points between those plotted have meaning.

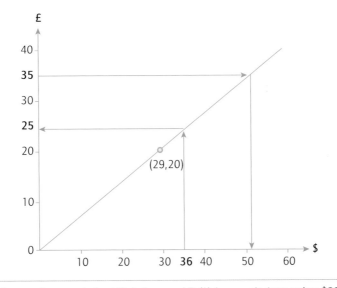

Figure 20.6 Conversion graph for US dollars and British pounds (assuming $29 = £20)

HOW DO YOU INTERPRET DISTANCE-TIME GRAPHS?

Particularly important in mathematics and science are **distance–time graphs**. They connect strongly with line graphs used for representing statistical data over time (see Chapter 26). Simple examples of distance–time graphs, such as those consisting of a few straight-line sections, can be introduced at the primary school level. Figure 20.7 is an example showing a journey of 8 kilometres completed by a walker in three stages in 120 minutes. The x-axis here represents the time that passes from the beginning of the journey; and the y-axis represents 'the distance from the starting point' at any given time. Below are examples of observations that may be made from the graph in Figure 20.7:

- After 40 minutes, the walker has completed 4 km.
- The walker has a 20-minute rest from walking after 60 minutes.
- After 100 minutes, the walker is 7 km from the start.
- The walker covers 2 km in 20 minutes in the first part of the walk.
- The walker covers only 1 km in 20 minutes in the last part of the walk.

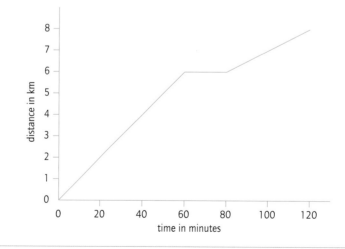

Figure 20.7   Distance-time graph for an 8-kilometre walk

The fact that the walker walks faster for the first part of the journey compared to the last is shown by the steeper slope for the first section of the graph. Notice also that I chose my words carefully for the variable represented by the vertical axis, deliberately not saying 'the distance walked'. Calling this 'the distance from the starting point' allows for the graph to slope downwards if at some point the walker decides to turn round and start walking back towards the starting point.

RESEARCH FOCUS: UNDERSTANDING LINEAR FUNCTIONS

The axes used in the Cartesian coordinate system are number lines. When the axes are used in plotting graphs, the position on the number line represents a variable, the fundamental idea of algebra. In a fascinating study, Carraher et al. (2006) provide evidence that primary school children can make use of algebraic representations like this that we might imagine to be beyond their reach. Their classroom-based research started with the assumptions that generalizing lies at the heart of algebraic reasoning, that arithmetic operations can be viewed as functions and that algebraic notation (letters used to represent variables and unknown numbers) can lend support to mathematical reasoning even amongst younger students. They show how the children they worked with in this research worked successfully with variables, functions and equations. For example, some children aged 9–10 years were told a story that involved the difference in height between three children. They were able to relate these differences to a number line with a central point labelled N, to represent the unknown height of one of the children, and other points labelled $N - 1$, $N - 2$, $N - 3$, …, to the left, and $N + 1$, $N + 2$, $N + 3$, …, to the right. The children were able to locate correctly the positions on this number line representing the other two children. In fact, most of them could successfully complete this task using N to represent the unknown height of any of the three children in the story. Children might use this number line to answer questions such as: if A has 3 dollars less than B and is given 5 dollars, how much more does A now have than B? Being able to answer this without knowing how much money either A or B has shows genuine algebraic reasoning. The authors also give instances of children using this 'algebraic' number line to reason, for example, that $n + 3 - 5 + 4$ is equal to $n + 2$, whatever the number n might be.

🧩 LEARNING RESOURCES

Access activities for your **lesson plans** at: https://study.sagepub.com/haylock7e

Before trying the self-assessment questions below, you should complete the **self-assessment questions** for this chapter at: https://study.sagepub.com/haylock7e

20.1: A knight's move in chess is two units horizontally (left or right) and one unit vertically (up or down), or two units vertically and one unit horizontally. Starting at (3, 3), which points can be reached by a knight's move?

20.2: On squared paper, plot some of the points corresponding to pairs of values of A and B in each of the tables (b), (c) and (d) in Figure 19.4 in Chapter 19. In each case, decide whether the relationship between A and B is linear.

20.3: Give an example of a variable which is directly proportional to each of the following independent variables: (a) the number of boxes of six eggs bought by a shopper; (b) the number of bars of a piece of music in waltz time that someone plays on the piano; and (c) the bottom number in a set of equivalent fractions.

20.4: Use Figure 20.4(b) to solve the equation $2x + 1 = 6$.

20.5: How would you use the fact that 11 stone is about the same as 70 kilograms to draw a conversion graph for stones and kilograms? How would you then use this to convert your personal weight (mass) from one to the other?

20.6: Refer to Figure 20.7: (a) How far is the walker from the start after 70 minutes? (b) How long does it take the walker to get to a point 7.5 km from the start? (c) What are the average speeds of walking (in kilometres per hour) for the first and last stages of the walk?

FURTHER PRACTICE

Access the website material for **Knowledge check 31: Conversion graphs** at: https://study.sagepub.com/haylock7e

FROM THE STUDENT WORKBOOK

Questions 20.01–20.20: Checking understanding (coordinates and linear relationships)

Questions 20.21–20.32: Reasoning and problem solving (coordinates and linear relationships)

Questions 20.33–20.41: Learning and teaching (coordinates and linear relationships)

GLOSSARY OF KEY TERMS INTRODUCED IN CHAPTER 20

Scatter diagram (scatter plot, scattergram, scattergraph)   A graphical representation of data for two variables for a given set, with horizontal and vertical axes representing the two variables, and the values of the two variables for each individual in the set plotted as points.

Axes (plural of axis)   In a two-dimensional coordinate system, two number lines drawn at right angles to represent the variables x and y. Conventionally, the horizontal axis represents the independent variable (x) and the vertical axis the dependent variable (y).

Origin   The point where the axes in a coordinate system cross; the point with co-ordinates (0, 0).

Coordinates   Starting from the origin, the distance moved in the x-direction followed by the distance moved in the y-direction to reach a particular point; recorded as (x, y).

Quadrant   One of the four regions into which the plane is divided by the two axes in a coordinate system.

First quadrant   The quadrant consisting of all those points with positive coordinates.

Linear relationship   A relationship between two variables that produces a straight-line graph. If the two variables are directly proportional, the straight line passes through the origin.

Conversion graph A straight-line graph used to convert a measurement from one unit to another (for example, volume in litres to volume in pints). If zero units in one unit is equal to zero units in the other then the line will pass through the origin and the two variables will be directly proportional.

Distance–time graph A graphical representation of the distance from the starting point of a moving object over time; the vertical axis usually represents the distance from the starting point and the horizontal axis the time that has passed. The steeper the slope, the faster the object is travelling.

SUGGESTIONS FOR FURTHER READING FOR SECTION E

1 Chapter 7 on algebra in Brown (2003) will provide the reader with comprehensive coverage of all the algebra they will ever need and more. It is written from the perspective of primary school teaching and raises important issues about teaching this area of mathematics. The reader is enabled to see how formal algebra evolves developmentally from early experiences of sorting and patterns.

2 Borthwick et al. (2021) provide a very practical guide to support teachers in recognizing opportunities to lay the foundation for algebraic thinking by making the most of children's early work with pattern recognition and analysis.

3 Chapter 7 of Orton (2004), written by Tony and Jean Orton, is entitled 'Pattern and the approach to algebra'. They discuss the use of number patterns as a route into algebra and explore children's approaches to identifying and articulating simple patterns. The chapter includes a study of the performance of 10–13-year-old children on generalizations from number patterns.

4 Chapter 4 of Nickson (2004) provides an insightful analysis of research into children's understanding of algebra.

5 To take the material in Chapter 20 further, you could look at Chapter 10 of Suggate, Davis and Goulding (2017), which deals with graphs and functions.

6 An extensive review of research into children's learning of early algebraic ideas is provided by Carraher and Schliemann (2007).

7 For a stimulating and comprehensive collection of scholarly articles related to younger children's learning of algebra, dip into Kaput et al. (2008).

8 Germia and Panorkou (2020) provide some interesting ideas for promoting the understanding of coordinates through the programming language, Scratch, which is widely used in schools.

SECTION F

MEASUREMENT

21   Concepts and Principles of Measurement   369
22   Perimeter, Area and Volume   393

WATCH THE SECTION OPENER VIDEO AT: HTTPS://STUDY.SAGEPUB.COM/HAYLOCK7E

CONCEPTS AND PRINCIPLES OF MEASUREMENT

IN THIS CHAPTER, THERE ARE EXPLANATIONS OF

- what children learn to measure;
- the different kinds of situations in which units of length are used;
- the distinction between volume and capacity;
- the distinction between mass and weight;
- two aspects of the concept of time;
- the role of comparison and ordering as a foundation for measurement;
- the principle of transitivity in the context of measurement;
- some principles of inequalities, using the signs < and >;
- conservation of length, mass and liquid volume;
- non-standard and standard units;
- the idea that all measurement is approximate;
- the difference between a ratio scale and an interval scale;
- SI and other metric units of length, mass and time, including the use of prefixes;
- how to convert between different metric units of measurement;
- the importance of estimation and the use of reference items;
- imperial units still in use and their relationship to metric units.

READ THIS CHAPTER'S CURRICULUM LINKS AT: HTTPS://STUDY.SAGEPUB.COM/HAYLOCK7E

WHAT DO CHILDREN LEARN TO MEASURE?

Children in primary school learn to measure:

- length and distance;
- liquid volume and capacity;
- mass (weight);
- time intervals and recorded time;
- temperature;
- area;
- solid volume;
- angle.

Measurement of area and solid volume are considered in Chapter 22, along with perimeter, which is an application of measurement of length. Measurement of temperature has been considered in Chapter 14 as a key context for exploring positive and negative integers. Measurement of angle is discussed in Chapter 23. In a sense, learning about the value of coins and notes and how to handle amounts of money could also be considered an aspect of measurement. Since this is such a significant context for learning about number and number

> **LEARNING AND TEACHING POINT**
>
> Children should develop their skills and understanding of measurement through practical, purposeful activities. They should learn to choose and use appropriate measuring devices, discussing the ideas of accuracy and approximation.

operations, money has been integrated into Chapters 6–12 and 16–17. Indeed, all measuring contexts are important for the application of number concepts and skills, so we have already used them extensively in previous chapters in this book. In this chapter, we focus therefore on some of the underlying principles of measurement, as well as clarifying some of the concepts and language involved.

IN WHAT DIFFERENT SITUATIONS ARE UNITS OF LENGTH APPLIED?

We tend to think of length as a very straightforward measuring concept. But, as so often with mathematics, for understanding of this concept the symbols and language have to be connected with a range of different pictures and real-life contexts. Consider a length of 1 metre (1 m). To what different kinds of situations might we connect this? Here are some possibilities:

1   1 m could be the length of a straight piece of wood (like the edge of a metre rule).
2   1 m could be the length of a line drawn on a piece of paper (using a metre rule).
3   1 m could be a step along a number line (for example, from 2.4 m to 3.4 m).
4   1 m could be the length of the diagonal of a rectangle (television screen sizes are measured like this).
5   1 m could be the height of a shelf.
6   1 m could be the size of the gap between two rows of chairs.
7   1 m could be the height of a four-year-old child (about 80% of children reach this height by the age of four years).
8   1 m could be the distance from one point to another.
9   1 m could be the perimeter of a rectangle.
10 1 m could be the distance along a curved path from one point to another.
11 1 m could be the circumference of a circle.

The reader will be able to add other examples to this list. The simple idea of a unit of length, as a measurement of finding how long is a straight line, becomes a surprisingly sophisticated notion, embracing lengths of imaginary lines, distance and lengths of non-straight paths. In statement 3 above, the 1 metre is a section of a straight line. In statement 4, there may not even be a line drawn, but we can still measure it. In statement 5, the understanding is that we are talking about an imaginary vertical line from the floor to the shelf. But in statement 6, the understanding is that we require an imaginary horizontal line drawn from the edge of one chair to another, but not any old horizontal line; it must be at right angles to the edges of the rows. In statement 7, we impose a vertical straight line on an object (a child!) that has no straight edges at all. In statement 8, we probably imagine a piece of string stretched from one point to the other. In statement 9, the thing we are measuring is composed of four straight sections, added together. In statements 10 and 11, we are applying the idea of length to a curved path, presumably by imagining the path stretched out into a straight line.

> **LEARNING AND TEACHING POINT**
>
> Children should have experiences of applying measurements of length not just to straight edges and straight lines, but also to sections of lines, imaginary lines drawn through objects, horizontal and vertical distances, and distances along non-straight paths.

CAN YOU EXPLAIN THE DISTINCTION BETWEEN VOLUME AND CAPACITY?

The **volume** of an object is the amount of three-dimensional space that it occupies. By historical accident, liquid volume and solid volume are conventionally measured in different units, although the concepts are exactly the same. Using standard units of measurement, liquid volume is measured in litres and millilitres and so on, whereas solid volume is measured in units such as cubic metres and cubic centimetres. A cubic metre is simply the volume of a cube with length of side 1 metre; and a cubic centimetre is the volume of a cube with side 1 centimetre. (Units are considered later in this chapter; solid volume is considered in the next.) In the metric system, the units for liquid and solid volume are related in a very simple way: 1 millilitre is the same volume as 1 cubic centimetre; or 1 litre is the same volume as 1000 cubic centimetres (see Figure 21.1).

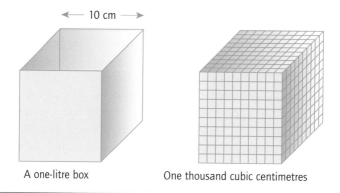

A one-litre box   One thousand cubic centimetres

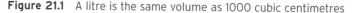

Figure 21.1 A litre is the same volume as 1000 cubic centimetres

Only containers have **capacity**. The capacity of a container is the maximum volume of liquid that it can hold. Hence, capacity is measured in the same units as liquid volume. For example, if a wine glass holds 180 millilitres of wine when filled to the brim then its capacity is 180 millilitres. In real-life situations, there is usually a distinction between 'brimful capacity' (for example, if you were to fill a medicine bottle to the brim) and 'nominal capacity' (which is what the bottle is intended to hold). A bottle of wine usually has a nominal capacity of 750 ml, but the brimful capacity would be around 780 ml.

WHAT IS THE DIFFERENCE BETWEEN MASS AND WEIGHT?

There is a problem here about the language we use to describe what we are measuring when, for example, we put a book in one pan of a balance and equalize it with, say,

200 grams in the other. Colloquially, most people say that what we have found out is that the book *weighs* 200 grams, or that its **weight** is 200 grams. This is technically incorrect. What we have discovered is that the book *weighs the same as* a mass of 200 grams, or that the **mass** of the book is 200 grams. This conflict between everyday language usage and the scientifically correct usage is not resolved simply by using the two words, mass and weight, interchangeably.

The units we use for weighing, such as grams and kilograms, or pounds and ounces, are actually units for measuring the mass of an object, not its weight. The mass is a measurement of the quantity of matter there is in the object. Note that this is not the same thing as the amount of space it takes up – that is the volume of the object. A small lump of lead might have the same mass as the 200-gram book, but it would take up much less space, because the molecules making up the piece of lead are much more tightly packed together than those in the book.

The problem with the concept of mass is that we cannot actually experience mass directly. I cannot see the mass of the book, feel it or perceive it in any way. When I hold the book in my hand, what I experience is the weight of the book. The weight is the force exerted on the book by the pull of gravity. I can feel this because I have to exert a force myself to hold the book up.

Of course, the weight and the mass are directly related: the greater the mass, the greater the weight, and therefore the heavier the object feels when I hold it in my hand. However, the big difference between the two is that, whereas the mass of an object is invariant, the weight changes depending on how far you are from the centre of the Earth (or whatever it is that is exerting the gravitational pull on the object).

We are all familiar with the idea that an astronaut's weight changes in space, or on the Moon, because the gravitational pull being exerted on the astronaut is less than it is on the Earth's surface. In some circumstances, for example when in orbit, this gravitational pull can effectively be cancelled out and the astronaut experiences 'weightlessness'. The astronaut can then place a book on the palm of his or her hand and it weighs nothing. On the Moon's surface, the force exerted on the book by gravity, that is, the weight of the book, is about one-sixth of what it was back on the Earth's surface. But throughout all this, the mass of the astronaut and the mass of the book remain unchanged. The book is still 200 grams, as it was on Earth, even though its weight has been changing constantly. (So a good way of losing weight is to go to the Moon, but this does not affect your waist size because what you really need to do is lose mass.)

An important point to note is that the balance-type weighing devices do actually measure mass. We put the book in one pan, balance it with a mass of 200 grams in the other pan, and because the book 'weighs the same as' a mass of 200 grams, we conclude that it also has a mass of 200 grams. Note that we would get the same result using the balance on the Moon. However, the pointers on spring-type weighing devices, such as many kitchen scales and bathroom scales, actually respond directly to weight. This means that they would give a different reading if we took them to the Moon, for example. But, of course, they are calibrated for use on the Earth's surface, so when I stand on the bathroom scales and the pointer indicates 68 kilograms I can rely on that as a measurement of my mass. On the Moon, it would point to around 11 kilograms; this would just be wrong.

Because weight is a force, it is measured in the units of force. The standard unit of force in the metric system is the **newton**, appropriately named after Sir Isaac Newton (1642–1727), the mathematical and scientific genius who first articulated this distinction between mass and weight. A newton is defined as the force required to increase the speed of a mass of 1 kilogram by 1 metre per second every second. A newton is actually about the weight of a small apple (on Earth) and a mass of a kilogram has a weight of nearly 10 newtons. You may not need to know this, although you could come across spring-type weighing devices with a scale graduated in newtons.

> **LEARNING AND TEACHING POINT**
>
> With older children in the primary range, you can discuss the effect of gravity and space travel on weight and the idea that mass does not change.

> **LEARNING AND TEACHING POINT**
>
> My approach is to refer to the things we use for weighing objects on a balance as *masses* and use the language *weighs the same as a mass of so many grams*. Then encourage children to say 'the mass is so many grams', whilst acknowledging that most people incorrectly say 'the weight is so many grams'.

One way to introduce the word 'mass' to primary school children is to refer to those plastic or metal things we use for weighing objects in a balance as 'masses' (rather than 'weights'). So we would have a box of 10-gram masses and a box of 100-gram masses and so on. Then when we have balanced an object against some masses, we can say that the object weighs the same as a mass of so many grams, as a step towards using the correct language, that the mass of the object is so many grams.

WHAT ABOUT MEASURING TIME?

There are two quite different aspects of time that children have to learn to handle. First, there is the idea of a *time interval*. This refers to the length of time occupied by an activity, or the time that passes from one instant to another. Time intervals are measured in units such as seconds, minutes, hours, days, weeks, years, decades, centuries and millennia.

Then there is the idea of *recorded time*, the time at which an event occurs. To handle recorded time, we use the various conventions for reading the time of day, such as o'clock, a.m. and p.m., the 24-hour system, together with the different ways of recording the date, including reference to the day of the week, the day in the month and the year. So, for example, we might say that the meeting starts at 15:30 on Monday, 17 October 2016, using the concept of 'recorded time', and that it is expected to last for 90 minutes, using the concept of 'time interval'.

Time is one aspect of measurement that has not gone metric, so the relationships between the units (60 seconds in a minute, 60 minutes in an hour, 24 hours in a day and so on) are particularly challenging. This makes it difficult, for example, to use a subtraction algorithm for finding the time intervals from one time to another. I strongly recommend that problems of this kind are done by an ad hoc process of adding on. For example, to find the length of time of a journey starting at 10:45 a.m. and finishing at 1:30 p.m., reason like this:

From 10:45 a.m.:
15 minutes to 11 o'clock,
then 2 hours to 1 o'clock,
then a further 30 minutes to 1:30 p.m.,
making a total of 2 hours and 45 minutes.

Setting this out as a formal subtraction would be inadvisable. But representing it as a calculation on a number line, as shown in Figure 21.2, would be highly advisable. A number line used like this is also referred to as a time-line.

Figure 21.2   The time interval from 10:45 a.m. to 1:30 p.m. on a number line (time-line)

Learning about time is also complicated by the fact that the hands on a conventional dial-clock go round twice in a day; it would have been so much more sensible to go round once a day. Because of the association of a circle with 12 hours on a clock face, I always avoid using a circle to represent a day. For example,

I would avoid a pie chart for 'how I spend a day' or a circular diagram showing the events of a day. For a picture of a day and the hours that pass until the day begins, again I would recommend a diagram like that shown in Figure 21.3. Children can add to this pictures or verbal descriptions of what they are doing at various times of day.

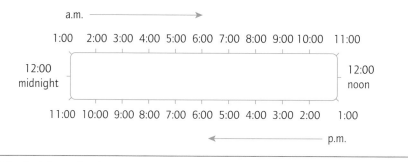

Figure 21.3 A picture of a day

Then there are the added complications related to the variety of watches and clocks that children may use, as well as the range of ways of saying the same time. For example, as well as being able to read a conventional dial-clock and a digital display in both 12-hour and 24-hour versions, children have to learn that the following all represent the same time of day: twenty to four in the afternoon, 3:40 p.m., 15:40 (also written sometimes as 1540 or 15.40). Incidentally, the colloquial use of, for example, 'fifteen hundred' to refer to the time 15:00 in the 24-hour system is an unhelpful abuse of mathematical language. It reinforces the misunderstanding, mentioned in Chapter 6, of thinking that '00' is an abbreviation for 'hundred'. I prefer the BBC World Service convention: 'The time is fifteen hours.'

A couple of further small points relate to noon and midnight. First, note that 'a.m.' and 'p.m.' are abbreviations for **ante meridiem** and **post meridiem**, meaning 'before noon' and 'after noon', respectively. This means that 12 noon is neither a.m. nor p.m. It is just 12 noon. Similarly, 12 o'clock midnight is neither a.m. nor p.m. Then, in the 24-hour system, midnight is the moment when the recorded time of day starts again, so it is not 24:00, but 00:00, '**zero hours**'.

WHAT PRINCIPLES ARE CENTRAL TO TEACHING MEASUREMENT IN THE PRIMARY AGE RANGE?

Some of the central principles in teaching for mastery of the mathematics of measurement relate to the following headings:

- comparison and ordering;
- transitivity;
- conservation;
- non-standard and standard units;
- approximation;
- a context for developing number concepts;
- the meaning of zero.

These are encountered in learning about all aspects of measurement.

WHAT HAVE COMPARISON AND ORDERING TO DO WITH MEASUREMENT?

The foundation of all aspects of measurement is direct comparison, putting two and then more than two objects (or events) in order, according to the attribute in question. The language of comparison, discussed in relation to subtraction in Figure 7.6 (see Chapter 7) is of central importance here. Two objects are placed side by side and children determine which is the longer, which is the shorter. Two items are placed in the pans of a balance and children determine which is the heavier, which is the lighter. Water is poured from one container to another to determine which holds more, which holds less. Two children perform specified tasks, starting simultaneously, and observe which takes a longer time, which takes a shorter time. No units are involved at this stage, simply direct comparison leading to putting two or more objects or events in order.

Recording the results of comparison and ordering can be an opportunity to develop the use of the *inequality signs* (see Chapter 6). So, for example, 'A is longer than B' can be recorded as A > B, and 'B is shorter than A' as B < A. This introduces in a practical context an important principle of inequalities that can be expressed formally as follows:

If A > B, then B < A (if A is greater than B then B is less than A).
If A < B, then B > A (if A is less than B then B is greater than A).

HOW DOES TRANSITIVITY APPLY TO MEASUREMENT?

In Chapter 13 we saw how the *transitive* property applies to the relationships 'is a multiple of' (illustrated in Figure 13.1) and 'is a factor of' (illustrated in Figure 13.4). The principle of transitivity is shown in Figure 21.4. If we know that A is related to B (indicated by an arrow) and B is related to C, the question is whether A is related to C as a logical consequence. With some relations (such as 'is a factor of'), it does follow logically and we can draw in the arrow connecting A to C. In other cases, it does not. For example, the relationship 'is a mirror image of' can be applied to a set of shapes: if shape A is a mirror image of shape B, and B is a mirror image of shape C, then it is *not* true that A must automatically be a mirror image of C. So that would not be a transitive relationship.

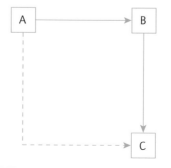

Figure 21.4 The transitive property

We can now see that whenever we compare and order three or more objects (or events) using a *measuring* attribute such as their lengths, their masses, their capacities or the length of time (for events), then we are again making use of a transitive relationship.

The arrow used in Figure 21.4 could represent any one of the measuring relationships used to compare two objects or events, such as: 'is longer than', 'is lighter than', 'holds more than' or 'takes less time than'. In each case, because A is related to B and B is related to C, it follows logically that A is related to C. This principle is fundamental to ordering a set of more than two objects or events: once we know A is greater than B and B is greater than C, for example, it is this principle which allows us not to have to check A against C. Grasping this is a significant step in the development of a child's understanding of measuring.

The transitive property of measurement can be expressed formally using inequality signs as follows:

If $A > B$ and $B > C$ then $A > C$.
If $A < B$ and $B < C$ then $A < C$.

WHAT IS CONSERVATION IN MEASUREMENT?

Next, we should note the principle of **conservation**, another fundamental idea in learning about measurement of length, mass and liquid volume. Children meet this principle first in the context of conservation of number, as discussed in Chapter 3 (see Figure 3.3 and accompanying text). They have to learn, for example, that if you rearrange a set of counters in different ways, you do not alter the number of counters. Similarly, if two objects are the same length, they remain the same length when one is moved to a new position: this is the principle of conservation of length. Conservation of mass is experienced when, for example, children balance two lumps of dough, then rearrange each lump in some way, such as breaking one up into small pieces and moulding the other into some shape or other, and then check that they still balance.

Conservation of liquid volume is the one that catches many children out. When they empty the water from one container into another, differently shaped container, as shown in Figure 21.5, by focusing their attention on the heights of the water in the containers, children often lose their grip on the principle that the volume of water has actually been unchanged by the transformation.

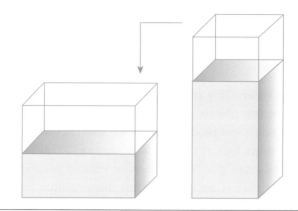

Figure 21.5 Conservation of liquid volume

HOW MIGHT CHILDREN LEARN ABOUT USING UNITS IN MEASUREMENT?

Fundamental to the idea of measuring is the use of a 'unit'. The use of non-standard units to introduce to children the idea of measuring in units is a well-established tra-dition in primary school mathematics teaching. For example, they might measure the length of items of furniture in spans, the length of a wall in cubits (a cubit is the length of your forearm), the mass of a book in conkers, the capacity of a con-tainer in egg-cups. Many adults make use of non-standard units of length in everyday life, especially, for example, when making rough-and-ready measurements for practical jobs around the house and garden. And we often hear references to a

length as being equivalent to so many London buses or a large area as being so many times the size of Wales and so on. The value of this approach is that children get experience of the idea of measuring in units through familiar, unthreatening objects first, rather than going straight into the use of mysterious things called millilitres and grams and so on.

Also, it is often the case that the non-standard unit is a more appropriate size of unit for early measuring experiences. For example, most of the things around the classroom the children might want to weigh will have a mass of several hundred grams. The gram is a very small mass for practical purposes to begin with, and the kilogram is far too large; conkers and glue-sticks are much more appropriate sizes of units for measuring mass in the early stages. Eventually, of course, we must learn that there is a need for a standard unit. The experience of working with non-standard units often makes this need explicit, as, for example, when two children measure the length of the hall in paces and get different answers.

WHAT IS THERE TO KNOW ABOUT APPROXIMATION?

The next major principle of measurement is that nearly all measurements are *approximate*. If you are measuring length, mass, time or capacity, all you can ever achieve is to make the measurement to the nearest something, depending on the level of accuracy of your measuring device. The reader may have noticed that against the stated mass or volume on many food and drink packages there is a large 'e'. This is a European symbol (known as the e-mark) indicating that the stated measurement is only an estimate within mandatory limits. So, when we say that a bottle contains 750 ml of wine, it has to be understood that this statement is an approximation. It might mean, for example, that it contains 750 ml if measured to the nearest 5 ml, in which case the volume would lie between 747.5 ml and 752.5 ml. When I say that a child is 90 cm tall, you have to realize that what I mean is that the child is 90 cm tall if measured to the nearest centimetre. This is particularly important when we do calculations with measurements, because of the problems of compounding errors that arise from rounding (see the discussion of rounding in Chapter 16). For example, when we calculate that 10 bottles containing 750 ml (to the nearest 5 ml) will provide 7.5 litres in total, by multiplying 750 ml by 10, this answer is correct only to the nearest 50 ml.

LEARNING AND TEACHING POINT

Make collections and displays of packages or labels, discussing which items are sold by mass or by volume, and note the units used. Challenge children to find as many different masses and volumes as they can and display these in order. Discuss the significance of the e-mark that occurs next to most measurements on food packages and labels.

The principle of measuring to the nearest something and the associated language should be introduced to primary school children from the earliest stages. Even when measuring in non-standard units, they will encounter this idea, as, for example, when determining that the length of the table is 'nearly 9 spans' or '8 spans and a bit' or 'about 9 spans' or 'between 8 and 9 spans'.

HOW IS MEASUREMENT A CONTEXT FOR DEVELOPING NUMBER CONCEPTS?

In discussing the principles of measurement, we should stress the central importance of measurement experiences as a context for developing number concepts. Throughout this book, I have used measurement problems and situations to reinforce ideas such as place value (Chapter 6), the various structures for the four operations (Chapters 7 and 10) and calculations with decimals (Chapter 17). For example, the key idea of comparison in understanding subtraction is best experienced in practical activities comparing heights, masses, capacities and time; and the key structure of division as the inverse of multiplication can be connected with finding out how many small containers can be filled from a large container.

DOESN'T ZERO JUST MEAN 'NOTHING'?

Finally in this section on principles of measurement, I comment on the meaning of zero. Measurements such as length, mass, liquid volume and capacity, and time intervals, are examples of what are called **ratio scales**. These are scales where the ratio of two quantities has a real meaning. For example, if a child is 90 cm tall and an adult is 180 cm tall, we can legitimately compare the two heights by means of ratio, stating that the adult is twice as tall as the child. Similarly, we can compare masses, capacities and time intervals by ratio.

However, recorded time, for example, is not like this: it would make no sense to compare, say, 6 o'clock with 2 o'clock by saying that one is three times the other. This is an example of what is called an **interval scale**. Comparisons can only be made by reference to the difference (the interval) between two measurements, for example saying that 6 o'clock is 4 hours later than 2 o'clock. Of course, you can compare the measurements in a ratio scale by reference to difference (for example, the adult is 90 cm taller than the child), but the point about an interval scale is that you *cannot* do it by ratio, you can only use difference. Temperature measured in °C (the **Celsius scale**) or °F (the **Fahrenheit scale**) is another example of something measured by using an interval scale. It would be meaningless to assert that 15 °C is three times as hot as 5 °C; the two temperatures should be compared by their difference.

The interesting mathematical point here is that the thing that really distinguishes a ratio scale from an interval scale is that in a ratio scale the zero means *nothing*, but in an interval scale it does not. When the recorded time is 'zero hours', time has not disappeared. When the temperature is 'zero degrees', there is still a temperature out there and we can feel it. But a length of 'zero metres' is no length; a mass of 'zero grams' is nothing; a bottle holding 'zero millilitres' of wine is empty; a time interval of 'zero seconds' is no time at all.

WHAT METRIC UNITS AND PREFIXES DO I NEED TO KNOW ABOUT?

There is an internationally accepted system of metric units called **SI units** (Système International). This system specifies one base unit for each aspect of measurement. For length, the SI unit is the **metre**, for mass it is the **kilogram** (not the gram), for time it is the *second*. There is no specific SI unit for liquid volume, since this would be measured in the same units as solid volume, namely **cubic metres**. However, the **litre** is a standard unit for liquid volume and capacity that is used internationally.

Other units can be obtained by various prefixes being attached to these base units. There is a preference for those related to a thousand: for primary school use, this would be just **kilo** (k), meaning 'a thousand', and **milli** (m), meaning 'a thousandth'. So, for example, throughout this book, we have used kilograms (kg) and grams (g), where 1 kg = 1000 g, and litres (l) and millilitres (ml), where 1 litre = 1000 ml.

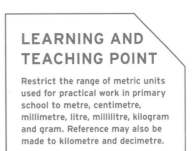

LEARNING AND TEACHING POINT

Note that the symbol used for litre (l) can be confused in print with the numeral 1, so it is often better to write or type the word *litre* in full.

Similarly, we can have kilometres (km) and metres (m), where 1 km = 1000 m, and metres and millimetres (mm), where 1 m = 1000 mm. These are the only uses of the prefixes kilo and milli likely to be needed in primary school.

For practical purposes, we will need other prefixes, especially **centi** (c), meaning 'a hundredth'. This gives us the really useful unit of length, the **centimetre** (cm), where 1 m = 100 cm. We might also find it helpful, for example when explaining place value and decimals with length (see Chapter 6),

LEARNING AND TEACHING POINT

Restrict the range of metric units used for practical work in primary school to metre, centimetre, millimetre, litre, millilitre, kilogram and gram. Reference may also be made to kilometre and decimetre.

to use the prefix **deci** (d), meaning 'a tenth', as in decimetre (dm), where 1 m = 10 dm. We might also note that some wine bottles are labelled 0.75 litres, others are labelled 7.5 dl (decilitres) and still others 75 cl (centilitres) or 750 ml (millilitres). These are all the same volume of wine.

HOW DO YOU CONVERT BETWEEN DIFFERENT METRIC UNITS OF MEASUREMENT?

Converting between different units of metric measurement requires the application of the procedures for multiplying and dividing by 10, 100 and 1000 explained in Chapter 17, plus understanding which of these is required.

Converting to and from the unit that does not have a prefix is the most straightforward. For example, a centimetre is one hundredth of a metre and 100 cm = 1 m; so, to convert metres to centimetres, multiply by 100; and to convert centimetres to metres, divide by 100. Similarly, a gram is one thousandth of a kilogram and 1 kg = 1000 g; so, to convert kilograms to grams, multiply by 1000; and to convert grams to kilograms, divide by 1000. Here are some examples:

8.60 m	= 860 cm	(multiplying by 100)
0.05 m	= 5 cm	(multiplying by 100)
505 cm	= 5.05 m	(dividing by 100)
24.5 cm	= 0.245 m	(dividing by 100)
8.60 kg	= 8600 g	(multiplying by 1000)
0.5 kg	= 500 g	(multiplying by 1000)
750 ml	= 0.75 litres	(dividing by 1000)
2.5 litre	= 2500 ml	(multiplying by 1000)
5 m	= 50 dm	(multiplying by 10)
28 dm	= 2.8 m	(dividing by 10)
8 km	= 8000 m	(multiplying by 1000)
400 m	= 0.4 km	(dividing by 1000)
2.5 g	= 2500 mg	(multiplying by 1000)
40 mg	= 0.040 g	(dividing by 1000)

Just a little trickier is converting between units when both of them have prefixes. For example, how do you convert between millimetres and centimetres? Or between decimetres and millimetres? Since there are 1000 mm in a metre and 100 cm in a metre, then clearly there are 10 mm in a centimetre. Likewise, since there are 100 cm in a metre and 10 dm in a metre, then 1 dm is equal to 10 cm. This is the basis of Figure 21.6, which shows how you move along from millimetres to centimetres to decimetres to metres, each time multiplying by 10; and, in the reverse direction, by dividing by 10. So, we have, for example, these equivalences:

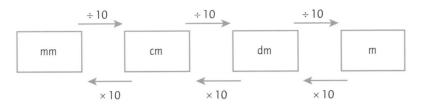

Figure 21.6 Converting between metric units of length

250 mm = 25 cm = 2.5 dm = 0.25 m   (each time dividing by 10)
1.75 m = 17.5 dm = 175 cm = 1750 mm   (each time multiplying by 10)

It follows then, as shown in Figure 21.6, that to convert between, say, millimetres and decimetres, we would multiply or divide by 10 and by 10 again, in other words by 100. So, for example, 1.25 dm = 125 mm (multiplying by 100); and 400 mm = 4 dm (dividing by 100). Clearly, the same principles would apply, for example, in converting between millilitres, centilitres, decilitres and litres.

WATCH THE PROBLEM SOLVED! VIDEO AT: HTTPS://STUDY.SAGEPUB.COM/HAYLOCK7E

HOW CAN I GET BETTER AT HANDLING UNITS OF MEASUREMENT?

Take every opportunity to practise *estimation* of lengths, heights, widths and distances, liquid volume and capacity, and mass. In the supermarket, take note of which items are sold by mass (although they will call it weight) and which by volume; estimate the mass or volume of items you are purchasing and check your estimate against what it says on the packet or the scales. This all helps significantly to build up confidence in handling less familiar units.

> **LEARNING AND TEACHING POINT**
>
> Make considerable use of estimation as a class activity, encouraging children to learn by heart the measurements of specific reference items such as those given here.

One way of becoming a better estimator is to learn by heart the sizes of some specific **reference items**. Children should be encouraged to do this for length, mass and capacity, and then to relate other estimates to these. Here are some that I personally use:

- A child's finger is about 1 cm wide.
- The children's rulers are 30 cm long.
- A sheet of A4 paper is about 21 cm by 30 cm.
- The distance from my nose to my outstretched fingertip is about 1 metre (100 cm).
- The classroom door is about 200 cm or 2 m high.
- The mass of an individual packet of crisps is 30 g.
- The mass of a standard-size tin of baked beans is about 500 g (including the tin).
- A standard can of drink has a capacity of 330 ml.
- A wine bottle holds 750 ml of wine.
- Standard cartons of milk or fruit juice are 1 litre (1000 ml).
- A litre of water has a mass of a kilogram (1000 g).

WHAT IMPERIAL UNITS ARE STILL IMPORTANT?

The attempt to turn the UK into a fully metricated country has not been entirely successful. A number of popular imperial units of measurement stubbornly refuse to go away. For example, some people still find temperatures given in degrees Fahrenheit to be more meaningful than those in degrees Celsius (also known as centigrade).

The prime candidate for survival would seem to be the *mile*: somehow I cannot imagine in the foreseeable future all the road signs and speed limits in the UK being converted to kilometres. Unfortunately, children cannot possibly have practical experience of measuring distances in kilometres in the classroom to compensate for the lack of experience of this metric unit in everyday life.

To relate miles to kilometres, the most common equivalence used is that 5 miles is about 8 kilometres. A simple method for doing conversions is to read off the corresponding speeds on the speedometer in a car, most of which are given in both miles per hour and kilometres per hour. For example, you can read off that 30 miles is about 50 km, 50 miles is about 80 km, and 70 miles is about 110 km.

> **LEARNING AND TEACHING POINT**
>
> To be realistic, work on journey distances and average speeds (see Chapter 27) is most appropriately done in miles and miles per hour.

> **LEARNING AND TEACHING POINT**
>
> Children could usefully memorize a few key approximate conversions between miles and kilometres. For example, 30 miles is about 50 km; 50 miles is about 80 km. They could also work out and memorize significant distances in these units. For example, children in Norwich might know that Norwich to London is about 120 miles or about 190 km.

There is also an intriguing connection between this conversion and a sequence of numbers called the **Fibonacci sequence**, named after Leonardo Fibonacci, a twelfth/ thirteenth-century Italian mathematician: 1, 1, 2, 3, 5, 8, 13, 21, 34, 55, 89 and so on. The sequential generalization here is to add the two previous numbers to get the next one. So, for example, the next number after 89 is 144 (55 + 89). Now, purely coincidentally, it happens that one of these numbers in miles is approximately the same as the next one in kilometres. For example, 2 miles is about 3 km, 3 miles is about 5 km, 5 miles is about 8 km, 8 miles is about 13 km and so on. This works remarkably well to the nearest whole number for quite some way. (See self-assessment question 21.1 below.)

Other common measurements of length still surviving in everyday usage, together with an indication of the kinds of equivalences that it might be useful to learn, are:

- the inch (about the width of an adult thumb, about 2.5 cm);
- the foot (about the length of a standard class ruler or a sheet of A4 paper, that is, about 30 cm);
- the yard (about 10% less than a metre, approximately 91 cm).

Imperial units of mass still used occasionally and unofficially in some markets are:

- the ounce (about the same as a small packet of crisps, that is, about 30 g);
- the pound (getting on for half a kilogram, about 450 g).

Many people still like to weigh themselves in stones: a stone is a bit more than 6 kg (about 6.35 kg). I find it useful to remember that 11 stone is about 70 kg: see self-assessment question 20.5 in the previous chapter. And, for those who enjoy this kind of thing, a hundredweight is just over 50 kg, and a metric tonne (1000 kg) is therefore just a bit more than an imperial ton (20 hundredweight).

LEARNING AND TEACHING POINT

Do not ignore imperial units still in everyday use. Discuss them and bring them into practical experience in the classroom. Give children opportunities to relate them to metric units.

Gallons have disappeared from the petrol station, but still manage to survive in common usage; for example, people still tell me how many miles per gallon their car does, even though petrol is almost universally sold in litres. My (very economical) car does 11 miles to the litre. A gallon is about 4.5 litres. Given the drinking habits of the British public, another contender for long-term survival must be the pint. I have noticed that some primary school children (and their parents) refer to any carton of milk as 'a pint of milk', regardless of the actual volume involved. A pint is actually just over half a litre (568 ml).

Conversion between metric and imperial units can be done using the methods described in Chapter 18 for direct proportion problems, or using conversion graphs as described in Chapter 20, so it provides a realistic context for some genuine mathematics for older children in primary school.

RESEARCH FOCUS: CHILDREN'S UNDERSTANDING OF LENGTH

What are the different stages of development that children go through in learning about the measurement of length? By evaluating a theoretical learning trajectory for length measurement ideas with children aged 4 to 7 years, Szilágyi et al. (2013) were able to identify at least seven stages through which a young child's understanding progresses. Here I can summarize just three of the stages involved. The reader should access the original article for details of the complete learning trajectory identified.

At the lowest level of development, for example, is the child who is what the researchers call a 'length quantity recognizer'. A child at this level can correctly acknowledge length relationships between pairs of objects that are already aligned – parallel along their lengths with endpoints on a line perpendicular to the lengths – but does not recognize the need for such alignment when asked to determine which is the longer or shorter of the two objects.

The fourth of the seven levels identified in this research is the child described as a 'serial orderer'. At this level of thinking, the child shows that he or she understands the complete ordering of six or more objects as a series in which any one object is longer than each of the preceding ones. This understanding involves the successive application of transitivity. Children at this level were able, for example, to place six towers in order of height correctly and to see the series of towers as a whole.

The highest level identified was the 'path measurer'. Children at this level have a sophisticated understanding of length. For example, they are able to verbalize their understanding of the conservation of length with respect to parts put together to form a whole. They will recognize that a zigzag road can be longer than a straight road, even if the distance between the endpoints is shorter. These children clearly understand that the length of the zigzag road is not the distance between its endpoints, but the sum of the lengths of its parts. They understand that a ruler is an object composed of equal units, and can therefore use a broken ruler or a ruler without numbers. These children understand that an object of length 3 m is also 300 cm long, and even 2 m plus 100 cm and so on.

ㄷㄷ LEARNING RESOURCES

Access activities for your **lesson plans** at: https://study.sagepub.com/haylock7e

Before trying the self-assessment questions below, you should complete the **self-assessment questions** for this chapter at: https://study.sagepub.com/haylock7e

21.1: Given that 1 mile is 1.6093 km (to four decimal places), use a calculator to find out how far the rule for changing miles to kilometres based on the Fibonacci sequence is correct to the nearest whole number.

21.2: Are these relationships transitive?

　　a 'Is earlier than' applied to times of the day;

　　b 'Is half of' applied to lengths.

21.3: Measure the length of a sheet of A4 paper to the nearest millimetre. Give the answer:

　　a in millimetres;

　　b in centimetres;

　　c in decimetres;

　　d in metres.

21.4: Classify each of these statements as possible or impossible:

　　a An elephant has a mass of about 7000 kg.

　　b A standard domestic bath filled to the brim contains about 40 litres of water.

　　c Yesterday I put exactly 20 litres of petrol in the tank of my car.

　　d A normal cup of coffee is about 2 decilitres.

　　e It takes me about a week to complete a walk of 1000 km.

　　f An envelope containing 8 sheets of standard A4 photocopier paper does not exceed 100 g in mass.

FURTHER PRACTICE

Access the website material for

- Knowledge check 32: Knowledge of metric units of length and distance
- Knowledge check 33: Knowledge of other metric units

at: https://study.sagepub.com/haylock7e

FROM THE STUDENT WORKBOOK

Questions 21.01–21.20: Checking understanding (concepts and principles of measurement)

Questions 21.21–21.30: Reasoning and problem solving (concepts and principles of measurement)

Questions 21.31–21.44: Learning and teaching (concepts and principles of measurement)

GLOSSARY OF KEY TERMS INTRODUCED IN CHAPTER 21

Volume   The amount of three-dimensional space taken up by an object; measured in cubic units, such as cubic centimetres or cubic metres.

Capacity   The volume of liquid that a container can hold; usually measured in litres and millilitres; only containers have capacity.

Weight   The force of gravity acting upon an object and therefore properly measured in newtons; colloquially used incorrectly as a synonym for mass.

Mass   A measurement of the quantity of matter in an object, measured, for example, in grams and kilograms; technically not the same thing as weight.

Newton   The SI unit of force (and therefore of weight); a newton is the force required to make a mass of 1 kg accelerate at the rate of one metre per second per second; named after Sir Isaac Newton (1642–1727), an English scientist and mathematician.

Ante meridiem and post meridiem   Abbreviated to a.m. and p.m., before noon and after noon respectively.

Zero hours   Midnight on the 24-hour clock.

Conservation (in measurement) The principle that a measurement remains the same under certain transformations. For example, the length of an object is conserved when its position is altered; the volume of water is conserved when it is poured from one container to another.

Ratio scale A measuring scale in which two measurements can be meaningfully compared by ratio; for example, if the mass of one object is 30 kg and the mass of another is 10 kg then it makes sense to say that one is 3 times heavier than the other.

Interval scale A measuring scale in which two measurements can be meaningfully compared only by their difference, not by their ratio; for example, if the temperature outside is –3 °C and the temperature inside is +15 °C then the temperature difference of 18 °C is the only sensible comparison to make.

Celsius scale (°C) A metric scale for measuring temperature, also called the centigrade scale, where water freezes at 0 degrees and boils at 100 degrees under standard conditions; named after Anders Celsius (1701–44), a Swedish astronomer, physicist and mathematician, who devised the scale.

Fahrenheit scale (°F) A non-metric scale for measuring temperature, where water freezes at 32 degrees and boils at 212 degrees under standard conditions; named after the inventor, Gabriel Fahrenheit (1686–1736), a German physicist.

SI units An agreed international system of units for measurement, based on one standard unit for each aspect of measurement.

Metre (m) The SI unit of length; about the distance from my nose to my fingertip when my arm is outstretched.

Kilogram (kg) The SI unit of mass; equal to 1000 grams.

Cubic metre (m³) The SI unit of volume; the volume of a cube of side 1 metre; written 1m³ but read as 'one cubic metre'.

Litre A unit used to measure liquid volume and capacity; equal to 1000 cubic centimetres. The mass of a litre of water is 1 kilogram.

Kilo A prefix (k) denoting a thousand; for example, a kilometre (km) is 1000 metres.

Milli A prefix (m) denoting one thousandth; for example, a millilitre (ml) is one-thousandth of a litre.

Centi A prefix (c) denoting one hundredth; for example, a centilitre (cl) is one hundredth of a litre.

Centimetre (cm) One-hundredth of a metre; 100 cm = 1 m; about the width of a child's little finger.

Deci   A prefix (d) denoting one tenth; for example, a decilitre (dl) is one-tenth of a litre.

Reference item   A measurement that is memorized and used as a reference point for estimating other measurements; for example, the capacity of a wine bottle is 750 ml.

Imperial units   Units of measurement that were, at one time, statutory in the UK, most of which have now been officially replaced by metric units.

Fibonacci sequence   A sequence of numbers in which each term is obtained by the sum of the two previous terms. Starting with 1, the sequence is 1, 1, 2, 3, 5, 8, 13, 21 ...

PERIMETER, AREA AND VOLUME

- the concept of perimeter;

- area measured in square units;

- how to find the perimeter and area of rectilinear two-dimensional shapes;

- the ideas of varying the area for a fixed perimeter, and varying the perimeter for a fixed area;

- measuring the volume of cuboids in cubic units;

- how surface area might vary for shapes with the same volume;

- ways of investigating areas of parallelograms, triangles and trapeziums;

- the units used for measuring area and the relationships between them;

- the units used for measuring volume and the relationships between them;

- the number π and its relationship to the circumference and diameter of a circle.

READ THIS CHAPTER'S CURRICULUM LINKS AT: HTTPS://STUDY.SAGEPUB.COM/HAYLOCK7E

WHAT IS THE PERIMETER OF A SHAPE?

The **perimeter** is the length of the boundary around the edge of a two-dimensional shape. Because perimeter is a length, it is measured in units of length (such as centimetres or metres). The rectangle in Figure 22.1(a) has two sides of length 3 units and two sides of length 5 units. So the length of the boundary, the perimeter, is $3 + 3 + 5 + 5 = 16$ units. It is not difficult to generalize from a few examples like this and to formulate the rule that the perimeter of a rectangle with adjacent sides of length a units and b units is $2(a + b)$ units.

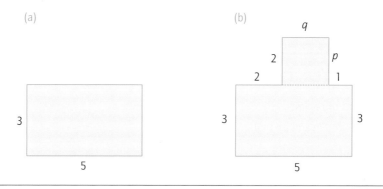

Figure 22.1 What are the perimeters of these shapes? What are their areas?

The shape in Figure 22.1(b) is an example of what is sometimes called a *rectilinear* shape. In mathematics, this word is usually applied to a shape that is bounded by straight lines that meet at right angles (see Chapter 23). This means that the shape can be subdivided into rectangles (as shown in Figure 22.1(b) by the dotted line). You can then use the fact that the opposite sides of any rectangle are equal to work out the lengths of the two sections of the perimeter that are not given, labelled p and q. In

this simple example, p is clearly equal to 2 units and q is $5 - 2 - 1 = 2$ units. The perimeter is therefore $5 + 3 + 2 + 2 + 2 + 2 + 1 + 3 = 20$ units.

WHAT IS AREA?

Area is a measure of the amount of two-dimensional space inside a boundary. To put it another way, it is a measure of the amount of two-dimensional space covered by a two-dimensional shape. We have already articulated in Chapter 10, in the context of understanding multiplication, the simple rule for finding the area of a rectangle: that you just multiply together the lengths of two adjacent sides. We then exploited this further as an image to support multiplication calculations in Chapter 12, and investigated it further in Chapter 19.

The units for measuring area are always *square units*. So, for example, the rectangle in Figure 22.1(a) has an area of 15 square units. Similarly, a rectangle 4 cm by 6 cm has an area of 24 **square centimetres**. A square centimetre is the space occupied by a square of side 1 cm. The area of 24 square centimetres is abbreviated to 24 cm^2, but should still be read as 24 *square centimetres*. If it is read as '24 centimetres squared', it could be confused with the area of a 24 cm square, which actually has an area of 576 square centimetres. The shape in Figure 22.1(b) is made up of a 5 by 3 rectangle with area 15 square units, and a 2 by 2 square with area 4 square units, giving a total area of 19 square units.

Larger areas, such as classroom floor space, would be measured in square metres. A square metre is the amount of space occupied by a square of side 1 metre. So, a classroom 5 m by 6 m would have a floor area of 30 m^2 (30 square metres). Much smaller areas can be measured in square millimetres (mm^2); for example, the area of a square of side 0.5 cm (5 mm) would be $5 \times 5 = 25$ mm^2. Larger land areas might be measured in square kilometres (km^2); for example, the area occupied by London Heathrow Airport is approximately 12 km^2. This means it is about the same area as a rectangle 4 km by 3 km.

> **LEARNING AND TEACHING POINT**
>
> Make sure that children are aware that the rule for finding the area of a rectangle applies only to rectangles. Using the phrase 'area is length times width' is sloppy and wrong. Other two-dimensional shapes have areas as well, but these are not found by applying that rule.

> **LEARNING AND TEACHING POINT**
>
> Children should learn how to find the area of a 'rectilinear' shape, by dividing it into rectangles, finding the areas of these and adding them up.

HOW DO YOU EXPLAIN THE IDEAS OF PERIMETER AND AREA SO THAT CHILDREN DO NOT GET THEM CONFUSED?

I like to use *fields* and *fences* to differentiate between area and perimeter. The area is the size of the field and the perimeter is the amount of fencing around the edge. Children can draw pictures of various fields on squared paper, such as the one shown in Figure 22.2. They can then count up the number of units of fencing around the edge, to determine the perimeter, which in this case is 18 units. Make sure that they count the units of fencing and not the squares around the edge, being especially careful going round corners not to miss out any units of fencing. To determine the area, they can fill the field with 'sheep', using unit-cubes to represent sheep; the number of sheep they can get into the field is a measure of the area. This is, of course, the same as the number of square units inside the boundary, in this case 16 square units.

LEARNING AND TEACHING POINT

To avoid confusion between area and perimeter, use the illustration of fields and fences to differentiate between these concepts and pose problems about area and perimeter in these terms. Area is the size of the field. Perimeter is the length of the fencing.

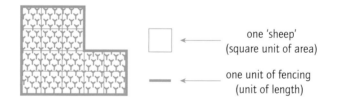

one 'sheep'
(square unit of area)

one unit of fencing
(unit of length)

Figure 22.2 Fields and fences

WHAT IS THE RELATIONSHIP BETWEEN PERIMETER AND AREA?

In general, there is no direct relationship between perimeter and area. This is something of a surprise for many people. It provides us with an interesting counter-example of the principle of conservation in measurement (see Chapter 21). When, for example, you rearrange the fencing around a field into a different shape, the perimeter is conserved, but the area is *not* conserved. For any given perimeter, there is a range of possible areas. Exploration of these ideas can provide a deep and rich mathematical experience for children, involving conjecture and generalization. Again, this is helpfully couched in terms of fields and fences.

The first challenge is to find as many different fields as possible that can be enclosed within a given amount of fencing. Figure 22.3 shows a collection of shapes, drawn on squared paper, all of which represent fields made up from rearranging 12 units of fencing. They all have different areas. An important discovery is the generalization that the largest area is obtained with a square field. This is the best use of the farmer's fencing material. (If we were not restricted to the grid lines on squared paper, the largest area would actually be provided by a circle: imagine the fencing to be totally flexible and push it out as far as you can in all directions in order to enclose the maximum area.)

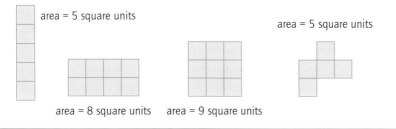

area = 5 square units

area = 5 square units

area = 8 square units   area = 9 square units

Figure 22.3 All these shapes have perimeters of 12 units

The second challenge is the reverse problem: keep the area fixed and find the different perimeters. In other words, what amounts of fencing would be required to enclose differently shaped fields all with the same area? Figure 22.4 shows a collection of fields all with the same area, 16 square units. As the 16 square units are rearranged here to make the different shapes, the area is conserved, but the perimeter is not conserved. Once again, we find the square to be the superior solution, requiring the minimum amount of fencing for the given area.

If it is not actually possible to make a square, using only the grid lines drawn on the squared paper (for example, with a fixed perimeter of 14 units, or a fixed area of 48 square units), we still find that the shape that is 'nearest' to a square

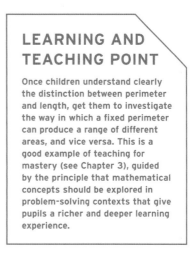

LEARNING AND TEACHING POINT

Once children understand clearly the distinction between perimeter and length, get them to investigate the way in which a fixed perimeter can produce a range of different areas, and vice versa. This is a good example of teaching for mastery (see Chapter 3), guided by the principle that mathematical concepts should be explored in problem-solving contexts that give pupils a richer and deeper learning experience.

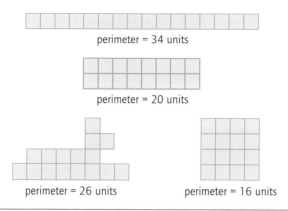

perimeter = 34 units

perimeter = 20 units

perimeter = 26 units

perimeter = 16 units

Figure 22.4 All these shapes have an area of 16 square units

gives the best solution. (The more general result, getting away from squared paper, is that the minimum perimeter for a given area is provided by a circle.) It is interesting to note that in some ancient civilizations land was priced by counting the number of paces around the boundary, that is, by the perimeter. A shrewd operator in such an arrangement could make a good profit by buying square pieces of land and selling them off in long thin strips.

WHAT ABOUT VOLUME AND SURFACE AREA?

In Chapter 21 we saw that the volume of an object is a measure of the amount of three-dimensional space it occupies. This is measured in 'cubic units'. In Figure 22.5(a), for example, the cuboid (a solid shape with six rectangular faces: see glossary for Chapter 25) illustrated has a volume of 12 cubic units, made up of 2 layers, each of which is made up of 2 rows of 3 cubic units. This illustrates how the volume of a cuboid is the product of the height, the length and the width: in this case, $2 \times 3 \times 2 = 12$ cubic units. This result generalizes to the formula 'volume of a cuboid = abc' for a cuboid with length a, width b and height c. Remember (see Chapter 19) that abc is an abbreviation used in algebra for $a \times b \times c$.

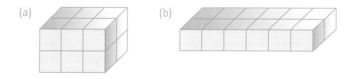

(a)   (b)

Figure 22.5 Two cuboids with the same volume, 12 cubic units

Solid volume can be measured in cubic centimetres (cm^3). One cubic centimetre ($1\ cm^3$) is the volume of a cube of side 1 cm – which is the standard size for the units (ones) in base-ten materials (see Chapter 6). Larger volumes, such as the volume of the classroom, might be measured in cubic metres (m^3). For example, a room 5 m long, 4 m wide and 2.5 m high has a volume of $5 \times 4 \times 2.5 = 50\ m^3$ (50 cubic metres). These ideas extend naturally to much smaller volumes measured in cubic millimetres (mm^3). For example, a cube of side 0.5 cm (5 mm) has a volume of $5 \times 5 \times 5 = 125\ mm^3$. Finding the volumes of solids other than a cuboid or a shape constructed from cuboids, where children can count the individual unit cubes, is beyond the curriculum for this age range.

The **surface area** of a solid object is the sum of the areas of all its surfaces, measured in square units. The cuboid in Figure 22.5(a) has four surfaces with areas of 6 square units and two with areas of 4 square units, giving a total surface area of 32 square units. Figure 22.5(b) illustrates another cuboid with the same volume, 12 cubic units. The two cuboids in Figure 22.5 can be made by arranging 12 unit cubes in different ways. But notice that although the volume is conserved when you do this, surprisingly, the surface area is

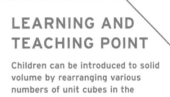

LEARNING AND TEACHING POINT

Children can be introduced to solid volume by rearranging various numbers of unit cubes in the shape of cuboids; use numbers with plenty of factors, such as 12 and 24. Older primary children could investigate the way the surface area changes.

not. In cuboid (a) we see that the total surface area is 32 square units, but in cuboid (b) it is 40 square units. This means that to cover (b) with paper you would need 40 square units of paper, but to cover (a) you would need only 32 square units. The fact that rearranging the volume changes the surface area actually explains why you sometimes need less wrapping paper if you arrange the contents of your parcel in a different way. The closer you get to a cube (or more generally to a sphere), the smaller the surface area.

HOW CAN YOU FIND AREAS OF SHAPES OTHER THAN RECTANGLES?

The area of a right-angled triangle (see Chapter 25) is easily found, because it can be thought of as half of a rectangle, as shown in Figure 22.6. In this example, the area of the rectangle is 24 square units, so the area of the triangle is 12 square units.

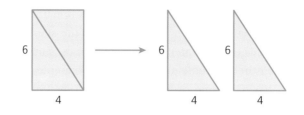

Figure 22.6   A right-angled triangle is half of a rectangle

Older children in some primary schools might explore the areas of some other geometric shapes, so their teachers should be confident with the following material. For example, the area of a parallelogram (see Chapter 25) is found by multiplying its height by the length of its base (any one of the sides can be called the base). Figure 22.7 shows how this can be demonstrated rather nicely, by transforming the parallelogram into a rectangle with the same height and the same base. For example, if the parallelogram has height 4 cm and base 3 cm, it has an area of 12 cm².

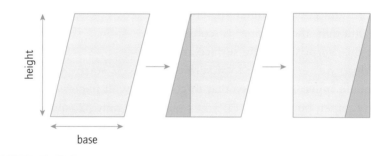

Figure 22.7   A parallelogram transformed into a rectangle

This then gives us a way of finding the area of any triangle. Just as any right-angled triangle can be thought of as half of a rectangle (Figure 22.6), so can any triangle be thought of as half of a parallelogram, as shown in Figure 22.8. If you make a copy of the triangle, and rotate it through 180°, the two triangles can be fitted together to form a parallelogram, with the same base and the same height as the original triangle. Since the area of the parallelogram is the base multiplied by the height, the area of the triangle is half the product of its base and height. For example, if the triangle has a height of 4 cm and a base of 3 cm, it has an area of 6 cm². Again, note that any one of the sides of the triangle can be taken as the base.

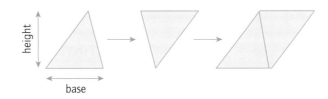

Figure 22.8 A triangle seen as half a parallelogram

Figure 22.9 shows a quadrilateral (see glossary for Chapter 23) with just one pair of sides that are parallel (see Chapter 25). In the UK, this is called a **trapezium**. This particular trapezium has a height of 3 cm and the parallel sides have lengths of 4 cm and 6 cm respectively. The area of a trapezium can be found by cutting it up into a parallelogram and a triangle, as shown in Figure 22.9. In this case, we produce a parallelogram with height

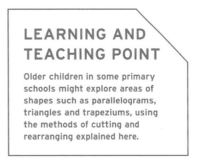

3 cm and base 4 cm (area 12 cm²), and a triangle of height 3 cm and base 2 cm (area 3 cm²). So, the area of this trapezium is 12 cm² + 3 cm² = 15 cm².

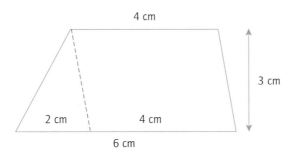

Figure 22.9 A trapezium transformed into a triangle and a parallelogram

WHAT SHOULD I KNOW ABOUT THE RELATIONSHIPS BETWEEN UNITS OF AREA?

Students often get confused when changing between areas measured in square centimetres and **square metres**. This is because they fail to recognize that there are actually 10,000 cm² in 1 m². That does seem an awful lot, doesn't it? But there are 100 centimetres in 1 metre, and a square metre can be made from 100 rows of 100 centimetre squares, each with an area of 1 cm². Imagine using four metre rulers to make a square metre on the floor. You really would need 10 thousand (100 × 100) centimetre-square tiles to fill this area. So, an area of 1 square metre (1 m²) is 10,000 cm². And an area of 1 square centimetre (1 cm²) is 0.0001 m². Thus, converting areas between cm² and m² involves shifting the digits four places in relation to the decimal point. For example, 12 cm² = 0.0012 m², and 0.5 m² = 5000 cm².

So, for example, what would be the area of a square of side 50 cm, expressed in square centimetres and in square metres? Using centimetres, the area is 50 × 50 = 2500 cm². Using metres, the area is 0.5 × 0.5 = 0.25 m². The reader should not be surprised at this result, because they should now recognize that 2500 cm² and 0.25 m² are the same area. We might also note that a square of side 50 cm would be a quarter of a metre square, and $^1/_4$ expressed as a decimal is 0.25.

The potential for bewilderment is even greater in converting between **square millimetres** and square metres, since there are a million square millimetres (1000 × 1000) in a square metre.

WHAT SHOULD I KNOW ABOUT THE RELATIONSHIPS BETWEEN UNITS OF VOLUME?

In considering the cuboids in Figure 22.5, we noted that the volume of a cuboid, measured in cubic units, is found by multiplying together the length, width and height. Now, a metre cube is 100 layers of 100 rows of 100 centimetre cubes. So 1 m³ (a cubic metre) must be equal to 100 × 100 × 100 = 1,000,000 cm³ (a million **cubic centimetres**). Thus, converting volumes between cm³ and m³ involves shifting the digits six places in relation to the decimal point. For example, 12 cm³ = 0.000012 m³, and 0.5 m³ = 500,000 cm³. If all this leads the reader to feel the need to brush up on their calculations with decimals, they should now return to Chapter 17.

WHAT IS Π?

A **circle** is the shape consisting of all the points at a fixed distance from a given point. The given point is called the *centre*. A line from the centre to any point on the circle is called a **radius**. A line from one point through the centre to the opposite point on the circle is called a **diameter**. The perimeter of a circle is also called the **circumference** (see Figure 22.10).

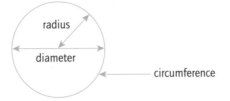

Figure 22.10   Terms used with circles

A point about terminology should be made here. Strictly, the word 'circle' refers to just the continuous line made up of the set of points lying on the circumference. But, in practice, we use the word loosely to refer to both this continuous line and the space it encloses. This enables us to talk about 'the area of a circle', meaning strictly 'the area enclosed by a circle'. Also note that we often use the words *radius*, *diameter* and *circumference* to mean not the actual lines but the lengths of these lines. So, for example, we could say, 'the radius is half the diameter', meaning 'the length of the radius is half the length of the diameter'.

One of the most amazing facts in all mathematics relates to circles. If you measure the circumference of a circle and divide it by the length of the diameter, you always get the same answer. This can be done in the classroom practically, using lots of differently sized circular objects, such as tin lids and crockery. To measure the diameter, place the circular object on a piece of paper between two blocks of wood, remove the object, mark the edges of the blocks on the paper and measure the distance between the marks. To measure the circumference, wrap some tape carefully around the object, mark where a complete circuit begins and ends, unwind the tape and measure the distance between the marks. Allowing for experimental error (which can be considerable with such crude approaches to measuring the diameter and the circumference), you should still find that in most cases the circumference divided by the diameter gives an answer of between about 3 and 3.3. Any results that are way out should be checked and measured again, if necessary.

This result, that the circumference is always about three times the diameter, was clearly known in ancient civilizations and used in construction. It occurs in a number of places in the Bible; for example, in I Kings 7.23, we read of a construction, 'circular in shape, measuring ten cubits from rim to rim … taking a line of thirty cubits to measure around it'. If we were able to measure more accurately, we might be able to determine that the ratio of the circumference to the diameter of a circle is about 3.14. This ratio is so important it is given a special symbol, the Greek letter π **(pi)**. The value of π can be found theoretically to any number of decimal places. It begins like this: 3.14159265358979323846264$3\ldots$ and actually goes on forever, without ever recurring (see Chapters 13 and 17). For practical purposes, rounding this to two decimal places, as 3.14, is sufficient. Because it cannot be written down as an exact or recurring decimal, π is an example of an **irrational number** (see Chapter 6).

Once we have this value, we can find the circumference of a circle, given the diameter, by multiplying the diameter by π, using a calculator if necessary. For example, a circle with radius 5 cm has a diameter of 10 cm and therefore a circumference of approximately $10 \times 3.14 = 31.4$ cm. And we can find the diameter of a circle, given the circumference, by dividing the circumference by π. For example, a metre trundle-wheel is a circle with a circumference of 100 cm, so the diameter will have to be approximately $100 \div 3.14 = 31.8$ cm. So the radius will have to be half this, namely 15.9 cm.

There is a common misunderstanding, perpetrated, I fear, by some mathematics teachers who should know better, that π is *equal* to $^{22}/_7$ or $3^1/_7$. This is not true. Three-and-one-seventh is an *approximation* for the value of π in the form of a rational number. As a decimal, $3^1/_7$ is equal to $3.142857142857\ldots$, with the 142857 recurring forever. Comparing this with the value of π given above, we can see that it is only correct to two decimal places anyway, and therefore no better an approximation than 3.14. In the days of calculators, 3.14 is bound to be a more useful approximation for π than $^{22}/_7$, which is probably best forgotten.

RESEARCH FOCUS: TWO-DIMENSIONAL SPACE CONCEPTS

Sarama et al. (2003) investigated the development of two-dimensional space concepts within the teaching of a mathematics unit on two-dimensional grids and coordinates to children aged 8 to 9 years. Data from case studies, interviews, paper-and-pencil tests and whole-class observations revealed that learners had to overcome substantial hurdles in learning to structure spatially two-dimensional grids, including 'interpreting the grid's components as line segments rather than regions; appreciating the precision of location of the lines required, rather than treating them as fuzzy boundaries or indicators of intervals; and learning to trace vertical or horizontal lines that were not axes' (Sarama et al., 2003: 285).

To measure space and shape in two dimensions, children have to co-ordinate their numerical and spatial knowledge. Children often confuse area and perimeter, because they are still wedded to a model of measuring length based on the idea of counting discrete items, like squares. This is the basis of the error that occurs when, given a rectangle drawn on squared paper, children count the squares along the inside edges of the perimeter. To grasp the concept of perimeter, children need an understanding of length as a one-dimensional continuous quantity, and then to be able to extend this from one dimension (the length of a line) to two dimensions (the length of a path). Barrett and Clements (1998) reported how one nine-year-old child in a teaching experiment restructured his strategic knowledge of length to incorporate measures of perimeter. The breakthrough occurred through his response to a problem about putting a fringe on a carpet placed on a floor covered in square tiles (compare putting a fence round a field). This is consistent with the findings of Sarama et al. (2003) that real-world contexts serve an important scaffolding role for most children at the early phases of learning – although they also found indications that these can impede further mathematical abstraction in some learners at later phases. These researchers also report that computer representations are significant in aiding the learning of both grid and coordinate systems.

LEARNING RESOURCES

Access activities for your **lesson plans** at: https://study.sagepub.com/haylock7e

Before trying the self-assessment questions below, you should complete the **self-assessment questions** for this chapter at: https://study.sagepub.com/haylock7e

22.1: Using 36 centimetre-squares on squared paper to make a rectangle, what is the largest perimeter you can get?

22.2: How can (a) 27 and (b) 48 unit-cubes be arranged in the shape of a cuboid to produce the minimum surface area?

22.3: What is the area of a square of side 5 mm, expressed in square millimetres, in square centimetres and in square metres?

22.4: What would be the volume of a cube of side 5 cm? Give your answer in both cm^3 and m^3.

22.5: How many cuboids 5 cm by 4 cm by 10 cm would be needed to build a metre cube?

22.6: How much ribbon will I need to go once round a circular cake with a diameter of 25 cm?

22.7: Roughly, what is the diameter of a circular running track which is 400 metres in circumference?

FURTHER PRACTICE

Access the website material for **Knowledge check 34: Knowledge of metric units of area and solid volume** at: https://study.sagepub.com/haylock7e

FROM THE STUDENT WORKBOOK

Questions 22.01–22.17: Checking understanding (perimeter, area and volume)

Questions 22.18–22.29: Reasoning and problem solving (perimeter, area and volume)

Questions 22.30–22.40: Learning and teaching (perimeter, area and volume)

GLOSSARY OF KEY TERMS INTRODUCED IN CHAPTER 22

Perimeter   The total length all the way round a boundary enclosing an area; like the length of fencing enclosing a field.

Area   The amount of two-dimensional space enclosed by a boundary; like the size of a field enclosed by a fence.

Square centimetre (cm^2)   The area of a square of side 1 centimetre; written 1 cm^2 but read as 'one square centimetre'. There are ten thousand square centimetres in a square metre.

Surface area   The sum of the areas of all the surfaces of a solid object.

Trapezium   A quadrilateral (four-sided shape) with two sides parallel.

Square metre (m^2)   The SI unit of area; the area of a square of side 1 metre; written 1 m^2 but read as 'one square metre'.

Square millimetre (mm^2)   The area of a square of side 1 millimetre; written 1 mm^2 but read as 'one square millimetre'. There are a million square millimetres in a square metre.

Cubic centimetre (cm^3)   The volume of a cube of side 1 centimetre; written 1 cm^3 but read as 'one cubic centimetre'.

Circle   A two-dimensional shape consisting of all the points that are a given distance from a fixed point, called the centre of the circle.

Radius   A line from the centre of a circle to any point on the circle; also the length of such a line; half the diameter.

Diameter   A line from any point on a circle, passing through the centre to the point opposite; also the length of such a line; twice the radius.

Circumference   The perimeter of a circle.

Pi (π)   A number equal to the ratio of the circumference of any circle to its diameter; about 3.14.

SUGGESTIONS FOR FURTHER READING FOR SECTION F

1   For those who teach younger children, there is a good chapter (Chapter 7) on learning about measurement through play in Tucker (2014).

2   Chapter 7 of Hansen (2020) provides an interesting analysis of children's errors and misconceptions in measurement.

3   Fenna (2002) is a comprehensive and authoritative dictionary, providing clear definitions of units, prefixes and styles of weights and measures within the Système International (SI), as well as traditional and industry-specific units. It also includes fascinating material on the historical and scientific background of units of measurement.

4   Chapter 8 of Haylock and Cockburn (2017) is on understanding measurement. We cover in this chapter the material on measurement, particularly from the perspective of teaching the topic to young children. The chapter concludes with a number of suggestions for classroom activities designed to promote understanding of the key ideas involved.

5   Have a look at Chapter 12 on area and Chapter 13 on capacity and volume in Suggate, Davis and Goulding (2017). These chapters explore thoroughly the ideas of area and volume, with some interesting problems to deepen the reader's grasp of these concepts and their applications.

6   A useful 'teaching for mastery' booklet on measurement of length, weight, area and volume has been produced by the TES Resource Team (2016), in partnership with Mathematics Mastery and The White Rose Hub.

7   Some excellent ideas for classroom measurement activities specifically related to early years mathematics are provided by NCETM (2019).

SECTION G

GEOMETRY

23   Angle   411
24   Transformations and Symmetry   423
25   Classifying Shapes   437

WATCH THE SECTION OPENER VIDEO AT: HTTPS://STUDY.SAGEPUB.COM/HAYLOCK7E

ANGLE

IN THIS CHAPTER, THERE ARE EXPLANATIONS OF

- the dynamic and static views of angle;
- comparison and ordering of angles;
- the use of turns and fractions of a turn for measuring angle;
- degrees;
- acute, right, obtuse, straight, reflex angles;
- some basic properties of angles;
- the sum of the angles in a triangle, a quadrilateral and so on.
- internal and external angles of a triangle

READ THIS CHAPTER'S CURRICULUM LINKS AT: HTTPS://STUDY.SAGEPUB.COM/HAYLOCK7E

WHAT IS THE DYNAMIC VIEW OF ANGLE?

An **angle** is a measurement. When we talk about the angle between two lines, we are not referring to the shape formed by the two lines, nor to the point where they meet, nor to the space between the lines, but to a particular kind of measurement.

There are two ways we can think of what it is that is being measured. First, there is the *dynamic* view of angle: the angle between the lines is a measurement of the size of the *rotation* involved when you point along one line and then turn to point along the other. This is the most useful way of introducing the concept of angle to children, because it lends itself to practical experience, with the children themselves pointing in one direction and then turning themselves through various angles to point in other directions. Also, when it comes to measuring angles between lines drawn on paper, children can physically point something, such as a pencil or a finger, along one line and rotate it about the intersection of the lines to point along the other line.

We should note that there are always two angles involved when turning from one direction to another, depending on whether you choose to rotate clockwise or anticlockwise, as shown in Figure 23.1.

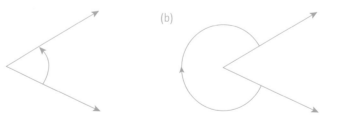

(a)   (b)

anticlockwise rotation   clockwise rotation

Figure 23.1   Angle as a measure of rotation

WHAT IS THE STATIC VIEW OF ANGLE?

As well as the idea of an angle as a rotation, there is also the *static* view of angle. This is where we focus our attention on how pointed is the shape formed by the two lines.

But the angle is still a measurement: we can think of it as a measurement of the difference in direction between the two lines. So, for example, in Figure 23.2, the angle marked in (a) is greater than that marked in (b). This is because the two lines in (b) are pointing in nearly the same direction, whereas the difference in the direction of the two lines in (a) is much greater. Thinking of angle as the difference in direction between two lines helps to link the static view of angle with the dynamic one of rotation, because the obvious way to measure the difference in two directions is to turn from one to the other and measure the amount of turn. Note that the convention used to indicate the angle between two lines is a small arc joining the lines, close to the point at which they meet, as shown in Figure 23.2.

Figure 23.2 Angle as a measure of the difference in direction

HOW ARE ANGLES MEASURED?

Figure 23.2 illustrates that, like any aspect of measurement, the concept of angle enables us to make comparisons and to order (see Chapter 21). This can be dynamically, by physically doing the rotations involved (for example, with a pencil) and judging which is the greater rotation, which the smaller. It can also be experienced more statically, by cutting out one angle and placing it over another to determine which is the more pointed (the smaller angle). Figure 23.3 shows a set of angles, (a)–(f), ordered from smallest to largest.

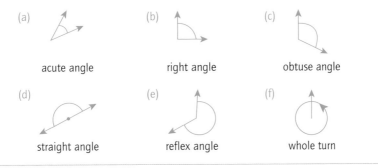

Figure 23.3 A set of angles in order from smallest (a) to largest (f)

The equivalent to making measurements of length, mass and capacity in non-standard units is to measure angle in turns and fractions of a turn, using the dynamic view. Figure 23.3(f) shows a whole turn, pointing in one direction and rotating until you point again in the same direction. Then, for example, if a child points north and then turns to point south, they have moved through an angle that can be called a half-turn. Similarly, rotating from north, clockwise, to east is a quarter-turn. Figure 23.3(b) shows a quarter-turn from a horizontal position to a vertical position. This explains why a quarter-turn is also called a **right angle**: it is an *upright* angle. Figure 23.3(d) illustrates why a half-turn, formed by two lines pointing in opposite directions, is also called a **straight angle**.

The next stage of development is to introduce a standard unit for measuring angles. For primary school use, this unit is the **degree**, where 360 degrees (360°) is equal to a complete turn. Hence, a right angle (quarter-turn) is 90° and a straight angle (half-turn) is 180°. There is evidence that this system of measuring angles in degrees based on 360 was used as far back as 2000 BC by the Babylonians, and it is thought that it may be related to the Babylonian year being 360 days.

So, we have here another example of a non-metric measurement scale in common use. There is actually a metric system for measuring angle, used in some European countries, in which 100 'grades' (pronounced as in French) is equal to a right angle, but this has never caught on in the UK. Interestingly, it is probably because 'centigrade' would then be one-hundredth of a grade, and therefore a measure of angle, that the 'degree centigrade' as a measure of temperature is officially called the 'degree Celsius', to avoid confusion.

The device used to measure angles in degrees is the **protractor**. I like to give children the opportunity to use a 360° protractor, preferably marked with only one scale and with a pointer which can be rotated from one line of the angle being measured to the other, thus emphasizing the dynamic view of angle. Even if there is not an actual pointer, children can still be encouraged to imagine the rotation always starting at zero on one line and rotating through 10°, 20°, 30° ... to reach the other.

CAN YOU REMIND ME ABOUT ACUTE, OBTUSE AND REFLEX ANGLES?

Mathematicians can never resist the temptation to put things in categories, thus giving them the opportunity to invent a new collection of terms. Angles are classified, in order of size, as: acute, right, obtuse, straight, reflex, as illustrated in Figure 23.3.

An **acute angle** is an angle less than a right angle.

An **obtuse angle** is an angle between a right angle and a straight angle.

A **reflex angle** is an angle greater than a straight angle, but less than a whole turn.

WHAT ELSE IS THERE TO KNOW ABOUT ANGLES?

Figure 23.4 shows some basic properties of angles that children may learn about in primary school and use to calculate angles that they do not know from those they do. First, Figure 23.4(a) shows the two angles that arise when one line meets another line. These two angles, labelled here as p and q, together make a straight angle. So $p + q = 180°$; or $q = 180° - p$. So, for example, if we know that p is equal to 70° it follows that q must equal 110°. Two angles that sum to 180° are known as *supplementary angles*.

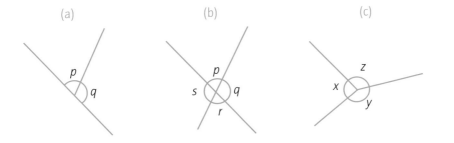

Figure 23.4   (a) p + q = 180° (b) q = s and p = r (c) x + y + z = 360°

Second, in Figure 23.4(b) there are two lines intersecting each other. This produces four angles meeting at the point of intersection (sometimes called a **vertex**). We already know from (a) that $p + q = 180°$. But p and s together also make a straight angle, so $p + s = 180°$ as well. It follows that q must equal s. By a similar argument, p must equal r. The general result here is that when two lines intersect, the opposite angles are equal. These are usually called *vertically opposite angles*; 'vertically' here refers to the fact that the two opposite angles are either side of a common vertex.

Third, Figure 23.4(c) shows three lines meeting at a point; this results in three angles around the point, labelled here as x, y and z. Imagine standing at the point where the lines meet and facing along one of the lines; then turn clockwise through the three angles in turn. Clearly, you finish up facing in the same direction as you did to start with. You have rotated through one whole turn, namely 360°. Hence, we deduce that $x + y + z = 360°$. This would, of course, be true for any number of lines and angles meeting at a point: that the angles at the point sum to 360°.

HOW CAN YOU SHOW THAT THE THREE ANGLES IN A TRIANGLE ADD UP TO 180 DEGREES?

A **triangle** is a two-dimensional shape with three sides and three vertices (plural of vertex). The three angles in a triangle always add up to 180 degrees. A popular way of seeing this property is to draw a triangle on paper, mark the angles, tear off the three corners and fit them together, as shown in Figure 23.5, to discover that together they form a straight angle, or two right angles (180°).

Figure 23.5   The three angles of a triangle make a straight angle

This illuminative experience uses the static view of angle. It is also possible to illustrate the same principle using the dynamic view, as shown in Figure 23.6, by taking an arrow (or, say, a pencil) for a walk round a triangle. Step 1 is to place an arrow (or the pencil) along one side of the triangle, for example on AC. Step 2 is to slide the arrow along until it reaches the vertex A; then rotate it through the angle at A. It will now lie along AB,

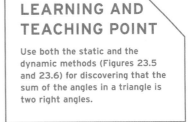

LEARNING AND TEACHING POINT

Use both the static and the dynamic methods (Figures 23.5 and 23.6) for discovering that the sum of the angles in a triangle is two right angles.

pointing towards A. Now, for step 3, slide it up to the vertex B and rotate through that angle. Finally, step 4, slide it down to the vertex C and rotate through that angle. The arrow has now rotated through the sum of the three angles and is facing in the opposite direction to which it started. Hence, the three angles together make a half-turn, or two right angles. Clearly, this will work for any triangle, not just the one shown here.

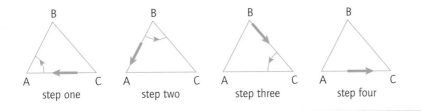

Figure 23.6   The three angles of a triangle together make a half-turn

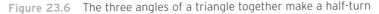

WHAT ABOUT THE SUM OF THE ANGLES IN SHAPES WITH MORE THAN THREE SIDES?

This is going beyond primary school mathematics, but primary teachers should know where the mathematics is heading. The four corners of any **quadrilateral** (a plane shape with four straight sides and four vertices) can be torn off and fitted together in the same way as was done with a triangle in Figure 23.5. It is pleasing to discover this way that they always fit together to make a whole turn, or four right angles (360°).

But, also, the procedure used in Figure 23.6 can be applied to a four-sided figure, such as that shown in Figure 23.7. Now we find that the arrow does a complete rotation, finishing up pointing in the same direction as it started, so we conclude that the sum of the four angles in a four-sided figure is four right angles.

WATCH THE PROBLEM SOLVED! VIDEO AT: HTTPS://STUDY.SAGEPUB.COM/HAYLOCK7E

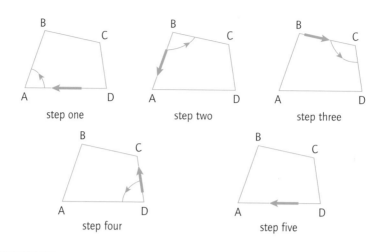

B   C

A   D

step one

B   C

A   D

step two

B   C

A   D

step three

B   C

A   D

step four

B   C

A   D

step five

Figure 23.7 The four angles of a quadrilateral together make a whole turn

The delight of this experience of taking an arrow for a walk round a triangle or quadrilateral is that it can easily be extended to five-sided shapes, six-sided, seven-sided and so on. The results can then be tabulated, using the approach given in Chapter 19, to formulate a sequential generalization and a global generalization. This is left as an exercise for the reader in self-assessment question 23.3 below.

WHAT ARE EXTERNAL ANGLES IN A TRIANGLE?

So far in this chapter, such as in Figure 23.5, when I have referred to the 'angles of a triangle' I have actually been talking about the three *internal* angles. Figure 23.8 shows what are called the three *external* angles of a triangle. In Figure 23.8(a), if you continue the line XY a little beyond Y, as shown, then the external angle at Y is the angle between the extension of XY and the line YZ. The first thing to note is that the internal and external angles at any vertex add up to 180°, because together they make a straight angle. In Figure 23.8, if the internal angle at X is 60° then the external angle will be 120°. Figure 23.8(b) shows the three external angles if you continue the lines beyond the vertices in the opposite directions to those used in Figure 23.8(a). Of course, the two ways of showing the external angle at any vertex produce the same angle: it is just the change in direction at a vertex as you travel clockwise or anticlockwise around the triangle.

If, for example, children are programming robotic devices (such as a Bee-Bot) around a triangle, either on the floor or on a computer, it is the external angle that the device has to turn through at each vertex, not the internal angle. If the device is travelling around the triangle shown in Figure 23.8(a), starting at Z and heading towards X, when it gets to X it has to turn anticlockwise through 120° (not 60°) to continue its journey along XY. (I am assuming here that the device moves *forward* along each side of the triangle, unlike the pencil in Figure 23.6, which travels backwards along one side.) When the device has travelled round the triangle and turned through the three external angles shown it has done one complete rotation through 360°. So the sum of the three external angles of a triangle must be 360°. Self-Assessment Question 23.4 at the end of this chapter invites you to consider the sum of the external angles of quadrilaterals.

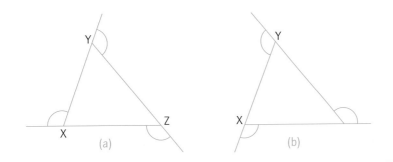

Figure 23.8 External angles of a triangle

RESEARCH FOCUS: THE CHALLENGING CONCEPT OF ANGLE

Developing a sound concept of angle presents serious challenges for the learner, because so many diverse situations have to be connected together under this umbrella term. Davey and Pegg (1991) report that when children are asked to define an angle, their responses tend to be of four kinds: (a) a corner which is pointy or sharp; (b) a place where two lines meet; (c) the distance or area between two lines; and (d) the difference between the slope of the two lines. Only the last of these is anywhere near an appropriate (static) concept of angle. Mitchelmore (1998) found that less than 10% of 8–9-year-olds mentioned turning when asked to give examples of angles. Children also do not easily associate a bend in a path with the idea of turning, and they find it difficult to perceive a slope as an angle. In a later study, Mitchelmore and White (2000) analysed the range of situations that children – not surprisingly – find difficulty in connecting together under this concept. These include: (1) intersections (road junctions, scissors, hands of a clock); (2) corners (on a table top, a pencil point); (3) bends (on a road, an arm or a leg); (4) slopes (ramp, railway signal, roof, hillside); (5) limited rotation (door, windscreen wiper, tap); (6) unlimited rotation (ceiling fan, revolving door). Their research indicates that children need extensive experience within each of these aspects and the opportunity to recognize deeper and deeper similarities between them, so that their conceptual understanding can progress from the specific to the general and then to the abstract. Smith, King and Hoyte (2014) explored how children's skills with angles could be developed through practical experience of body movement. Using a Kinect Camera children created angles using their arms and received feedback. Following some instruction and guidance there were significant gains in the children's ability to estimate and draw angles.

LEARNING RESOURCES

Access activities for your **lesson plans** at: https://study.sagepub.com/haylock7e

Before trying the self-assessment questions below, you should complete the **self-assessment questions** for this chapter at: https://study.sagepub.com/haylock7e

23.1: (a) Through what angle would you turn if you faced west and then turned clockwise to face north-east? (b) What if you turned anticlockwise? (c) Approximately, what is the largest angle you can make between your index finger and your middle finger?

23.2: Put these angles in order and classify them as acute, right, obtuse, straight or reflex: 89°, $\frac{1}{8}$ of a turn, 150°, 90°, $\frac{3}{4}$ of a turn, 200°, 2 right angles, 95°.

23.3: Use the idea of taking an arrow round a shape, rotating through each of the internal angles in turn, to find the sum of the angles in: (a) a five-sided two-dimensional shape (a pentagon); (b) a six-sided two-dimensional shape (a hexagon); and (c) a seven-sided two-dimensional shape (a heptagon). Give the answers in right angles, tabulate them and formulate both the sequential and the global generalizations. What would be the sum of the angles in a two-dimensional shape with 100 sides?

23.4: Draw any quadrilateral with vertices A, B, C, D and continue the sides to show four external angles (as shown for the triangle in Figure 23.8(a)). Imagine a device being programmed to travel forwards around the quadrilateral. When it has done this and turned through all four external angles what will be the total angle it has turned through? What can you deduce about the sum of the four external angles of a quadrilateral?

FURTHER PRACTICE

FROM THE STUDENT WORKBOOK

Questions 23.01–23.18: Checking understanding (angle)

Questions 23.19–23.27: Reasoning and problem solving (angle)

Questions 23.28–23.34: Learning and teaching (angle)

GLOSSARY OF KEY TERMS INTRODUCED IN CHAPTER 23

Angle   Dynamically, a measure of the amount of turn (rotation) from one direction to another; statically, the difference in direction between two lines meeting at a point.

Right angle   An upright angle, a quarter-turn, 90°.

Straight angle   A half-turn, 180°.

Degree   A measure of angle; 360 degrees (360°) is a complete turn.

Protractor   A device for measuring angles.

Acute angle   An angle between 0° and 90°.

Obtuse angle   An angle between 90° and 180°.

Reflex angle   An angle between 180° and 360°.

Vertex (plural: vertices)   The point where two lines meet at an angle; in a plane geometric shape with straight sides, a point where two sides meet; similarly, for a three-dimensional shape, a point where three or more edges meet.

Triangle   A plane shape with three straight sides, three vertices and three interior angles. The three angles of any triangle add up to 180°.

Quadrilateral   A plane shape with four straight sides, four vertices and four interior angles. The four angles of any quadrilateral add up to 360°.

TRANSFORMATIONS AND SYMMETRY

IN THIS CHAPTER, THERE ARE EXPLANATIONS OF

- transformation, equivalence and congruence in the context of shape;
- translation, reflection and rotation as types of congruence;
- scaling up and down by a scale factor in the context of shape;
- similar shapes;
- reflective and rotational symmetry for two-dimensional shapes.

READ THIS CHAPTER'S CURRICULUM LINKS AT: HTTPS://STUDY.SAGEPUB.COM/HAYLOCK7E

HOW ARE THE IDEAS OF TRANSFORMATION AND EQUIVALENCE IMPORTANT IN UNDERSTANDING SHAPE AND SPACE?

The two basic processes in geometry are (a) moving or changing shapes, and (b) classifying shapes. These involve the fundamental concepts of transformation and equivalence. In Chapter 3 we saw that these two concepts are key processes in understanding mathematics in general. In this chapter, we shall see how important they are in terms of understanding shapes in particular. The reader may recall from Chapter 3 (see, for example, Figure 3.2) that there are two fundamental questions when considering the relationship between two mathematical entities: How are they the same? How are they different? The first question directs our attention to an *equivalence*, the second to a *transformation*. Much of what we have to understand in learning geometry comes down to recognizing the equivalences that exist within various transformations of shapes, which transformations preserve which equivalences, and how shapes can be different yet the same.

For example, if I draw a large rectangle on the board and ask the class to copy it onto their paper, they all dutifully do this, even though none of them has a piece of paper anything like large enough to produce a diagram as big as mine. Their rectangles are different from mine, that is, different in size, whilst in many respects they might be the same as mine. They have transformed my diagram,

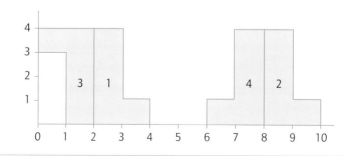

Figure 24.1   The same but different

but produced something, which, I hope, is in some way equivalent. The diagrams are the same, but different.

In a mathematics lesson, Cathy was asked to draw on squared paper as many different shapes as possible that are made up of five square units. Figure 24.1 shows four of the shapes that she drew. Her teacher told her that they were all the same shape. Cathy insisted they were all different. Who was right? The answer is, of course, that they are both right. The shapes are all the same in some senses and different in others.

WHAT IS A CONGRUENCE?

Consider how the shapes in Figure 24.1 are the same. All four shapes are made up of five square units; they have the same number of sides (six); each shape has one side of length four units, one side of length three units, one side of length two units and three sides of length one unit; each shape has five 90° angles and one 270° angle; and the sides and the angles are arranged in the same way in each of the shapes to make what we might recognize as a letter 'L'. Surely they are identical – the same in every respect?

Well, they certainly are **congruent shapes**. This word, congruent, describes the relationship between two shapes that have sides of exactly the same length, angles of exactly the same size, with all the sides and angles arranged in exactly the same way, as in the four shapes shown in Figure 24.1. A practical definition would be that you could cut out one shape and fit it exactly over the other one. The pages in this book are congruent: as you can see, one page fits exactly over the next.

A transformation of a shape that changes it into a congruent shape is called a **congruence**. Three types of congruence are explained below: translation, rotation and reflection.

WHAT IS A TRANSLATION?

So, how are the shapes in Figure 24.1 different? Figure 24.2 shows just the two shapes, 1 and 2. How is shape 2 different from shape 1? Cathy's argument is that they are in different positions on the paper: shape 1 is here and shape 2 is over there, so they are not the same shape. Surely every time I draw the shape in a different position I have drawn a different shape, in a sense. So, the shapes are different if you decide to take *position* into account.

In order to do this, we need a system of coordinates, as outlined in Chapter 20. The transformation that has been applied to shape 1 to produce shape 2 is called a **translation**. We saw in Chapter 20 that we can use the coordinate system to describe

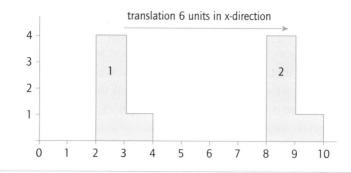

Figure 24.2   A translation

movements. So, this translation can be specified by saying, for example, that to get from shape 1 to shape 2 you have to move 6 units in the *x*-direction and 0 units in the *y*-direction. Any movement of our shape like this, so many units in the *x*-direction and so many in the *y*-direction, without turning, is a translation. Note that every point in shape 1 moves the same distance in the same direction to get to the corresponding point in shape 2.

WHAT IS A ROTATION?

If we now decide that, for the time being, we will not count translations as producing shapes which are different, what about shape 3? Is that different from shape 1? These two shapes are shown in Figure 24.3. Cathy's idea is that shape 1 is an L-shape the right way up, but shape 3 is upside down, so they are different. So, the shapes are different if you decide to take their orientation on the page into account. In order to take this into account, we need the concept of direction, and, as we saw in Chapter 23, to describe a difference in direction we need the concept of *angle*. So, to transform shape 1 into shape 3 we can apply a **rotation**.

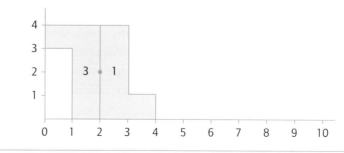

Figure 24.3   A rotation

To specify a rotation, we have to indicate the point about which the rotation occurs and the angle through which the shape is rotated. If shape 1 is rotated through an angle of 180° (either clockwise or anticlockwise), about the middle of its left-hand side – the point with coordinates (2, 2) – then it is transformed into shape 3. Imagine copying shape 1 on to tracing paper, placing a pin in the point (2, 2) and rotating the tracing paper through 180°. The shape would land directly on top of shape 3. Any movement of our shape like this, turning through some angle about a given centre, is a rotation. Note that every line in shape 1 is rotated through this angle of 180° to get to the corresponding line in shape 3.

WHAT IS A REFLECTION?

Now, if we decide that, for the time being, we will not count rotations or translations as producing shapes which are different, what about shape 4? How is that different from shape 1? These two shapes are shown in Figure 24.4. Cathy's idea is that they are actually mirror images of each other and this makes them different. This difference can be made explicit by colouring the shapes, say, red, before cutting them out. If we did that with the shapes in Figure 24.1, we would find that shapes 1, 2 and 3 can all be placed on top of each other and match exactly, with the red faces uppermost. But shape 4 only matches if we turn it over so that it is red face down. This surely makes it different from all the others? The transformation that has been applied now to shape 1 to produce shape 4 is a **reflection**.

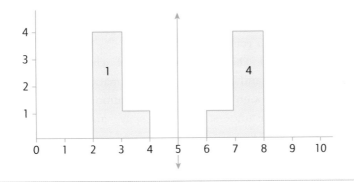

Figure 24.4   A reflection

To specify a reflection, all you have to do is to identify the mirror line. In the case of shapes 1 and 4, the mirror line is the vertical line passing through 5 on the *x*-axis, shown as a double-headed arrow in Figure 24.4. Each point in shape 1 is matched by a corresponding point in shape 4, the other side of the mirror line and the same distance from it. For example, the point (3, 1) in shape 1 is 2 units to the left of the mirror line, and the corresponding point in shape 4 (7, 1) is 2 units to the right of the mirror line. Any transformation of our shape like this, obtained by producing a mirror image in any given mirror line, is a reflection.

ARE THERE OTHER TRANSFORMATIONS OF SHAPES I SHOULD KNOW ABOUT?

There are many ways in which a two-dimensional shape can be transformed yet still remain in some sense 'the same'. Perhaps the most extreme example you will come across is the kind of transformation that changes a network of railway lines into the familiar London Underground map. This is an example of what is called a *topological* transformation, in which all the lengths and angles can change, and curved lines can become straight lines, or vice versa; but the map still retains significant features of the original network to enable you to determine a route from one station to another. Then there is the kind of *perspective* transformation that we make when we draw, for example, the side of a building as seen at an angle, in which a rectangle might be transformed into a trapezium (see Chapter 23). Other transformations we might encounter are those that change a square into an oblong rectangle or a rectangle into a par-

> **LEARNING AND TEACHING POINT**
>
> Make use of some of the computer-based resources available that can give primary children practical experience of transforming a shape by translating it, rotating it, reflecting it, scaling it up or down and so on.

allelogram (see Chapter 25). These are examples of what are called *affinities*. The mathematics of topological and perspective transformations and affinities is beyond the scope of this book, although all these transformations are used by children intuitively as they develop their understanding of space and shape (see Chapter 9 of Haylock and Cockburn, 2017).

But there is one other way in which shape can be transformed and yet remain in some sense 'the same', which is taught directly in the primary curriculum. This is by **scaling a shape** up or down by a scale factor. This idea is illustrated in Figure 24.5.

Shape P is scaled up into shape Q by applying a scale factor of 3. This means that the lengths of all the lines in shape P are multiplied by 3 to produce shape Q. Each length in shape Q is three times the corresponding length in shape P. It is significant that we multiply here, because, as we saw in Chapter 10, scaling is one of the most important structures of multiplication. We should also note the connection with ratio. In Chapter 15 we discussed scale drawings and maps as an application of ratio. So, we could express the relationship between shapes P and Q in Figure 24.5 by saying that the ratio of any length in P to the corresponding length in Q is 1:3; or that P is a scale drawing of Q using a scale of 1:3.

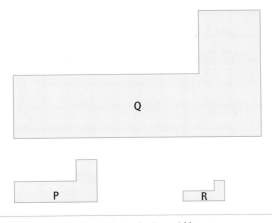

Figure 24.5   Scaling by a factor of 3 and by a factor of $^1/_2$

One of the best practical experiences of scaling is using a photocopier. Here you can scale something by factors such as 141% (which is a scale factor of 1.41, enlarging A4 to A3) or by 71% (which is scaling by a factor of 0.71, reducing A3 to A4). Similarly, as I write this chapter my computer is telling me that what is on the screen has been scaled by a factor of 125% (1.25), so I can see only about two-thirds of a page at a time, but it's comfortably large print. At the click of a key, I can change this to a scaling of 50% (0.5) and I get six pages on the screen at the same time, but it's now too small to read. **Scaling up** is achieved by a scale factor greater than 1. When the scale factor is less than 1, a shape is subject to a **scaling down** and so made smaller. Scaling by a factor of 1 leaves a shape unchanged, of course. Shape R is a scaling of shape P; this is clearly a scaling down, rather than a scaling up, because R is smaller than P. The scale factor here is $^1/_2$ (or 0.5, or 50%). In this case, each length in R is a half of the corresponding length in P. The ratio of a length in P to the corresponding length in R is 1:$^1/_2$, although we should note that this could be expressed by the equivalent ratio 2:1.

Figure 24.6 shows a rectangle with sides of lengths 2 cm and 3 cm scaled up by a factor of 3 to produce a rectangle with sides 6 cm by 9 cm. Notice what has happened to the area of the rectangle as a result of this transformation: it has been scaled up by a factor of 9, from 6 cm² to 54 cm². This illustrates a general result, that when the lengths in a shape are scaled by a particular factor the area is scaled by the *square* of the scale factor. You should be able to see easily in Figure 24.6 that the larger rectangle is three times longer and three times wider than the smaller rectangle, but that you would need *nine* of the smaller rectangles to cover the area of the larger one.

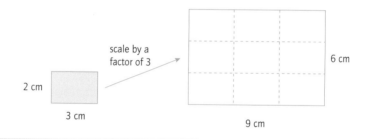

Figure 24.6   Scaling by a factor of 3 increases the area by a factor of 9

Notice also how the concept of *inverse processes* (see the glossary in Chapter 7) applies to scalings. In Figure 24.5, shape P is transformed into shape Q by scaling by a factor of 3; shape Q is transformed into shape P by scaling by a factor of $^1/_3$. Shape P is transformed into shape R by scaling by a factor of $^1/_2$; shape R is transformed into shape P by scaling by a factor of 2. In general, scalings by factors of n and $^1/_n$ are inverse transformations: one undoes the effect of the other. So, for example, on a photocopier, enlarging something by a factor of 1.25 and then reducing this by a factor of 0.8 would get you back to what you started with. Note that 1.25 × 0.8 = 1. Scaling by a factor of 1 leaves a shape unchanged.

WHAT IS MEANT BY 'SIMILAR' SHAPES?

A scaling certainly changes a shape. It changes it by scaling up or down all the lengths by a given scale factor. But shapes P, Q and R in Figure 24.5 are still 'the same shape' in many ways, even though they are not congruent. In technical mathematical language, we say they are **similar shapes**. This does not just mean that they look a bit like each other. The word 'similar' has a very precise meaning in this context. If shape P is similar to shape Q, then:

- for each line, vertex and angle in P, there is a corresponding line, vertex and angle in Q;
- the length of each line in Q is in the same ratio to the length of the corresponding line in P;
- each angle in P is equal to the corresponding angle in Q.

This last point is particularly significant. The lengths of the lines change – they are scaled up or down – but the angles do not change. This is why the shapes still look the same.

WHAT IS REFLECTIVE SYMMETRY?

Sometimes when we reflect a shape in a particular mirror line, it matches *itself* exactly, in the sense that the mirror image coincides precisely with the original shape. Shape A in Figure 24.7 is an example of this phenomenon. As in Figure 24.4 I have shown the mirror line as a double-headed arrow. This line divides the shape into two identical halves that are mirror images of each other. If we cut the shape out, we could fold it along the mirror line and the two halves would match exactly. The shape is said to have **reflective symmetry**

(sometimes called *line symmetry*) and the mirror line is called a **line of symmetry**.

Another approach is to colour the shape (so you can remember which face was up to start with), cut it out and turn it face down: we find that the shape turned face down could still fit exactly into the hole left in the paper.

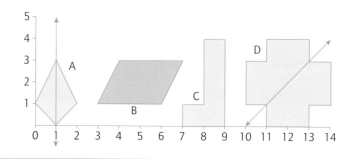

Figure 24.7   Are these shapes symmetrical?

Shape D in Figure 24.7 also has reflective symmetry. There are actually four possible lines that divide this shape into two matching halves, with one half the mirror image of the other, although only one of these is shown in the diagram. Finding the other three lines of symmetry is an exercise for the reader (self-assessment question 24.3). Again, notice that if you colour the shape, cut it out and turn it face down, it could fit exactly into the hole left in the paper.

Shape C in Figure 24.7 does not have reflective symmetry. The colouring, cutting out and turning face down routine demonstrates this nicely, since it is clear that we would not then be able to fit the shape into the hole left in the paper.

WHAT IS ROTATIONAL SYMMETRY?

Shape B (a parallelogram) in Figure 24.7 is perhaps a surprise, because this too does *not* have reflective symmetry. If you think, for example, that one of the diagonals is a line of symmetry, copy the shape onto paper and use the colouring, cutting out and turning face down procedure; or try folding it in half along the diagonal. But it does have a different kind of symmetry. To see this, trace the shape onto tracing paper and then rotate it around the centre point through a half-turn. The shape matches the original shape exactly. A shape that can be rotated on to itself like this is said to have **rotational symmetry**. The point about which we rotate it is called the **centre of rotational symmetry**.

Another practical way of exploring rotational symmetry is to cut out a shape carefully and see how many ways it can be fitted into the hole left in the paper, by rotation. For example, if we colour shape B in Figure 24.7 and cut it out, there would be two ways in which we could fit it into the hole left in the paper, by rotating it, without turning the shape face down. We therefore say that the **order of rotational symmetry** for this shape is *two*. Shape D also has rotational symmetry (see self-assessment question 24.3).

Shapes A and C do not have rotational symmetry. Well, not really. I suppose you could say that they have rotational symmetry of order one, since there is *one* way in which their cut-outs could fit into the hole left in the paper, without turning them face down. In this sense, all two-dimensional shapes would have rotational symmetry of at least order one. While recognizing this, it is usual to say that shapes like A and C do *not* have rotational symmetry.

The ideas of reflective and rotational symmetry are fundamental to the creation of attractive designs and patterns, and are employed effectively in a number of cultural traditions, particularly Islam. Children can learn first to recognize these kinds of symmetry in the world around them, and gradually learn to analyse them and to employ them in creating designs of their own.

LEARNING AND TEACHING POINT

Use the tracing paper approach to explore the ideas of rotation and rotational symmetry.

LEARNING AND TEACHING POINT

An interesting project is to make a display of pictures cut from magazines that illustrate different aspects of symmetry.

LEARNING AND TEACHING POINT

Use geometrical designs from different cultural traditions, such as Islamic patterns, to provide a rich experience of transformations and symmetry.

RESEARCH FOCUS: YOUNG CHILDREN AND SPATIAL REASONING

According to the classic model proposed by Piaget and Inhelder (1956), young children around the age of 3–5 years will reason from what is termed an egocentric point of view, using themselves as the point of reference for identifying positions and the orientation of objects – they are apparently unable to see position and orientation from someone else's point of view. This finding was challenged by Hughes and Donaldson (1979) who argued that the task used by Piaget and Inhelder did not assess the child's ability adequately, simply because it was neither meaningful nor purposeful from the child's perspective. Hughes and Donaldson investigated the spatial reasoning of 3–5-year-olds in the context of a game in which children had to hide a boy doll from a policeman doll. Their work showed that when young children encounter spatial questions in a context that makes sense to them, they can indeed see things from someone else's point of view. Zvonkin (1992), for example, observed how four-year-olds could engage in deep exploration of the concept of reflective symmetry in the context of a game. Working in pairs, one child constructed half a pattern with coloured counters, while the other child matched this to make a pattern with line symmetry. The level of reasoning shown, in which the second child has to consider the position of their counters in relation to the positioning of the first child's counters, would again seem to suggest that Piaget and Inhelder's model is an inadequate description of what young children can understand and achieve.

LEARNING RESOURCES

Access activities for your **lesson plans** at: https://study.sagepub.com/haylock7e

Before trying the self-assessment questions below, you should complete the **interactive self-assessment questions** for this chapter at: https://study.sagepub.com/haylock7e

24.1: Describe the congruences that transform shape 2 in Figure 24.1 into: (a) shape 1; (b) shape 4; (c) shape 3.

24.2: These questions refer to Figure 24.5:

   a   Imagine shape R constructed from 5 square units. How many of these square units would be needed to construct shapes P and Q?

   b   Scaling by what factor transforms shape Q into shape R?

24.3: In shape D in Figure 24.7, where are the other three lines of symmetry? What is the order of rotational symmetry of this shape?

24.4: Describe all the symmetries of shapes E, F and G in Figure 24.8.

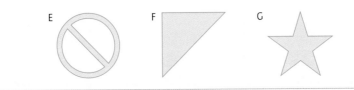

Figure 24.8 Identify the symmetries in these shapes

FURTHER PRACTICE

FROM THE STUDENT WORKBOOK

Questions 24.01–24.20: Checking understanding (transformations and symmetry)

Questions 24.21–24.30: Reasoning and problem solving (transformations and symmetry)

Questions 24.31–24.39: Learning and teaching (transformations and symmetry)

GLOSSARY OF KEY TERMS INTRODUCED IN CHAPTER 24

Congruent shapes Two or more shapes that can be transformed into each other by congruences.

Congruence Any transformation of a shape that leaves unchanged the lengths and angles; congruences are translations, rotations, reflections and combinations of these.

Translation A transformation in which a shape is slid from one position to another, without turning.

Rotation A transformation in which a shape is rotated through an angle about a centre of rotation; every line in the shape turns through the same angle.

Reflection   A transformation in which a shape is reflected in a mirror line and changed into its mirror image.

Scaling a shape   A transformation in which all the lengths in a shape are multiplied by the same factor, called the scale factor; the angles remain unchanged, but the area of the shape is multiplied by the square of the scale factor.

Scaling up   Scaling by a factor greater than 1.

Scaling down   Scaling by a factor less than 1 (but greater than zero).

Similar shapes   Two shapes either one of which is a scaling of the other. In similar shapes, corresponding lines are in the same ratio and corresponding angles are equal.

Reflective symmetry   The property possessed by a shape that is its own mirror image; also called line symmetry.

Line of symmetry   The mirror line in which a shape with reflective symmetry is reflected onto itself.

Rotational symmetry   The property possessed by a shape that can be mapped exactly onto itself by a rotation (other than through a multiple of 360°).

Centre of rotational symmetry   The point about which a shape with rotational symmetry is rotated in order to map onto itself.

Order of rotational symmetry   The number of ways in which a shape can be mapped onto itself by rotations of up to 360°. For example, a square has rotational symmetry of order four.

CLASSIFYING SHAPES

IN THIS CHAPTER, THERE ARE EXPLANATIONS OF

- the importance of classification as a process for making sense of the shapes in the world around us;
- polygons, including the meaning of 'regular polygon';
- different kinds of triangles;
- different kinds of quadrilaterals;
- tessellations;
- polyhedra, including the meaning of 'regular polyhedron';
- two-dimensional nets for three-dimensional shapes;
- various three-dimensional shapes, including prisms and pyramids;
- reflective symmetry applied to three-dimensional shapes.

READ THIS CHAPTER'S CURRICULUM LINKS AT: HTTPS://STUDY.SAGEPUB.COM/HAYLOCK7E

WHY ARE THERE SO MANY TECHNICAL TERMS TO LEARN IN GEOMETRY?

We need the special language of geometry in order to classify shapes into categories. In Chapter 3 I explained how classification is a key intellectual process that helps us to make sense of our experiences and one that is central to understanding mathematics. By coding information into categories, we condense it and gain some control over it. We form categories in mathematics by recognizing attributes shared by various elements (such as numbers or shapes). These elements are then formed into a set. Although the elements in the set are different, they have something the same about them, they are in some sense equivalent. When it is a particularly interesting or significant set, we give it a name. Because it is important that we should be able to determine definitely whether or not a particular element is in the set, the next stage of the process is often to formulate a precise definition. This whole process of classification is particularly significant in making sense of shapes and developing geometric concepts.

The learner will recognize an attribute common to certain shapes (such as having three sides), form them into a category, give the set a name (for example, the set of triangles) and then, if necessary, make the classification more explicit with a precise definition. This process of classifying and naming leads to a greater confidence in handling shapes and a better awareness of the shapes that make up the world around us.

So, to participate in this important process of classification of shapes, we need first a whole batch of mathematical ideas related to the significant properties of shapes that are used to put them in various categories. This will include, for example, reference to whether the edges of the shape are straight, the number of sides and angles, whether various angles are equal or right angled, and whether sides are equal in length or parallel. Second, we need to know the various terms used to name the sets, supported where necessary by a definition. My experience is that many primary teachers and trainee teachers have a degree of uncertainty about some of these terms that undermines

> ## LEARNING AND TEACHING POINT
>
> The role of a definition in teaching and learning is not to enable children to formulate a concept, but to sharpen it up once the concept has been formed informally through experience and discussion; and to deal with doubtful cases.

> ## LEARNING AND TEACHING POINT
>
> Children will develop geometric concepts, such as those discussed in this chapter, by experience of classifying, using various attributes of shapes, informally in the first instance, looking for exemplars and non-exemplars, and discussing the relationships between shapes in terms of sameness and difference.

their confidence in teaching mathematics. For their sake, the following material is provided for reference purposes.

WHAT ARE THE MAIN CLASSES OF TWO-DIMENSIONAL SHAPES?

The first classification of two-dimensional shapes we should note separates out those with only straight edges from those, such as circles, semicircles and ellipses, which have curved edges. A two-dimensional closed shape made up entirely of straight edges is called a **polygon**. The straight edges are called *sides*. In discussing shapes, we should restrict the use of the word 'side' to the straight edges of a polygon. It is not appropriate, for example, to refer to the circumference of a circle as a 'side'; I am perplexed when a circle is referred to in some texts as a shape 'with one side'.

Polygons can then be further classified depending on the number of sides: *triangles* with three sides, *quadrilaterals* with four, **pentagons** with five, **hexagons** with six, **heptagons** with seven, **octagons** with eight, **nonagons** with nine, **decagons** with ten and so on. Two sides in a polygon meet at a *vertex* (see glossary for Chapter 23). The number of vertices in a polygon is always the same as the number of sides. You may enjoy convincing yourself why this must be the case.

WHAT ARE REGULAR POLYGONS?

An important way in which we can categorize polygons is by recognizing those that are *regular* and those that are not, as shown in Figure 25.1. A **regular polygon** is one in which all the sides are the same length and all the internal angles are the same size. For example, a regular octagon has eight equal sides and eight angles, each of which is equal to 135°. To work out the angles in a regular polygon, use the rule ($2N - 4$ right angles) deduced in self-assessment question 23.3 (Chapter 23) to determine the total of the angles in the polygon (for example, for an octagon the sum is $2 \times 8 - 4 = 12$ right angles, that is, 1080°), then divide this by the number of angles (for example, for the octagon, 1080° ÷ 8 = 135°). Note that a regular triangle has three lines of reflective symmetry; a regular quadrilateral (a square) has four; a regular pentagon has five; and so on.

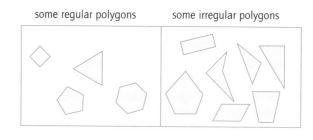

some regular polygons   some irregular polygons

Figure 25.1   Regular and irregular polygons

The word 'regular' is often misused when peo-ple talk about shapes, as though it were synony-mous with 'symmetric' or even 'geometric'. For example, the rectangular shape of the cover of this book is *not* a regular shape, because two of the sides are longer than the other two – unless my publisher has decided to surprise me and produce a square book. I'm also a bit disappointed when almost every time you see an example of, say, a pentagon (or a hexagon) used in material for pri-mary children, it seems to be a regular one. This seems to me to confuse the distinction between a pentagon in general and a regular pentagon.

WHAT ARE THE DIFFERENT CATEGORIES OF TRIANGLES?

There are basically two ways of categorizing triangles. The first is based on their angles, the second on their sides. Figure 25.2 shows examples of triangles categorized in these ways.

First, we can categorize a triangle as being *acute angled*, *right angled* or *obtuse angled*. (See Chapter 23 for the classification of angles.) An *acute-angled triangle* is one in which all three angles are acute, that is, less than 90°. A *right-angled triangle* is one in which one of the angles is a right angle; it is not possible, of course, to

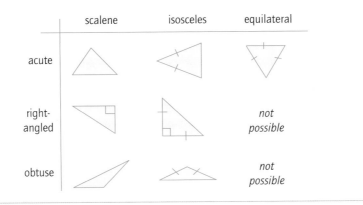

Figure 25.2   Categories of triangles

have two right angles since the sum of the three angles has to be 180°, and this would make the third angle zero. The right angles are indicated in the triangles in Figure 25.2 in the conventional fashion. An *obtuse-angled triangle* is one with an obtuse angle, that is, one angle greater than 90° and less than 180°; again, it is possible to have only one such angle because of the sum of the three angles having to equal 180°. This condition also makes it impossible to have a triangle containing a reflex angle (greater than 180°).

Second, looking at the sides, we can categorize triangles as being *equilateral*, *isosceles* or *scalene*. An **equilateral** (equal-sided) **triangle** is one in which all three sides are equal. Because of the rigid nature of triangles, the only possibility for an equilateral triangle is one in which the three angles are also equal (to 60°). So an equilateral triangle must be a regular triangle. This is only true of triangles. For example, you can have an equilateral octagon (with eight equal sides) in which the angles are not equal.

An **isosceles triangle** is one with two sides equal. In Figure 25.2, the equal sides are those marked with a small dash. An isosceles triangle has a line of symmetry passing through the middle of the angle formed by the two equal sides. If the triangle is cut out and folded in half along this line of symmetry, the two angles opposite the equal sides match each other. In this way, we can discover practically that a triangle with two equal sides always has two equal angles.

Finally, a **scalene triangle** is one with no equal sides.

Using these different categorizations, it is then possible to determine seven different kinds of triangle, as shown in Figure 25.2.

WHAT ARE THE DIFFERENT CATEGORIES OF QUADRILATERALS?

An important set of quadrilaterals is the set of **parallelograms**, that is, those with two pairs of opposite sides parallel. Figure 25.3 shows some examples of parallelograms. Two lines drawn in a two-dimensional plane are said to be **parallel lines** if theoretically they would never meet if continued indefinitely. This describes the relationship between the opposite sides in each of the shapes drawn in Figure 25.3. In Chapter 24 we saw that not all parallelograms have reflective symmetry; in Figure 25.3 only the special parallelograms A, B and C have reflective symmetry. But they do all have rotational symmetry at least of order two. This means that the opposite angles match onto each other, and the opposite sides match onto each other, when the shape is rotated through a half-turn. In other words, the opposite angles in a parallelogram are always equal and the opposite sides are always equal. There are then two main ways of classifying parallelograms. One of these is based on the angles, the other on the sides.

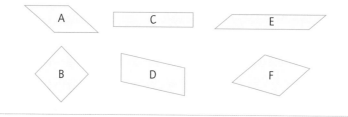

Figure 25.3   Examples of parallelograms

In primary school mathematics, the most significant aspect of the angles of a parallelogram is whether or not they are right-angled. If they are, as, for example, in shapes B and C in Figure 25.3, the shape is called a **rectangle**. Note that if one angle in a parallelogram is a right angle, because opposite angles are equal and the four angles add up to four right angles, then *all* the angles must be right angles. The rectangle is probably the most important four-sided shape from a practical perspective, simply because our artificial world is so much based on the rectangle. It is almost impossible to look anywhere that people live in this country and not see rectangles.

A **rhombus** is a parallelogram in which all four sides are equal, as, for example, in shapes A and B in Figure 25.3. A *square* (shape B) is therefore a rhombus that is also a rectangle, or a rectangle that is also a rhombus. It is, of course, a quadrilateral with all four sides equal and all four angles equal (to 90°), so 'square' is another name for a regular quadrilateral.

There is an important point about language to make here. A square is a rectangle (a special kind of rectangle) and a rectangle is a parallelogram (a special kind of parallelogram). Likewise, a square is a rhombus (a special kind of rhombus) and a rhombus is a parallelogram (a special kind of parallelogram). Sometimes you hear teachers talking about 'squares or rectangles', for example, as though they were different things, overlooking the fact that squares are a subset of rectangles. If you need to distinguish between rectangles that are squares and those that are not, then you can refer to *square rectangles* (such as B in Figure 25.3) and **oblong rectangles** (such as C). Figure 25.4 summarizes the relationships between different kinds of quadrilaterals, using an arrow to represent the phrase 'is a special kind of'.

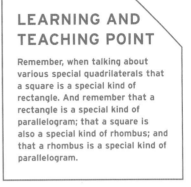

LEARNING AND TEACHING POINT

Remember, when talking about various special quadrilaterals that a square is a special kind of rectangle. And remember that a rectangle is a special kind of parallelogram; that a square is also a special kind of rhombus; and that a rhombus is a special kind of parallelogram.

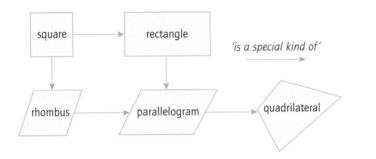

Figure 25.4   The relationships between different kinds of quadrilaterals

WHAT IS A TESSELLATION?

One further way of classifying two-dimensional shapes is to distinguish between those that *tessellate* and those that do not. A **tessellation** is a repeating pattern of tiles of one or more given shapes that fit together without gaps. A particular shape is said to tessellate if it can be used on its own to make a tessellation. This means that the shape can be used over and over again to cover a flat surface, the shapes fitting together without any gaps. In practical terms, for example, we might be asking whether the shape could be used as a tile to cover the kitchen floor (without worrying about what happens when we reach the edges).

The commonest shapes used for tiling are, of course, squares and other rectangles, which fit together so neatly without any gaps, as shown in Figure 25.5(a). This is no doubt part of the reason why the rectangle is such a popular shape in a technological world. Figure 25.5(b) demonstrates the remarkable fact that *any* triangle tessellates. If the three angles of the triangle are called A, B and C, then it is instructive to identify the six angles that come together at a point where six triangles meet in the tessellation, as shown. Because A, B and C add up to 180°, a straight angle, we find that they fit together at this point, neatly lying along straight lines. By repeating this arrangement in all directions, the triangle can clearly be used to form a tessellation.

LEARNING AND TEACHING POINT

Children can investigate which shapes tessellate and which do not, discovering, for example, that all triangles and all quadrilaterals do. They could use a plastic or card shape as a template, drawing round it in successive positions. Tessellation is a mathematical topic that directs the learner's attention to some of the key properties of shapes. It also has interesting cross-curricular links with aspects of art and design related to different cultural contexts.

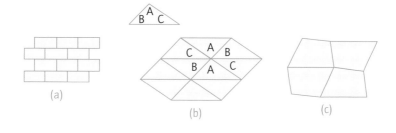

(a)

(b)

(c)

Figure 25.5 Tessellations

It is also true that *any* quadrilateral tessellates, as illustrated in Figure 25.5(c). Because the four angles add up to 360°, we can arrange for four quadrilaterals to meet at a point with the four different angles fitting together without any gaps. This pattern can then be continued indefinitely in all directions. Interestingly, apart from the equilateral triangle and the square, the only other regular polygon that tessellates is the regular hexagon, as seen in the familiar honeycomb pattern.

WHAT ARE THE MAIN CLASSES OF THREE-DIMENSIONAL SHAPES?

The first classification of three-dimensional shapes is to separate out those that have curved surfaces, such as a **sphere** (a perfectly round ball), a hemisphere (a sphere cut in half), a **cylinder** (like a baked-bean tin) and a **cone** (see the nearest motorway, or Figure 25.10). A three-dimensional shape that is made up entirely of flat surfaces (also called **plane surfaces**) is called a **polyhedron** (plural: *polyhedra*). How can you tell that a surface is a plane surface? Mathematically, the idea is that you can join up any two points on the surface by a straight line drawn on the surface. A spherical surface is not plane, for example, because two points can be joined only by drawing circular arcs.

To describe a polyhedron, we need to refer to the plane surfaces, which are called **faces** (not sides, note), the lines where two faces meet, called **edges**, and the points where edges meet, called *vertices* (plural of *vertex*). The term 'face' should only be used for plane surfaces, like the faces of polyhedra. It is not correct, for example, to refer to a sphere 'as a shape with one face'. A sphere has one continuous, smooth surface – but it is not 'a face'.

As with polygons, the word *regular* is used to identify those polyhedra in which all the faces are the same shape, all the edges are the same length, the same number of edges meet at each vertex in identical configurations, and all the angles between edges are equal. Whereas there are an infinite number of different kinds of regular polygons, there are, in fact, only five kinds of **regular polyhedra**. These are shown in Figure 25.6: (a) the regular **tetrahedron** (4 faces, each of which is an equilateral triangle); (b) the regular **hexahedron** (usually called a *cube*; 6 faces, each of which is a square); (c) the regular **octahedron** (8 faces, each of which is an equilateral triangle); (d) the regular **dodecahedron** (12 faces, each of which is a regular pentagon); and (e) the regular **icosahedron** (20 faces, each of which is an equilateral triangle).

Figure 25.6   The regular polyhedra

These and other solid shapes can be constructed by drawing a two-dimensional **net**, such as those shown for the regular tetrahedron and the cube in Figure 25.7, cutting these out, folding and sticking. It is advisable to incorporate some flaps for gluing in appropriate positions before cutting out.

A **prism** is a shape made up of two identical polygons at opposite ends, joined up by parallel lines. Figure 25.8 illustrates (a) a triangular prism, (b) a rectangular prism and (c) a hexagonal prism. I like to think of prisms as being made from cheese: a

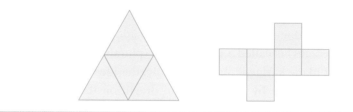

Figure 25.7   Nets for a regular tetrahedron and a cube

polyhedron is a prism if you can slice the cheese along its length in a way in which each slice is identical. Note that they are all called prisms, although colloquially the word is often used to refer just to the triangular prism.

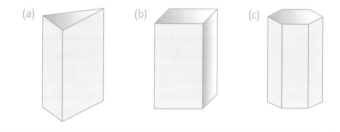

Figure 25.8   Prisms

Note also that another name for a rectangular prism (Figure 25.8b) is a **cuboid**: this is a three-dimensional shape in which all the faces are rectangles. A cube is, of course, a special kind of cuboid in which all the faces are squares. Note further that some 'sugar cubes' are cuboids but not cubes.

Another category of three-dimensional shapes to be mentioned here is the set called pyramids, illustrated in Figure 25.9. A **pyramid** is made up of a polygon forming the *base*, and then lines drawn from each of the vertices of this polygon to some point above, called the *apex*. The result of this is to form a series of triangular faces rising up from the edges of the base, meeting at the apex. Note that (a) a triangular-based pyramid is actually

LEARNING AND TEACHING POINT

Formal assessment of children's mastery of three-dimensional shapes often requires them to interpret two-dimensional representations of solid shapes. They have to learn to visualize drawings such as those in Figure 25.8 as three-dimensional objects. Construction of some simple three-dimensional shapes from nets is an excellent practical activity for primary school children that enhances their ability to visualize in this way, while drawing on a wide range of geometric concepts and practical skills.

a tetrahedron by another name, and that (b) a square-based pyramid is the kind we associate with ancient Egypt.

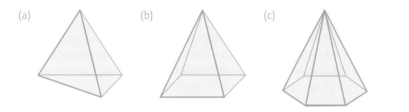

(a)   (b)   (c)

Figure 25.9   Pyramids

HOW DOES REFLECTIVE SYMMETRY WORK IN THREE DIMENSIONS?

To conclude this chapter, I will make a brief mention of *reflective symmetry* as it is applied to three-dimensional shapes. In Chapter 24 we saw how some two-dimensional shapes have reflective symmetry, with a line of symmetry dividing the shape into two matching halves, one a mirror image of the other. The same applies to three-dimensional shapes, except that it is now a **plane of symmetry** that divides the shape into the two halves.

This is like taking a broad, flat knife and slicing right through the shape, producing two bits that are mirror images of each other, as illustrated with a cone in Figure 25.10. A cone is a three-dimensional shape consisting of a circular base and one continuous curved surface tapering to a point (called the apex or the vertex) directly above the centre of the circular base. A cone actually has an infinite number of planes of symmetry, since any vertical slice through the apex of the cone can be used. All the three-dimensional shapes illustrated in the figures in this chapter have reflective symmetry. For example, the regular tetrahedron in Figure 25.6(a) has six planes of symmetry. Children can experience this idea by slicing various fruits in half, or by slicing through solid shapes made out of some moulding material.

Figure 25.10   A plane of symmetry in a cone

RESEARCH FOCUS: THE VAN HIELE LEVELS OF GEOMETRIC REASONING

In the field of research into children's geometric learning, an influential framework has been that developed by the Dutch husband and wife team of Pierre van Hiele and Dina van Hiele-Geldof: the Van Hiele levels of geometric reasoning (see, for example, Burger and Shaughnessy, 1986). There are five levels, of which only the first three are relevant to the primary school age range: (1) *visualization*, in which children can name and recognize shapes, by their appearance, but cannot specifically identify properties of shapes or use characteristics of shapes for recognition and sorting; (2) *analysis*, in which children begin to identify properties of shapes and learn to use appropriate vocabulary related to properties; and (3) *informal deduction*, in which children are able to recognize relationships between and among properties of shapes or classes of shapes and are able to follow logical arguments using these properties. The implication of this framework is that geometry taught in primary school should be informal and exploratory, aimed at moving children through visualization to analysis. Only the more able children will move to level 3. For example, a study of 12-year-olds in Malysia (Hok et al., 2015) found that none of the children in the sample were working at level 3. Van Hiele (1999) stresses that children's experience should begin with play, investigating plane and solid shapes, building and taking apart, drawing and talking about shapes in the world around them. This early informal experience is seen to be essential as the basis for more formal activities in secondary school geometry.

⊂⊃ LEARNING RESOURCES

Access activities for your **lesson plans** at: https://study.sagepub.com/haylock7e

Before trying the self-assessment questions below, you should complete the **self-assessment questions** for this chapter at: https://study.sagepub.com/haylock7e

25.1: Why is it not possible to have an equilateral, right-angled triangle or an equilateral, obtuse-angled triangle? (See Figure 25.2)

25.2: What is another name for: (a) a rectangular rhombus; (b) a regular quadrilateral; (c) a triangle with rotational symmetry; (d) a rectangular prism; and (e) a triangular-based pyramid?

25.3: Which of the following shapes tessellate? (a) a parallelogram; (b) a regular pentagon; and (c) a regular octagon.

25.4: How many planes of symmetry can you identify for a cube?

25.5: Figure 25.7 shows one net for a cube, using 6 squares joined edge to edge. Find the ten other such nets for a cube.

FURTHER PRACTICE

FROM THE STUDENT WORKBOOK

Questions 25.01–25.17: Checking understanding (classifying shapes)

Questions 25.18–25.32: Reasoning and problem solving (classifying shapes)

Questions 25.33–25.40: Learning and teaching (classifying shapes)

GLOSSARY OF KEY TERMS INTRODUCED IN CHAPTER 25

Polygon A two-dimensional closed shape, consisting of straight sides.

Pentagon, hexagon, heptagon, octagon, nonagon, decagon Polygons with, respectively, 5, 6, 7, 8, 9 and 10 sides (and angles).

Regular polygon A polygon in which all the sides are equal in length and all the angles are equal in size.

Equilateral triangle A triangle with all three sides equal in length; the three angles are also equal, and each one is therefore $60°$.

Isosceles triangle A triangle with two equal sides; the two angles opposite these two equal sides are also equal.

Scalene triangle   A triangle with all the three sides different in length.

Parallelogram   A quadrilateral with opposite sides parallel and equal in length.

Parallel lines   Two lines drawn in the same plane, which, if continued indefinitely, would never meet.

Rectangle   A parallelogram in which all four of the angles are right angles. A square is a rectangle with all sides equal in length.

Rhombus   A parallelogram in which all four sides are equal in length; a diamond. A square is a rhombus with all four angles equal.

Oblong rectangle   A rectangle that is not a square.

Tessellation   A repeating pattern of tiles of one or more given shapes that fit together without gaps; it must be possible to continue the pattern in all directions as far as you wish. A particular shape is said to *tessellate* if it can be used on its own to make a tessellation.

Sphere   A completely round ball; a solid shape with one continuous surface, in which every point on the surface is the same distance from a point inside the shape called the centre.

Cylinder   A three-dimensional shape, like a baked-bean tin, consisting of two identical circular ends joined by one continuous curved surface.

Cone   A solid shape consisting of a circular base and one continuous curved surface tapering to a point (the apex or vertex) directly above the centre of the circular base.

Plane surface   A completely flat surface. Any two points on the surface can be joined by a straight line drawn on the surface.

Polyhedron (plural: **polyhedra**)   A three-dimensional shape with only straight edges and plane surfaces.

Face   One of the plane surfaces of a polyhedron.

Edge   The intersection of two surfaces; in particular, the straight line where two faces of a polyhedron meet.

Regular polyhedron   A polyhedron in which all the faces are identical shapes, the same number of edges meet at each vertex in an identical configuration, and all the edges are equal in length.

Tetrahedron, hexahedron, octahedron, dodecahedron, icosahedron   Polyhedra with, respectively, 4, 6, 8, 12 and 20 faces. The regular forms of these five shapes are the only possible regular polyhedra. A cube is a regular hexahedron.

Net A two-dimensional arrangement of shapes that can be cut out and folded up to make a polyhedron.

Prism A polyhedron consisting of two opposite identical faces with their vertices joined by parallel lines.

Cuboid A rectangular prism; a six-faced polyhedron in which any two opposite faces are identical rectangles. A cube is a cuboid in which all the faces are square.

Pyramid A polyhedron consisting of a polygon as a base, with straight lines drawn from each of the vertices of the base to meet at one point, called the apex.

Plane of symmetry A plane that cuts a solid shape in two halves that are mirror images of each other.

SUGGESTIONS FOR FURTHER READING FOR SECTION G

1   A sound explanation of the key mathematical ideas of angle can be found in Chapter 15 ('Angles and compass directions') of Suggate, Davis and Goulding (2017).

2   Chapter 2 of Nickson (2004) provides a detailed summary of research into children's understanding of shape and space, with consideration of how this can be applied in the classroom.

3   Those who teach younger children will find plenty of good suggestions in Chapter 6 of Tucker (2014), where she writes about children learning concepts of shape and space through play.

4   Chapter 8 of Hansen (2020) provides an interesting analysis of children's errors and misconceptions in geometry.

5   Chapter 3 of Henderson and Taimina (2020) addresses the question 'What is an angle?' They describe angles as being of three types: dynamic, static (as in a geometric shape) and a measure. There are some interesting exercises in the chapter to help the reader understand these ideas.

6   Those who teach younger children will find it worth their while tracking down Davis et al. (2015), a book edited on behalf of the Spatial Reasoning Study Group. The first section addresses: (a) what is spatial reasoning and why should we care? (b) the development of spatial reasoning in young children and (c) developing spatial thinking in early years mathematics education.

7   An extensive review of research into the development of spatial and geometric thinking is provided by Battista (2007), one of the handbooks produced by the US National Council for Teaching of Mathematics (NCTM).

8   See if you can persuade your library to get hold of a copy of Chval et al. (2016); this book, also from the NCTM, provides sound practical guidance on teaching for understanding in geometry and some aspects of measurement.

SECTION H

STATISTICS AND PROBABILITY

26 Handling Data   455
27 Comparing Sets of Data   474
28 Probability   492

WATCH THE SECTION OPENER VIDEO AT: HTTPS://STUDY.SAGEPUB.COM/HAYLOCK7E

HANDLING DATA

IN THIS CHAPTER, THERE ARE EXPLANATIONS OF:

- sorting data according to various criteria and the use of Venn diagrams and Carroll diagrams;

- universal set, subset, complement of a set, intersection of sets;

- population, variable, and values of a variable in the context of statistical data;

- the four stages of handling data: collecting, organizing, representing, interpreting;

- the use of tallying and frequency tables for collecting and organizing data;

- the idea of sampling when undertaking a survey of a large population;

- the differences between discrete data, grouped discrete data and continuous data;

- the representation of discrete data in block graphs;

- the representation of discrete and grouped discrete data in bar charts;

- the misleading effect of suppressing zero in a frequency graph;

- other ways of representing data: pictograms, pie charts and line graphs.

READ THIS CHAPTER'S CURRICULUM LINKS AT: HTTPS://STUDY.SAGEPUB.COM/HAYLOCK7E

WHAT ARE VENN DIAGRAMS?

Sorting according to given criteria is one of the most fundamental processes in mathematics. For example, when a child in a reception class counts how many children have brought packed lunches, they have first to sort the children into two subsets: those who have packed lunches and those who do not.

Technically, sorting involves the concepts of a **population** (sometimes called the **universal set**), the values of a **variable**, and **subset**. For example, the 25 children in a Year 5 class in a rural school were sorted according to how they came to school that day: walking, by bicycle, by car or by bus. In this case, the children in the class constitute the *universal set*. This is the set containing all the things under consideration. In statistical language, this is called the *population*. The *variable* that distinguishes between the members of the population in this example is the way they came to school. There are four *values* of this variable: walk, bicycle, car and bus. The four different values of the variable sort the set of children into four *subsets*. The use of the word 'value' may seem a little strange here, but the concept is essentially the same as when we use a *numerical* variable to sort children, such as how many children in their family – when it seems more natural to talk about the various numbers of children in a family (1, 2, 3, 4, …) as being the values of the variable.

Set diagrams are visual structures that aid children's understanding of the process of sorting and the relationships between various sets and subsets. Figure 26.1 is an elementary example of a **Venn diagram**, where circles (or other closed shapes) are used to represent the various subsets of the 25 children. John Venn (1834–1923) was a Cambridge logician and philosopher who developed the use of such diagrams for representing various logical relationships between sets and subsets.

> ## LEARNING AND TEACHING POINT
>
> Learning to sort data according to given criteria is the foundation of counting and also of data handling. Give young children lots of experiences of sorting, first using the actual objects themselves.

> ## LEARNING AND TEACHING POINT
>
> Use the children themselves for sorting. For example, get children to move to different corners of the classroom depending on their answers to questions like: how did you travel to school today? Which quarter of the year is your birthday? What's your favourite fruit out of apple, banana, grapes or melon? Then do the same kind of sorting by getting the children to stand in circles drawn on the playground, before moving on to simple examples of set diagrams.

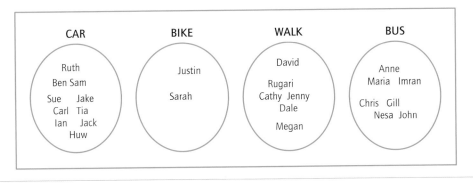

Figure 26.1   An elementary Venn diagram: how we travel to school

Figure 26.2 shows another kind of simple Venn diagram used for sorting. Here the question asked is 'Did you travel to school by car?' The universal set comprises the 25 children in the class, the names of all of which are to be placed somewhere within the rectangular box. The variable is again how they travel to school. But now the sorting uses only two values of the variable, namely 'by car' and 'not by car'. All those who travel by car are placed inside the circle and all the others go outside it. The set of children who do not travel by car – those lying outside the circle – is called the **complement** of the set of those who travel by car. This diagram is an important illustration of the partitioning structure of subtraction, linked with the question 'how many do not?' (see Chapter 7).

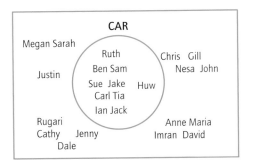

Figure 26.2   A Venn diagram showing the complement of a set

An interesting example of sorting using a Venn diagram occurs when two variables are used simultaneously. An example is shown in Figure 26.3, which arises from simultaneously sorting the children into those who travel by car and those who do not, and those who are girls and those who are not. This sorting generates four subsets: girls who travel by car, girls who do not travel by car, those who are not girls who travel by car, and those who are not girls who do not travel by car. If you think of each child answering the questions, 'are you a girl?' and 'did you travel to school by car?', the four subsets represent 'yes, yes', 'yes, no', 'no, yes' and 'no, no'. The 'yes, yes' subset – those whose names are placed in the section of the diagram where the two sets overlap – is called the **intersection** of the two sets.

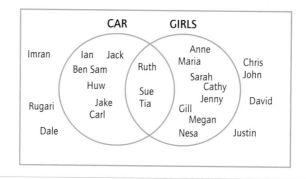

Figure 26.3   A Venn diagram showing the intersection of two sets

WHAT ARE CARROLL DIAGRAMS?

Lewis Carroll (1832–98) was not just the author of the *Alice* books but also a mathematician with a keen interest in logic. He devised what has become known as the

Figure 26.4   A Carroll diagram for sorting using two variables

Carroll diagram, another way of representing the subsets that occur when a set is sorted according to two variables. Figure 26.4 shows the data about girls and travel by car sorted in the same way as in Figure 26.3, but presented in a Carroll diagram. The four subsets generated by the sorting process are clearly identified by this simple diagrammatic device: girl, car; girl, not car; not girl, car; not girl, not car.

WHAT DO CHILDREN HAVE TO LEARN ABOUT HANDLING STATISTICAL DATA?

Essentially, there are four stages involved in handling statistical data: collecting it, organizing it, representing it and interpreting it. (*Note*: I adopt the current usage of *data* as a singular noun, meaning 'a collection of information'.)

Children should learn how to collect data as part of a purposeful enquiry, setting out to answer specific questions that they might raise. This might involve the skills associated with designing simple questionnaires. For example, the data used above about how children travel to school might arise as part of a geography-focused project on transport, and be used to make comparisons between, say, the children in this rural school and those in a city school. A useful technique here is that of **tallying**, based on counting in fives. Data should then be organized in a **frequency table**. Figure 26.5 shows both these processes for the information collected from our Year 5 class of 25 children.

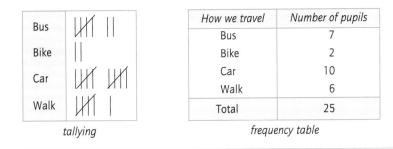

tallying   frequency table

Figure 26.5 Using tallying and a frequency table

In undertaking a survey, primary children can be introduced informally to the concept of *sampling*. With a large population, such as the population of Cardiff, for example, it is clearly not possible to gather data from all 365,000 residents. So, those who do surveys will collect data from a carefully selected **sample**. Primary children can appreciate the principle that the sample must be as far as possible a fair representation of the population. This principle resonates with their understanding of a fair test in science. Three factors can contribute to achieving a fair sample: (a) selecting members of the population at random; (b) ensuring that the proportions of significant categories of individuals (such as male/female, employed/unemployed, age bands, social class) in the sample are similar to the proportions in the population; and (c) making the sample as large as possible – in general, the larger the sample, the more reliable the findings.

Various kinds of graphs and diagrams – including Venn and Carroll diagrams, block graphs, bar charts, pictograms and pie charts – can then be used to represent the data, before the final step of interpreting it.

> **LEARNING AND TEACHING POINT**
>
> Children should experience all four stages in handling data: collecting it, organizing it, representing it and interpreting it. The last and important step of interpretation is often the least well-addressed in primary mathematics teaching.

> **LEARNING AND TEACHING POINT**
>
> Motivation for doing a statistical enquiry is higher when the data is collected by the children themselves, higher still when it is collected to answer some questions they have posed themselves, and even higher when it is about themselves.

WHAT IS DISCRETE DATA?

The word *discrete* means 'separate'. Discrete data is information about a particular population that automatically sorts the members of the population into quite distinct, separate subsets. The information about travelling to school, shown in Figures 26.1 and 26.5, is a good example of discrete data, since it sorts the children automatically into four separate, distinct subsets: those who come by bus, by car, by bicycle or on foot. Other examples of this kind of discrete data that children might collect, organize, display and interpret would include: their favourite television programme, chosen from

> **LEARNING AND TEACHING POINT**
>
> Three ways to promote mastery in interpreting graphs: (a) write about what the graph tells us, particularly in relation to the questions and issues that prompted the collection of the data; (b) write sentences about what the graph tells us, incorporating key words, such as *most, least, more than, less than*; and (c) make up a number of questions that can be answered from the graph, to pose to each other.

a list of six possible programmes; the daily news-paper taken at home, including 'none'; and the month in which they were born. We can also refer to the variable that gives rise to a set of discrete data as 'a **discrete variable**'. Separate subsets are formed for each individual value taken by the variable. The number in each subset is called the **frequency**.

We have seen that sometimes a discrete variable is numerical, rather than just descriptive. For example, children might be asked what size shoes they wear, or how many pets they have. The values of a discrete numerical variable will usually be a regular sequence of numbers across a particular range. For a particular class of children, possible values for shoe sizes, for example, might be 3, 3.5, 4, 4.5, 5, 5.5, or 6. The number of pets a child has might be 0, 1, 2, 3, 4, 5 or 6. Initially, we should use variables that have no more than a dozen values, otherwise we finish up with too many subsets to allow any meaningful interpretation. Frequency data collected and organized, as in Figure 26.5, can then be displayed in a conventional block graph or bar chart.

WHAT'S THE DIFFERENCE BETWEEN BLOCK GRAPHS AND BAR CHARTS?

Block graphs and bar charts are two important stages in the development of graphical representation of frequencies. In the **block graph** shown in Figure 26.6(a), each square is shaded individually, as though each square represents one child. In interpreting the graph, the child can count the number of squares, as though counting the number of children in each subset, so there is no need for a vertical axis. Block graphs are introduced

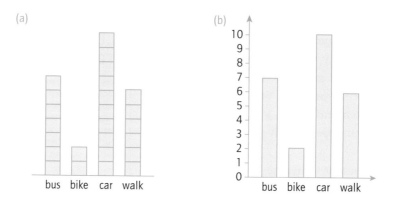

Figure 26.6 Two stages in the use of graphs: (a) block graph; (b) bar chart

first by sticking on or shading individual squares, the number of which represents the frequency.

Figure 26.6(b) is based on a more sophisticated idea. Now, the individual contributions are lost and it is the height of the column that indicates the frequency, rather than the number of squares in the column. In this **bar chart**, we read off the frequencies by relating the tops of the columns to the scale on the vertical axis. It should be noted that the numbers label the points on the vertical axis, not the spaces between them. So, in moving from the block graph to the bar chart, we have progressed from counting to measurement. This is an important step because, even though we may still use squared paper to draw these graphs, we now have the option of using different scales on the vertical axis, appropriate to the data: for example, with larger populations we might take one unit on the vertical scale to represent 10 people.

Notice that, in Figure 26.6, I have used the convention, sometimes used for discrete data, of leaving gaps between the columns; this is an appropriate procedure because it conveys pictorially the way in which the variable sorts the population into discrete subsets. In order to present an appropriate picture of the distribution of the data, it is essential that the columns in a block graph or bar chart be drawn with equal widths.

WHAT IS SUPPRESSION OF ZERO?

There is a further important point to make about using bar charts to represent frequencies, illustrated by the graphs shown in Figure 26.7. These were produced prior to a general election to show the numbers of votes gained by three political parties (which

I have called A, B and C) in the previous general election. The graph in Figure 26.7(a) was the version put out by our local party A candidate to persuade us that we would be wasting our vote by voting for party C. By not starting the frequency axis at zero, a totally false picture is presented of the relative standing of party C compared with A and B. Because the purpose of drawing a graph is to give us an instant overview of the relationships within the data, this procedure (called **suppression of zero**) is nearly always inappropriate or misleading and should be avoided. The graph in Figure 26.7(b), properly starting the vertical axis at zero, presents a much more honest picture of the relative share of the vote.

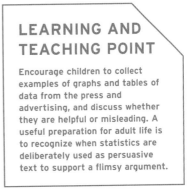

LEARNING AND TEACHING POINT

Encourage children to collect examples of graphs and tables of data from the press and advertising, and discuss whether they are helpful or misleading. A useful preparation for adult life is to recognize when statistics are deliberately used as persuasive text to support a flimsy argument.

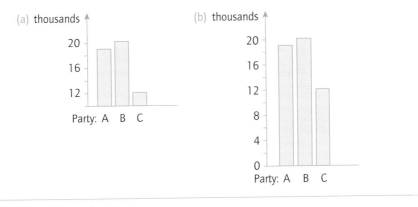

Figure 26.7   Suppression of zero

SO WHAT IS MEANT BY 'GROUPED DISCRETE DATA'?

Discrete data, like that in the examples above, is the simplest kind of data to handle. Sometimes, however, there are just too many values of the variable concerned for us to sort the population into an appropriate number of subsets. So the data must first be organized into groups.

For example, a group of Year 5 children was asked how many writing implements (pencils, pens, felt-tips and so on) they had with them one day at school. The responses were as follows: 1, 2, 2, 3, 4, 4, 5, 5, 5, 6, 6, 8, 8, 8, 9, 9, 10, 10, 11, 13, 14, 14, 14, 15, 15, 18, 19, 25, 26, 32. Clearly, there are just too many possibilities here to represent this data in a bar chart as it stands. The best procedure, therefore, is to group the data, for example, as shown in Figure 26.8.

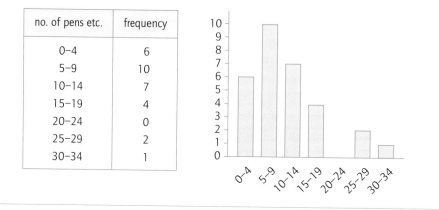

no. of pens etc.	frequency
0–4	6
5–9	10
10–14	7
15–19	4
20–24	0
25–29	2
30–34	1

Figure 26.8   Handling grouped discrete data

Of course, the data could have been grouped in other ways, producing either more subsets (for example, 0–1, 2–3, 4–5 and so on: 17 groups) or fewer (for example, 0–9, 10–19, 20–29, 30–39). When working with **grouped discrete data** like this, a rule of thumb is that we should aim to produce from five to twelve groups: more than twelve, we have too much information to take in; fewer than five, we have lost too much information. Note the following principles for grouping data like this:

- The range of values in the subsets (0–4, 5–9, 10–14, 15–19 and so on) should be the same in each case. (For 'range', see the glossary for Chapter 27.)
- The groups should not overlap.
- The groups must between them cover all the values of the variable.
- Groups with zero frequency (like 20–24 in Figure 26.8) should not be omitted from the table or from the graph.
- If possible, aim for the number of groups to be from five to twelve.

WHAT OTHER KIND OF DATA IS THERE APART FROM DISCRETE?

Discrete data contrasts with what is called *continuous data*. This is the kind of data produced by a variable that can theoretically take any value on a continuum. For example, a horticulturalist is collecting data about the heights of a group of 30 dahlia plants in a greenhouse. The measurements might come anywhere on a tape measure from, say, 50 cm to 100 cm. They are not just restricted to particular, distinct points on the scale. For example, the height of one plant at the beginning of a week is measured as 62.5 cm. At the end of the week this has increased to about 64.5 cm. Obviously, the height of the plant has increased continuously from one measurement to the next, on the way taking every possible value in between. The height would not suddenly jump from one value to the other. This is a characteristic of a **continuous variable**. The contrast with a discrete variable like, for example, the number of pets you own, is clear. If you have three pets and then get a fourth, you suddenly jump from three to four, without having to pass through 3.1 pets, 3.2 pets and so on. Measurements of length, mass, volume and time intervals are all examples of continuous data.

Having said that, we will always have to record a measurement 'to the nearest something' – see the discussions on rounding in Chapters 16 and 21. The effect of this is immediately to change the values of the continuous variable into a set of discrete data. For example, we might measure the length of running stride of a sample of 7–11 year olds *to the nearest centimetre*. This now means the data we handle is restricted to separate, distinct values: such as 70 cm, 71 cm, 72 cm and so on. So, in practice, the procedure for handling this kind of data – produced by recording a series of measurements to the nearest something – is no different from that for handling discrete data with a large number of potential values, by grouping it as explained above.

It is therefore an appropriate activity for primary school children. For example, children can collect data about: their heights (to the nearest centimetre); their masses (to the nearest tenth of a kilogram); the circumferences of their heads (to the nearest millimetre); the volume of water they can drink in one go (to the nearest tenth of a litre); the time taken to run 100 metres (to the nearest second); and so on. Each of these is technically a continuous variable, but by being measured to the nearest something it generates a set of discrete data, which can then be grouped appropriately and represented in a graph. I would then suggest that we might reflect the fact that the data originated from a continuous variable by drawing the columns in the graph with no gaps between them, as shown in Figure 26.9. Any further development of the handling of continuous data than this would be beyond the scope of primary school mathematics.

> **LEARNING AND TEACHING POINT**
>
> If primary children are collecting data arising from a continuous variable (such as their height), get them first to record the measurements to the nearest something (for example, to the nearest centimetre) and then group the results and handle them like grouped discrete data.

Stride measurements to nearest centimetre	frequency
50–54	1
55–59	2
60–64	5
65–69	5
70–74	3
75–79	8
80–84	3
85–89	2
90–94	1

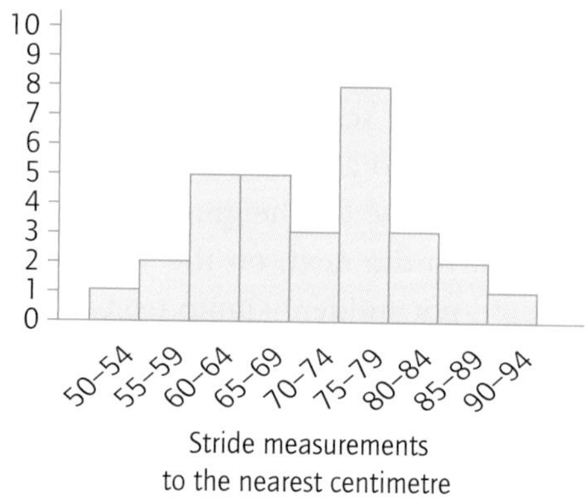

Figure 26.9   A graph derived from a continuous variable

WHAT ABOUT PICTOGRAMS?

Figure 26.10 shows how a **pictogram** can be used to represent the data given in Figure 26.5. Here the names of the children in various sets have been replaced by icons (pictures), organized in neat rows and columns. It is essential for a pictogram to work that the icons are lined up both horizontally and vertically. The pictogram is clearly only a small step from a block graph, where the icons are replaced by individual shaded squares. So, in terms of developing children's statistical understanding, it should be seen as an early stage of pictorial representation of data, coming between the representations used in Figures 26.1 and 26.6.

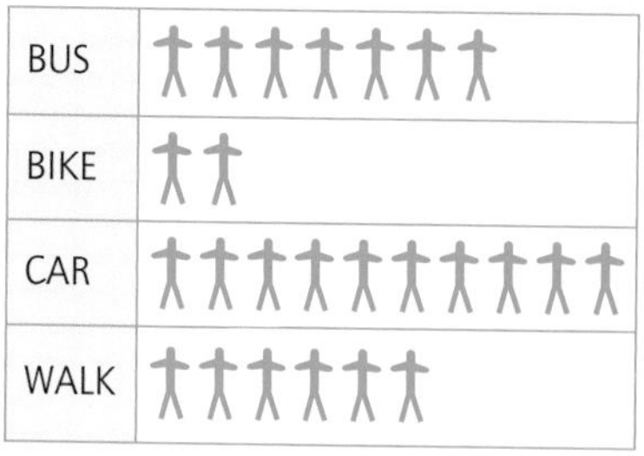

Figure 26.10   A pictogram

It is also possible to use pictograms where each icon represents a number of individuals rather than just one. So, for example, the data in Figure 26.10 might be presented with a key to indicate that each icon represents two people, in which case the numbers of icons needed would have to be one for bike, five for car, three for walk, and three and a half for bus! Yes, that does seem rather bizarre. But the Guidance for the English National Curriculum for mathematics is that children should compare information 'using many-to-one correspondence in pictograms with simple ratios 2, 5, 10' (DfE, 2013: 16). To be honest, these many-to-one correspondences are usually unnecessary, unhelpful and inappropriate with the small populations that would be used in primary mathematics. Many-to-one representations in pictogram form should really only be used with *larger* populations. We should remember that the point of a pictorial representation of data is to enable us to take in the overall relationships involved at a glance. A problem with this approach is that when an icon like the one used in Figure 26.10 is used to represent, say, 10 people, it just seems bizarre and misleading to use a picture of one person to represent 10 people and fractions of the icon to represent numbers of people less than 10. In a many-to-one representation, I would always advocate using an abstract icon, such as a circle or a square. The pictogram is a popular format in newspapers and advertising, because it is more eye-catching than just a plain bar chart, so children will have to learn how to interpret them. But, mathematically, a simple bar chart with an appropriate scale on the vertical axis is clearer and more accurate for representing the frequencies of various subsets within a larger population.

WHAT ARE PIE CHARTS USED FOR?

The **pie chart**, shown in Figure 26.11, is a much more sophisticated idea. Here, it is the angle of each slice of pie that represents the proportion of the population in each

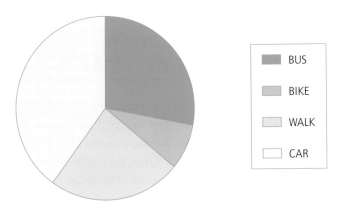

Figure 26.11 A pie chart

LEARNING AND TEACHING POINT

Only use pie charts with variables that have a small number of values. For example, girl or boy; their means of getting to school; their age in years; for each match this season the number of goals scored by the school football team.

LEARNING AND TEACHING POINT

Because computer software generates pie charts so easily, primary school children should learn how to interpret pie charts – but they do not need to be able to draw them for themselves.

subset. It is common practice to write these proportions as percentages (see Chapter 18) within the slice itself, if possible. A really important principle of the pie chart is that the whole pie must represent the whole population. Pie charts are really only appropriate for discrete data with a small number of subsets, say, six or fewer. In fact, the effectiveness of a pie chart as a way of displaying discrete data decreases as the number of subsets increases! Pie charts are often used to show what proportions of a budget are spent in various categories.

The mathematics for producing a pie chart can be difficult for primary school children, unless the data is chosen very carefully. For example, to determine the angle for the slice representing travel by bus, in Figure 26.11, we would have to divide 360 degrees by 25 to determine how many degrees per person in the population, and then multiply by 7. Using a calculator, the key sequence is: 360, ÷, 25, ×, 7, = (answer 100.8, which is about 101 degrees). This angle then has to be drawn using a protractor.

Fortunately, this can all be done nowadays by a computer. If the data in question is entered on a database or on a spreadsheet, then usually there is available the choice of a bar chart, a pie chart or a line graph (see below). Many simple versions of such data-handling software are available for use in primary school.

WHEN MIGHT A LINE GRAPH BE USED TO REPRESENT STATISTICAL DATA?

The other type of graph used sometimes for representing statistical data is the **line graph**. The mathematics involved in using and interpreting these is very similar to that for distance–time graphs (see Chapter 20). An appropriate example of the use of a line graph is shown in Figure 26.12. Like pie charts, line graphs are easily produced by entering the data in a computer spreadsheet or database. Figure 26.12 – showing the number of children on a primary school roll at the beginning of each school year for a number of years – was produced in this way. For statistical data, a line graph is really only appropriate where the variable along the horizontal axis is 'time'. In this example, the movement of the line, up and down, gives a picture of how the number on the roll is changing over time. A line graph would therefore be totally inappropriate as a means of presenting discrete data such as that relating to travelling to school in Figure 26.5, and similarly inappropriate for all the other examples of statistical data used in this chapter.

LEARNING AND TEACHING POINT

Use line graphs for statistical data only where the variable along the horizontal axis is 'time'. Possible examples would be: the midday temperature each day over a month; the number of children who walk or cycle to school each day over a month; the population of a local town each decade over a century.

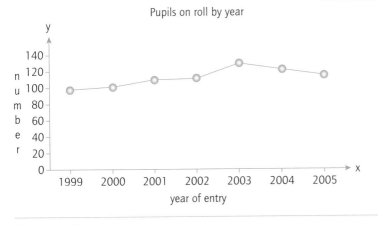

Figure 26.12   A line graph

RESEARCH FOCUS: STATISTICAL THINKING

Drawing on a bank of research and their experience with primary school children over the course of a whole year, Jones et al. (2000) have developed a useful framework for describing children's statistical thinking. This framework describes four key statistical processes. These are: (1) *describing data*, which involves extracting information from a set of data or graph and making connections between the data and the context from which it came; (2) *organizing and reducing data*, which includes ordering, grouping and summarizing data; (3) *representing data*, which refers to constructing various kinds of visual representation of the data; and (4) *analysing and interpreting data*, which involves recognizing patterns and trends and making inferences and predictions. For each of these four aspects, the researchers were able to identify four levels of thinking: *Level 1*, where the reasoning is idiosyncratic, often unrelated to the given data and tending to draw more on the individual's own experience and opinions; *Level 2*, where the child begins to use quantitative reasoning to comment on the data; *Level 3*, where quantitative reasoning is used consistently to formulate judgements; and *Level 4*, which uses a more analytical approach in exploring data and in making connections between the data and the context. The framework is a helpful tool both for the assessment of children's understanding and for informing teaching plans. This framework, along with others, is discussed in Sharma (2017) in which the importance of 'statistical literacy' is argued, as well as some of the thinking around it.

ᒍ ᓄ LEARNING RESOURCES

Access activities for your **lesson plans** at: https://study.sagepub.com/haylock7e

Before trying the self-assessment questions below, you should complete the **self-assessment questions** for this chapter at: https://study.sagepub.com/haylock7e

26.1: Children in a class answer 'yes' or 'no' to these two questions: (i) Are you 11 years old? (ii) Did you walk to school today? How do their answers to these questions put the children in four subsets? How could these four subsets be represented in (a) a Venn diagram, and (b) a Carroll diagram?

26.2: Make up two questions that can be answered from the graphs shown in each of: (a) Figure 26.6; (b) Figure 26.8; and (c) Figure 26.9.

26.3: For a sample of 50 trainees, the amount of money in coins in their possession on a given morning ranged from zero to £4.50. How would you choose to group this data in order to represent it in a bar chart?

26.4: (a) In a class of 36 children, 14 come to school by car. What angle would be needed in the slice of a pie chart to represent this information? (b) What would it be for 14 children out of a class of 33?

26.5: A survey is to be conducted of the opinions of the children in a large junior school (Years 3–6) about what time the school day should start. There is a total of 500 children in the school, so it is decided to get responses from a sample of 50. One suggestion is to interview the first 50 children arriving at school one morning. What's wrong with this method of sampling? How would you suggest the sample might be obtained?

FURTHER PRACTICE

Access the website material for:

Knowledge check 35: Bar charts and frequency tables for discrete data
Knowledge check 36: Bar charts for grouped discrete data
Knowledge check 37: Bar charts for continuous data
Knowledge check 38: Interpreting pie charts

at: https://study.sagepub.com/haylock7e

FROM THE STUDENT WORKBOOK

Questions 26.01–26.12: Checking understanding (handling data)

Questions 26.13–26.17: Reasoning and problem solving (handling data)

Questions 26.18–26.30: Learning and teaching (handling data)

GLOSSARY OF KEY TERMS INTRODUCED IN CHAPTER 26

Population   The term used in statistics for the complete set of all the people or other things for which some statistical data is being collected; synonymous with 'universal set' in set theory.

Universal set   The term used in set theory for the complete set of all things under consideration; synonymous with 'population' in statistics. In a Venn diagram, the universal set is usually represented by a rectangular box.

Variable (in statistics)   An attribute that can vary from one member of a population to another, the different values of which can be used to sort the population into subsets; variables may be non-numerical (such as choice of favourite fruit) or numerical (such as the mark achieved in a mathematics test).

Subset   A set of members within a given set that have some defined attribute, or that take a particular value of a variable.

Venn diagram   A way of representing the relationships between various sets and subsets using enclosed regions (such as circles); children can use these for sorting experiences by placing the members of various sets or subsets within the appropriate regions. (See Figures 26.1–26.3.)

Complement of a set   All the things in the universal set that are not within a given set. For example, the complement of the set of 7-year-olds in a class is the set of all those who are not 7 years old.

Intersection of two sets   The set of all those things that are common to the two sets. In a Venn diagram, the intersection is represented by the overlap between two enclosed regions. The intersection of the set of girls and the set of 7-year-olds is the set of 7-year-old girls.

Carroll diagram   A 2 by 2 grid used for sorting the members of a set according to whether or not they possess each of two attributes. The four cells of the grid correspond to 'yes, yes', 'yes, no', 'no, yes' and 'no, no'. (See Figure 26.4.)

Tallying   A simple way of counting, making a mark for each item counted, with every fifth mark used to make a group of five. (See Figure 26.5.)

Frequency table   A table recording the frequencies of each value of a variable. (See Figure 26.5.)

Sample   In a statistical survey, a representative selection of a large population for which data is collected; in general, the larger the sample, the more reliable are the results as a representation of the whole population.

Discrete variable   A variable that can take only specific, separate (discrete) values. For example, 'number of children in a family' is a discrete variable, because it can take only the values 0, 1, 2, 3, 4 and so on. When the value of this variable changes, it goes up in jumps.

Frequency   The number of times something occurs within a population.

Block graph   An introductory way of representing discrete data, in which each member of the population is represented by an individual square (stuck on or coloured in) arranged in columns. The frequency of a particular value of the variable is simply the number of squares in that column. (See Figure 26.6a.)

Bar chart   A graphical representation of data, where frequencies are represented by the heights of bars or columns. (See Figure 26.6b.)

Suppression of zero   The misleading practice of starting the vertical axis in a frequency graph at a number other than zero; this gives a false impression of the relative frequencies of various values of the variable.

Grouped discrete data   Data arising from a discrete (usually numerical) variable where the different values of the variable have been grouped into intervals, in order to reduce the number of subsets. For example, marks out of 100 in a mathematics test might be grouped into intervals 1–10, 11–20, 21–30, 31–40 and so on.

Continuous variable   A variable that can take any value on a continuum. For example, 'the height of the children in my class' is a continuous variable. When the value of this variable for a particular child has changed, it will have done so continuously, passing through every real number value on the way.

Pictogram   A way of representing discrete data, in which each member of the population is represented by an individual picture or icon arranged in rows or columns. (See Figure 26.10.) With larger populations, each picture or icon may represent a number of individuals rather than just one.

Pie chart   A way of representing statistical data where the population is represented by a circle (the pie) and each subset is represented by a sector of a circle (a slice of the circular pie), with the size of each sector indicating the frequency. (See Figure 26.11.)

Line graph   Mainly used for statistical data collected over time; the frequencies (or other measurements) are plotted as points and each point is joined to its neighbours by straight lines. (See Figure 26.12.) A line graph is therefore useful for showing trends over time.

COMPARING SETS OF DATA

READ THIS CHAPTER'S CURRICULUM LINKS AT: HTTPS://STUDY.SAGEPUB.COM/HAYLOCK7E

IN THIS CHAPTER, THERE ARE EXPLANATIONS OF:

- how two data sets using the same variable can be presented for comparison;

- the idea of an average as a representative value for a set of data;

- three measures of average: the mean, the median and the mode;

- how to calculate mode, median and mean from a frequency table;

- quartiles and the five-number summary of a distribution;

- range and inter-quartile range as measures of spread;

- percentiles and deciles;

- the concept of average speed.

HOW CAN TWO DATA SETS BE PRESENTED PICTORIALLY FOR COMPARISON?

This chapter focuses mainly on the professional needs of the teacher for understanding some basic statistics. Often, we will want to represent two sets of data side by side for comparison. This will usually be where the same variable is used for two different populations. For example, a school may wish to compare data for 25 boys and 30 girls in a mathematics assessment. The variable here is the level achieved by each child. Assume that this variable takes the values of 1 2, 3 or 4, as shown in this frequency table:

	Boys	Girls
Level 1	5	3
Level 2	10	18
Level 3	6	6
Level 4	4	3

What is the best way of representing this data in one diagram so that we can visually compare the achievements of the boys with that of the girls? The first thing to note is that because the sample sizes are different (25 boys, 30 girls), we cannot really use the raw data for comparison. For example, the 6 boys achieving level 3 is a greater proportion of the set of boys than the 6 girls achieving level 3. This would not be a problem if we had two sets of the same size. But with different-sized sets we need to compare the proportions of boys and girls achieving various levels. The obvious way to do this is to express the proportions as percentages:

	Boys	Girls
Level 1	20%	10%
Level 2	40%	60%
Level 3	24%	20%
Level 4	16%	10%

There are a number of ways of representing this data to make it possible to compare them at a glance. Figure 27.1 shows two of them. Figure 27.1(a) puts the columns side by side, making it easy to compare performances for each level. Figure 27.1(b) is a better representation if you want

LEARNING AND TEACHING POINT

Only a small portion of the mathematics in this chapter on comparing sets of data would be taught in primary school. But teachers themselves need to be confident with all the material here in order to make sense of the official statistics generated in the field of education and schooling.

to focus on comparing how the boys and girls were spread across the different levels. Note that the representation for each set in Figure 27.1(b) is essentially the same idea as a pie chart, but using a bar-model rather than a circle to represent the whole set.

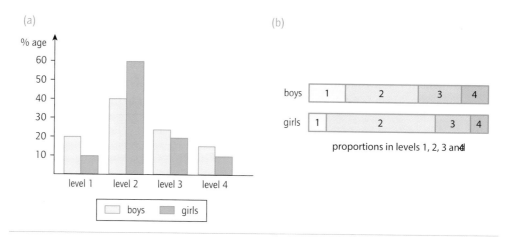

Figure 27.1   Comparing two sets of data for the same variable

WHAT ARE AVERAGES FOR?

The purpose of finding an **average** is to produce a *representative value* for a set of numerical data. There are three kinds of average to be considered: the **mean** (also called the **arithmetic mean**), the **median** and the **mode**. Although they are calculated in different ways, the important purpose shared by all three of these measures of average is to provide one number that can represent the whole set of numbers. This 'average' value will then enable us:

1   to make comparisons between different sets of data, by comparing their means, medians or modes; and
2   to make sense of individual numbers in a set by relating them to these averages.

In the discussion below, we use these different kinds of average to consider the marks out of 100 gained by two groups of children (Group A, 14 children; Group B, 11 children) in the same mathematics and English tests, as follows:

Group A: Mathematics	23, 25, 46, 48, 48, 49, 53, 60, 61, 61, 61, 62, 69, 85
Group B: Mathematics	36, 38, 43, 43, 45, 47, 60, 63, 69, 86, 95
Group A: English	45, 48, 49, 52, 53, 53, 53, 53, 54, 56, 57, 58, 59, 62
Group B: English	45, 52, 56, 57, 64, 71, 72, 76, 79, 81, 90

HOW DO YOU FIND THE MEAN?

To find the mean value of a set of numbers, three steps are involved:

1 Find the sum of all the numbers in the set.
2 Divide by the number of numbers in the set.
3 Round the answer appropriately, if necessary (see Chapter 16).

For example, to find the mean score of group A above in mathematics:

1 The sum of the scores is 751.
2 Divide 751 by 14, using a calculator to get 53.642857.
3 Rounding this to, say, one decimal place, the mean score is about 53.6.

The logic behind using this as a representative value is that the total marks obtained by the group would have been the same if all the children had scored the mean score (allowing for the possibility of a small error introduced by rounding). I imagine the process to be one of pooling. All the children put all their marks into a pool, which is then shared out equally between all 14 of them. This is an application of the concept involved in division structures associated with the word 'per' (see Chapter 10): we are finding the 'marks per student', assuming an equal sharing of all the marks awarded between them. An example that illustrates this well would be to find the mean amount of money that a group of

> **LEARNING AND TEACHING POINT**
>
> Explain the idea of a mean in terms of a group pooling some resources then sharing them out equally. For example, to find the mean number of writing implements in the possession of a group of six children, they could put them all on the table and then share them equally between the members of the group. How many do they each get?

people have in their possession. This could be done by putting all their money on the table and then sharing it out again equally between the members of the group. This is precisely the process that is modelled by the mathematical procedure for finding the mean.

We can now use this procedure to make comparisons. For example, to compare group A with group B in mathematics, we could compare their mean scores. Group B's mean score is about 56.8 (625 ÷ 11). This would lend some support to an assertion that, on the whole, group B (mean score 56.8) has done better in the test than group A (mean score 53.6).

We can also use average scores to help make sense of individual scores. For example, let us say that Luke, a child in group B, scored 60 in mathematics and 64 in English. Reacting naively to the raw scores, we might conclude that Luke did better in English than in mathematics. But comparing the marks with the mean scores for his group leads to a different interpretation: Luke's mark for mathematics (60) is above the mean (56.8), while the higher mark he obtained for English (64) is actually below the mean for his group (which works out to be 67.5). This would lend some support to the view that Luke has actually done better in mathematics than in English.

WHAT IS THE MEDIAN?

The *median* is simply the number that comes in the middle of the set when the numbers are arranged in numerical order. For example, the median mass of baby boys at age 22 weeks in the UK is given as 7.5 kg. To understand this, I imagine all the 22-week-old baby boys in the UK arranged in a long line from the lightest at one end to the heaviest at the other. If I walk halfway along the line and stop, the mass of the baby I am standing next to is 7.5 kg, the median mass.

It is very common, for example, for government education statistics to use medians as representative, average values. Finding this average value is much easier than calculating the mean, especially when you are dealing with a very large population with a large number of possible values for the variable being considered. The only small complication arises when there is an even number of elements in the set, because then there is not a middle one. The process for finding the median is as follows:

1   Arrange all the numbers in the set in order from smallest to largest.
2   If the number of numbers in the set is odd, the median is the number in the middle.
3   If the number of numbers in the set is even, the median is the mean of the two numbers in the middle, in other words, halfway between them.

If you are not sure how to decide where the middle of a list is situated, here's a simple rule for finding it: if there are n items in the list, the position of the middle one (the median) is 'half of $(n + 1)$'. For example, with 11 items in the list the position of the median is half of 12, which is the sixth item. With 83 items in the list, the position of the median would be half of 84, which is the 42nd item. If n is even, this formula still tells you where to find the median. For example, with 50 items, the formula gives the position of the median as half of 51, which is $25^1/_2$: we interpret this to mean 'halfway between the 25th and 26th items'.

So, for example, for group B mathematics, with a set of 11 children, the median is the sixth mark when the marks are arranged in order; hence the median is 47. For group A mathematics, with a set of 14, the median comes halfway between the seventh and eighth marks, which is halfway between 53 and 60; hence the median is 56.5.

Interestingly, if we use the median rather than the mean as our measure of average, we would draw a different conclusion altogether when comparing the two groups: that, on the whole, group A (median mark of 56.5) has done rather better than group B (median mark of 47)!

Although the median is often used for large sets of statistics, it sometimes has advantages over the mean when working with a small set of numbers, as in these examples. The reason for this is that the median is not affected by one or two extreme values, such as the 95 in group B. For a small set of data, a score much larger than the rest, like this one, can increase the value of the mean quite sig-nificantly and produce an average value that does

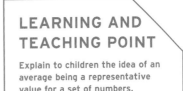

LEARNING AND TEACHING POINT

Explain to children the idea of an average being a representative value for a set of numbers, enabling us to make comparisons between different sets.

not represent the group in the most appropriate way. To take an extreme case, imagine that in a test nine children in a group of ten score 1 and the other child scores 100: this data produces a mean score of 10.9 and a median of 1! There is surely no argument here with the view that the median 'represents' the performance of the group as a whole more appropriately. All this simply serves to illustrate the fact that most sets of statistics are open to different interpretations – which is why I have used the phrase 'lends some support to ...' when drawing conclusions from the data in these examples.

Returning to Luke, who scored 60 for mathematics and 64 for English, we can compare his performance with the median scores for his group, which were 47 and 71 respectively.

These statistics again lend support to the assertion that he has done better in mathematics (well above the median) than in English (well below the median).

WHAT IS THE MODE AND WHEN WOULD YOU USE IT?

The mode is simply the value of the variable that occurs most frequently. For example, for group A mathematics, the mode (or the modal mark) is 61, because this occurs three times, which is more than any other number in the set. For group B mathematics, the mode is 43. This is actually a daft way of determining representative marks for these sets of data. The mode is really only of any use as a measure of average when you are dealing with a large set of data and when the number of different values in the set of data is quite small. A good example of the use of a mode would be when discussing an 'average' family. In the UK, the modal number of children in a family is two, because more families have two children than any other number. So, if I were to write a play featuring an 'average' family, there would be two children in it. Clearly, the mode is more use here than the mean, since 2.4 children would be difficult to cast.

Like the mean and the median, the mode enables us to make useful comparisons between different sets, when it is an appropriate and meaningful measure of average; for example, when comparing social factors or educational achievements in, say, some cities in China with some European states, the modal numbers of children per family could be very significant statistics to consider.

The mode can also be used with non-numerical variables and with grouped numerical data. For example, if we collected data about the colour of hair for the children in a class we might conclude that the modal colour is brown. If we collected the heights of children in a class, measured to the nearest centimetre, and then grouped these into intervals of 5 cm, we might conclude that the modal interval of heights is, say, 145–149 cm, meaning more children were in this interval of heights than any other.

HOW DO YOU CALCULATE THE MODE, MEDIAN AND MEAN FROM A FREQUENCY TABLE?

For the median and the mean, we can only do this with a numerical variable and where the data has not been grouped into intervals. So, these are procedures likely to be used for a numerical variable that takes a fairly small number of values. For example, a primary school recorded the following information regarding absence one term:

No. of days absent	No. of children
0	36
1	29
2	12
3	10
4	6
5	2
Total no. of children	95

Of the three different averages that we could use, the mode is the easiest to read off from a frequency table like this. It is simply the value with the largest frequency, in this case the mode is 0 days. More children were absent for 0 days than for any other number of days.

It would also be quite appropriate to use the median number of days as our representative value for this data. To find this, we need the number of days absent for the child who would come in the middle if we lined them all up in order of the number of days they were absent (assuming that none of them were absent when we did this!). From the left, we would have first the 36 children who were absent for 0 days, then the 29 who were absent for 1 day and so on. With a total of 95 children, the child in the middle would be the 48th child in the line. This would be one of the children in the group who were absent for 1 day. So the median number of days absent is 1 day.

The mean would also be an appropriate representative value for this set of data. Calculating this needs a bit more work. We must first add up all the numbers of days absent for all 95 children. That is 36 lots of 0 days, plus 29 lots of 1 days, plus 12 lots of 2 days and so on. The most convenient way of doing this is to add another column to the frequency table, showing the product of the number of days and the number of children.

No. of days absent	No. of children	Days × children
0	36	0
1	29	29
2	12	24
3	10	30
4	6	24
5	2	10
Total	95	117

Summing the numbers in the third column gives us a total of 117 days absent, which is shared between 95 children. So, the mean number of days absent is 117 ÷ 95 = 1.23 days approximately.

WATCH THE PROBLEM SOLVED! VIDEO AT: HTTPS://STUDY.SAGEPUB.COM/HAYLOCK7E

WHAT IS A FIVE-NUMBER SUMMARY?

To describe a set of numerical data and to get a feel for how the numbers in the set are distributed, a **five-number summary** is often used. First, we list all the numbers in the set in order from smallest to largest. This enables us to find five significant numbers that help us to describe the distribution and to compare it with another set of data. Two of these significant numbers are simply the minimum and maximum values, the first and last numbers in the list. The third one is the median, which has been explained above.

The other two are the **lower quartile (LQ)** and the **upper quartile (UQ)**. The lower quartile, the median and the upper quartile are three numbers that divide the list into four quarters. Just as the median is the midpoint of the set, the lower and upper quartiles are one-quarter and three-quarters of the way along the list respectively.

Here are the mathematics scores again for groups A and B considered above:

Group A: Mathematics   23, 25, 46, 48, 48, 49, 53, 60, 61, 61, 61, 62, 69, 85
Group B: Mathematics   36, 38, 43, 43, 45, 47, 60, 63, 69, 86, 95

For group B, the median is the sixth score (47), the lower quartile is the third score (43) and the upper quartile is the ninth score (69). Group B is a convenient size for discussing quartiles because it is fairly easy to decide where the quarter points of the list are situated. Group A is not so straightforward. There is a similar rule to that for the median for deciding where the lower and upper quartiles come. For completeness I will explain it, but you really do not have to be able to do this! The position of the lower quartile is one-quarter of $(n + 1)$ and that of the upper quartile is three-quarters of $(n + 1)$. So, for group B, with 11 items, the positions of the LQ and UQ are one-quarter and three-quarters of 12, namely positions 3 and 9 in the list. However, for group A, with 14 scores in the list, the position of the lower quartile would be one-quarter of 15, which is $3\frac{3}{4}$. This means that it comes three-quarters of the way between the third and fourth scores (46 and 48), which is 47.5. The position of the upper quartile is three-quarters of 15, which is $11\frac{1}{4}$. This means that it comes one-quarter of the way between the eleventh and twelfth scores (61 and 62), which is 61.25.

I should say that this kind of fiddling around, deciding precisely where the **quartiles** are located between particular items in a list, is definitely not necessary in practice when you are dealing with *large* sets of data. Anyway, the reader's requirements will be only to understand the idea of a quartile when it is met in government statistics, not to be able to calculate quartiles for awkward sets of data.

The five-number summaries for groups A and B for their scores for mathematics are therefore as follows:

	Group A	Group B
Min	23	36
LQ	47.5	43
Median	56.5	47
UQ	61.25	69
Max	85	95

This is a fairly standard way of presenting data from two populations for comparison. Teachers may well encounter government data about educational performance presented in this form. For example, the performance of primary schools in mathematics in two areas, X and Y, might be compared by the following five-number summaries based on data from the mathematics national assessments for 11-year-olds:

	X	Y
Min	26	23
LQ	47	48
Median	65	73
UQ	84	93
Max	96	100

The variable used here is the percentage of children in each school in the area gaining a specified level in the mathematics assessment. For example, the upper quartile of 84 for area X means this: if the schools in area X are listed in order from the school with the lowest percentage of children achieving the specified level to the school with the highest percentage, then the school that is three-quarters of the way along the list had 84% of children achieving the specified level. Glancing at these summaries, we can see that there is very little difference between the results of the bottom quarter (from the minimum to the lower quartile) of the schools in the two areas. But the results for area Y are markedly better than area X, with a higher median and a higher upper quartile. The comparison shows that the 'average' and higher-performing schools in area Y are doing better than those in area X.

WHAT IS THE RANGE?

We return to the test scores for group A for mathematics and English given earlier in this chapter:

Group A: Mathematics   23, 25, 46, 48, 48, 49, 53, 60, 61, 61, 61, 62, 69, 85
Group A: English   45, 48, 49, 52, 53, 53, 53, 53, 54, 56, 57, 58, 59, 62

We will compare group A's marks for mathematics with their marks for English. By looking just at the means (53.6 for mathematics and 53.7 for English), we might conclude that the sets of marks for the two subjects were very similar. Looking at the actual data, it is clear that they are not. The most striking feature is that the mathematics marks are more widely spread and the English marks are relatively closely clustered together.

Statisticians have various ways of measuring the degree of 'spread' (sometimes called 'dispersion') in a set of data. The reader may have heard, for example, of the 'standard deviation'. These measures of spread have a similar purpose to the measures of average: they enable us to compare sets of data and to make sense of individual items of data.

The simplest measure of spread is the **range**. This is just the difference between the largest and the smallest values in the set. So, for example, when comparing Group A's mathematics and English scores, we would note that, although they have about the same mean scores, the range for mathematics is 62 marks (the difference between 85 and 23), whereas the range for English is only 17 marks (the difference between 62 and 45). Clearly, the mathematics marks are more spread out.

WHAT IS THE INTER-QUARTILE RANGE AND WHY IS IT USED?

One or two exceptionally high or low scores in a set will result in the range not being a good indication of how spread out is *most* of the data in the set – it might therefore give a false impression when comparing two sets of data. So, it is better to use what is called the **inter-quartile range**. This is simply the difference between the quartiles. Since this measure excludes the top quarter and the bottom quarter of the data, the set is not affected by what happens at the extremes and a better indication is given of the spread of most of the data.

In the data given above for comparing area X and area Y, the inter-quartile range for the data for area X is 37 (that is, 84 – 47), whereas the inter-quartile range for area Y is 45 (that is, 93 – 48). This indicates that there is a greater spread of percentages of children achieving the specified level in mathematics in the schools in area Y than in area X.

WHAT ARE PERCENTILES AND DECILES?

The process of identifying quartiles and the median involves listing all the data in the set in numerical order and then dividing the set into four parts (quarters), with equal numbers of items in each part. With larger populations, it is common practice often to divide the list into a hundred equal parts. The values used to separate these hundred parts are called **percentiles**. Note that the word 'percentile' is sometimes abbreviated to '%ile'.

If you read that the 90th percentile score in a test administered to a large number of children is 58, this means that the bottom 90% of children scored 58 or less and the top 10% of children scored 58 or more. Similarly, if you read that the 20th percentile score was 26, this means that the bottom 20% of the children scored 26 or less and that the top 80% of children scored 26 or more. It follows therefore that the lower quartile can also be referred to as the 25th percentile, the median as the 50th percentile and the upper quartile as the 75th percentile.

Often, children's performances in standardized tests will be given in terms of percentiles. For example, a report on a nine-year-old states that 'Tom's standardized score for reading accuracy is 125, which is at the 95th percentile.' This puts Tom in the top 5% for reading accuracy for his age range. The raw percentage scores in a numeracy test administered to a sample of 450 trainee teachers are presented in terms of percentiles as follows:

10th %ile	58%
20th %ile	60%
30th %ile	63%
40th %ile	65%
50th %ile	70%
60th %ile	78%
70th %ile	84%
80th %ile	88%
90th %ile	92%

This example is included deliberately because it can be confusing when the numbers in the set of data are themselves percentages, as in this case. Readers should not get confused between the *percentiles*, which refer to percentages of the number of items in the set, and the *percentage scores*, which are the actual items of data in the set. The following are examples of observations that could be made from this data:

- A trainee scoring 94% on the numeracy test is in the top 10% in this sample.
- A trainee scoring 64% on the test is well below average, with more than 60% of others doing better than this.
- The median score on the test was 70%.
- The top half of the sample scored 70% or more on the test.
- The bottom 30% of the sample scored 63% or less on the test.
- The top 20% of trainees scored 88% or more on the test.
- The lower quartile was somewhere between 60% and 63%.

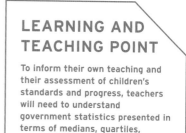

LEARNING AND TEACHING POINT

To inform their own teaching and their assessment of children's standards and progress, teachers will need to understand government statistics presented in terms of medians, quartiles, percentiles and deciles.

Sometimes reports will divide the set into 10 equal parts, using what are called the **deciles**. The 90th percentile, for example, can also be called the 9th decile and so on.

HOW DOES THE IDEA OF 'AVERAGE SPEED' FIT IN WITH THE CONCEPT OF AN AVERAGE?

In the UK, children's first experience of speed is usually the speed of a vehicle, measured in 'miles per hour'. Note that speed gives us another example of that important little word, 'per'.

The idea of **average speed** derives from the concept of a mean. Over the course of a journey in my car, my speed will be constantly changing; sometimes – at red traffic lights – it will be zero. When we talk about the average speed for a journey, it is as though we add up all the miles covered during various stages of the journey and then share them out equally 'per hour'. This uses the same idea of 'pooling' which was the basis for calculating the mean of a set of numbers. So, if my journey covers 400 miles in total and takes 8 hours, the average speed is 50 miles per hour (400 ÷ 8).

The logic here is that if I had been able to travel at a constant speed of 50 miles in each hour, the journey would have taken the same time (8 hours). So, the average speed (in miles per hour) is the total distance travelled (in miles) divided by the total time taken (in hours). We can then extend this definition of average speed to apply to journeys where the time is not a whole number of hours; for example, for a journey of 22 miles in 24 minutes (0.4 hours) the average speed is $22 \div 0.4$, which is 55 miles per hour. And, of course, the same principle applies whatever units are used for distance and time; for example, if the toy car takes 5 seconds to run down a ramp of 150 centimetres, the average speed is 30 centimetres per second ($150 \div 5$).

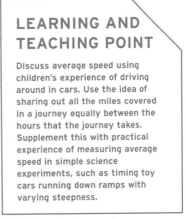

LEARNING AND TEACHING POINT

Discuss average speed using children's experience of driving around in cars. Use the idea of sharing out all the miles covered in a journey equally between the hours that the journey takes. Supplement this with practical experience of measuring average speed in simple science experiments, such as timing toy cars running down ramps with varying steepness.

RESEARCH FOCUS: SUBJECT KNOWLEDGE FOR TEACHING (2)

What do you need to know to be a good mathematics teacher? Ball, Thames and Phelps (2008) have developed a useful model for analysing teacher content knowledge in mathematics. They distinguish between *common subject knowledge*, which includes recognizing wrong answers and being able to do the mathematical tasks that the children are given, and the *special subject knowledge* that is required to be an effective teacher. This includes being able to analyse errors, evaluate alternative ideas, give mathematical explanations and choose appropriate mathematical representations. (Readers may find that elsewhere these two aspects of subject knowledge in mathematics teaching are referred to as *substantive* and *disciplinary knowledge*.) Ball, Thames and Phelps argue that teachers also need pedagogical subject knowledge. This has two strands. *Knowledge of content and learners* includes the ability to anticipate errors and misconceptions, to interpret learners' incomplete thinking and to predict their responses to mathematical tasks. *Knowledge of content and teaching* includes the ability to sequence content for teaching, to recognize the pros and cons of different representations and to handle novel approaches. Burgess (2009) showed how this model could be used to evaluate the teaching of statistics, through observing four teachers working with children aged 9–13 years on data-handling tasks involving more than one variable. The study revealed, for example, numerous instances where – because of inadequate special subject knowledge of statistics or pedagogical subject knowledge related to the learning and teaching of statistics – teachers missed opportunities to respond to and exploit the children's suggestions for processing the given data. Burgess concluded that these missed opportunities impacted negatively on the children's learning and understanding of statistical concepts and processes.

LEARNING RESOURCES

Activities for your **lesson plans** related to this chapter can be accessed at: https://study.sagepub.com/haylock7e.

Before trying the self-assessment questions below, you should complete the **self-assessment questions** for this chapter at: https://study.sagepub.com/haylock7e

27.1: In a survey in England, a sample of teenagers aged 13 to 17 years was asked to name up to three current UK government ministers. Figure 27.2 compares the proportions of boys and girls who could correctly name 0, 1, 2 or 3 ministers. From the diagram, what comparisons might be drawn between the boys and the girls?

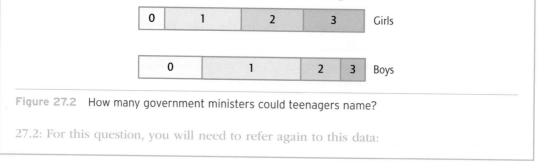

Figure 27.2   How many government ministers could teenagers name?

27.2: For this question, you will need to refer again to this data:

Group A: Mathematics	23, 25, 46, 48, 48, 49, 53, 60, 61, 61, 61, 62, 69, 85
Group B: Mathematics	36, 38, 43, 43, 45, 47, 60, 63, 69, 86, 95
Group A: English	45, 48, 49, 52, 53, 53, 53, 53, 54, 56, 57, 58, 59, 62
Group B: English	45, 52, 56, 57, 64, 71, 72, 76, 79, 81, 90

a Compare the mean and median scores for English for groups A and B. Which group on the whole did better?

b Find the mean score for English for the two groups combined. Is the mean score of the two groups combined equal to the mean of the two separate mean scores?

c Find the median scores and the ranges for English and mathematics for the two groups combined.

d John, in group A, scored 49 for mathematics. How does this score compare with the performance of group A as a whole?

27.3: Toy car P travels 410 centimetres in 6 seconds; toy car Q travels 325 centimetres in 5 seconds. Which has the greater average speed?

27.4: The following table shows the percentages of children reaching a specified level in a national assessment for reading, in schools with more than 20% and up to 35% of children eligible for free school meals:

95th %ile	UQ	60th %ile	median	40th %ile	LQ	5th %ile
94	83	78	76	72	67	52

St Anne's primary school has 24% of children eligible for free school meals, so it comes into this group. In the reading assessment, 69% of their children achieved the specified level. How well did they do compared with schools in this group?

FURTHER PRACTICE

Access the website material for:

Knowledge check 39: Calculating means
Knowledge check 40: Modes
Knowledge check 41: Medians
Knowledge check 42: Upper and lower quartiles

Knowledge check 43: Measures of spread, range and inter-quartile range
Knowledge check 44: Bar charts for comparing two sets of data

at: https://study.sagepub.com/haylock7e

FROM THE STUDENT WORKBOOK

Questions 27.01–27.17: Checking understanding (comparing sets of data)

Questions 27.18–27.28: Reasoning and problem solving (comparing sets of data)

Questions 27.29–27.36: Learning and teaching (comparing sets of data)

GLOSSARY OF KEY TERMS INTRODUCED IN CHAPTER 27

Average　A representative value for a set of numerical data, enabling comparisons to be made between sets; three types of average are the mean, the median and the mode.

Mean (arithmetic mean)　For a set of numerical data, the result of adding up all the numbers in the set and dividing by the number in the set.

Median　The value of the one in the middle when all the items in a set of numerical data are arranged in order of size. If the set has an even number of items, the median comes halfway between the two in the middle. In a set of n items arranged in order, the position of the median is $\frac{1}{2}$ of $(n + 1)$.

Mode　The value in a set of numerical data that occurs most often; a type of average only appropriate for large sets with a relatively small number of possible values.

Five-number summary　A way of summarizing a set of numerical data by giving the minimum, the lower quartile, the median, the upper quartile and the maximum.

Lower quartile (LQ)　If the items in a set of numerical data are arranged in order of size from smallest to largest, the position of the lower quartile is $\frac{1}{4}$ of $(n + 1)$.

Upper quartile (UQ)　If the items in a set of numerical data are arranged in order of size from smallest to largest, the position of the upper quartile is $\frac{3}{4}$ of $(n + 1)$.

Quartiles　The three items in a set of numerical data arranged in order of size, from smallest to largest, that come one quarter of the way along (the lower quartile), in the middle (the median) and three quarters of the way along (the upper quartile).

Range　In a set of numerical data, the difference between the largest and smallest value; a simple measure of spread that can be used to compare two sets of data.

Inter-quartile range   The difference between the upper and lower quartiles; a measure of spread, not affected by what happens at the extremes.

Percentile   The values that separate into 100 parts a large set of data arranged in order of size. To say that a child's score in a standardized test is at the 95th percentile, for example, means that the top 5% of children obtained this score or better.

Decile   The values that separate into ten parts a large set of data arranged in order of size. To say that a child's score in a standardized test is at the 7th decile, for example, means that the top three tenths of children obtained this score or better.

Average speed   The total distance travelled on a journey divided by the time taken; if the distance is measured in miles and the time in hours, the average speed is given in miles per hour.

PROBABILITY

IN THIS CHAPTER, THERE ARE EXPLANATIONS OF

- the meaning of probability as a measurement applied to events;
- some of the language we use to indicate probability subjectively;
- the use of a numerical scale from 0 to 100%, or from 0 to 1, for measuring probability;
- estimating probability from statistical data;
- estimating probability from data obtained by repeating an experiment a large number of times;
- estimating probability by using theoretical arguments based on symmetry and equally likely outcomes;
- the use of two-way tables for identifying all the possible equally likely outcomes from an experiment involving two independent events;
- mutually exclusive events;
- rules for combining probabilities for independent and mutually exclusive events;
- a simple model for assessing risk.

READ THIS CHAPTER'S CURRICULUM LINKS AT: HTTPS://STUDY.SAGEPUB.COM/HAYLOCK7E

WHAT IS PROBABILITY?

This interesting and relevant topic is not currently taught in primary schools in England or Wales. But I am including it here because there are many readers of this book who teach in primary schools in other countries, including Scotland and Northern Ireland, where probability is part of the primary school curriculum (Education Scotland, 2010a; CCEA, Northern Ireland, 2019b). It is also included in this book to help the reader to make sense of the many assertions in their professional life where the language of probability is used or, as is often the case, misused. For this reason I go well beyond any primary school curriculum in the material that follows.

LEARNING AND TEACHING POINT

Introduce probability by getting children in small groups to write down events that might occur in the next 12 months, and then to rank them in order from least likely to most likely. Focus on the language of comparison: *more likely than* and *less likely than*.

First, we should recognize that in mathematics **probability** is a measurement, just like any other measurement such as length or mass. Second, it is a measurement that is applied to *events*. But what it is about an event that is being measured is surprisingly elusive. My view is that what we are measuring is how strongly we believe that the event will happen. We describe this level of belief with words ranging from 'impossible' to 'certain', and compare our assessment of different events by talking about one being 'more likely' or 'less likely' than another. This strength of belief is determined by different kinds of *evidence* that we may assemble.

Sometimes this evidence is simply the accumulation of our experience, in which case our judgement about how likely one event may be, compared with others, is fairly subjective. For example, one group of students wrote down some events that might occur during the following 12 months and ranked them in order from the least likely to the most likely, as follows:

1   It will snow in Norwich during July.
2   Norwich City will win the FA Cup.
3   Steve will get a teaching post.
4   There will be a general election in the UK.
5   Someone will reach the summit of Mount Everest.

When they were then told that Steve had an interview at a school the following week for a post for which he was ideally suited, this extra piece of evidence had an immediate effect on their strength of belief in event (3) and they changed its position in the ranking.

By using 'more likely than' and 'less likely than', this activity is based on the ideas of comparison and ordering, always the first stages of the development of any aspect of measurement. The next stage would be to introduce some kind of measuring scale.

A **probability scale** can initially use everyday language, such as:

impossible;

almost impossible;

fairly unlikely;

evens;

fairly likely;

almost certain;

certain.

For example, we might judge that event (1) is 'almost impossible', event (2) is 'fairly unlikely' and event (5) is 'almost certain'. When we feel that an event is as likely to happen as not to happen, we say that 'the chances are **evens**'.

To introduce a numerical scale, we can think of awarding marks out of 100 for each event, with 0 marks for an event we believe to be impossible, 100 marks for an event we judge to be certain, and 50 marks for 'evens'. For example, purely subjectively, the students in the group awarded 1 mark for event (1), 5 marks for

event (2), 50 marks for event (4) and 99 marks for event (5). Event (3) started out at 40, but moved to 75 when the new evidence was obtained.

If these marks out of 100 are now thought of as percentages and converted to decimals (see Chapter 18 for how to do this), we have the standard scale used for measuring probability, ranging from 0 (impossible), through 0.5 (evens), to 1 (certain). For example, the **subjective probabilities** that we assigned to events (2) and (5) were 0.05 and 0.99 respectively.

HOW CAN YOU MEASURE PROBABILITY MORE OBJECTIVELY?

There are essentially three ways of collecting evidence that can be used for a more objective estimate of probability:

1   We can collect statistical data and use the idea of relative frequency.
2   We can perform an experiment a large number of times and use the relative frequency of different outcomes.
3   We can use theoretical arguments based on symmetry and equally likely outcomes.

HOW DOES RELATIVE FREQUENCY RELATE TO PROBABILITY?

The first of these three approaches, based on **relative frequency**, is used extensively in the world of business, such as insurance or marketing, where probabilities are often assessed by gathering statistical data.

For example, to determine an appropriate premium for a life insurance policy for a person such as a 60-year-old male university lecturer living in Suffolk, an insurance company would use the probabilities that he might live to 70, to 80, to 90 and so on. To determine these probabilities, they might collect statistical data about male academics of his age living in East Anglia and find what proportion of these academics survive to various ages. If it is found that out of 250 cases, 216 live to 70, the evidence would suggest that a reasonable estimate for the probability of his living to this age is 86.4% (216 ÷ 250) or, as a decimal, 0.864.

Since it is normally impractical to obtain data from the entire population, this application of probability is usually based on evidence collected from a sample (see Chapter 26). For example, what is the probability that a word chosen at random from a page of text in this book will have four letters in it? To answer this, we could use the data from a random sample of 100 words given in self-assessment question 28.1 at the end of this chapter. Since 14 of these words have four letters, the relative frequency of four-letter words in the sample is 14%. So an estimate for the probability, based on this evidence, would be 0.14. If we wanted to be more confident of this estimate, we would choose a larger sample than 100 words and make it more representative of the whole book by selecting the words from a number of different chapters.

HOW IS PROBABILITY MEASURED BY EXPERIMENT?

The second procedure for obtaining objective estimates for a probability applies the same idea, but to an experiment, often the kind of thing that can be experienced in a classroom. Now the 'event' in question is an *outcome* of the experiment.

For example, the experiment might be to throw three identical dice simultaneously. Let's say that the outcome we are interested in is that the score on one of them should be greater than the sum of the scores on the other two. What is the probability of this outcome? A useful experience for children is first to make a subjective estimate of the probability. This may be just based on intuition, in the sense that they have a feeling about how likely the various outcomes are but cannot state any evidence to support this. They then perform the experiment a large number of times, recording the numbers of successes and failures. For example, they might make a subjective estimate that the chances of this happening would be a bit less than evens, so the probability is, say, about 0.40. Then the dice are thrown, say, 200 times and it is found that the number of successes is 58. Hence, the

LEARNING AND TEACHING POINT

The material discussed here on experimental and theoretical probability would be excellent as extension material for children at the top end of primary school.

relative frequency of successes is 29% (58 ÷ 200) and so the best estimate for the probability, based on this evidence, would be 0.29. This is called **experimental probability**.

HOW IS PROBABILITY DETERMINED THEORETICALLY?

For some experiments, we can consider all the possible outcomes and make estimates of **theoretical probability** using an argument based on *symmetry*. Experiments with coins and dice lend themselves to this kind of argument.

The simplest argument would be about tossing one coin. There are only two possible outcomes: heads and tails. Given the symmetry of the coin, there is no reason to assume that one outcome is more or less likely than the other. So we would conclude that the probability of a head is 0.5 and the probability of a tail is 0.5. Notice that the sum of the probabilities of all the possible outcomes must be 1. This represents 'certainty': we are certain that the coin will come down on either heads or tails.

Similarly, if we throw a conventional, six-faced die, there are six possible outcomes, all of which, on the basis of symmetry, are equally likely. We therefore determine the probability of each number turning up to be one-sixth, or about 0.17. We can also determine the probability of events that are made up of various outcomes. For example, there are two scores on the die that are multiples of three, so the probability of throwing a multiple of three would be two-sixths, or about 0.33.

So, the procedure for determining the probability of a particular event by this theoretical approach is:

1   List all the possible equally likely outcomes from the experiment, being guided by symmetry, but also thinking carefully to ensure that the outcomes listed really are equally likely.

2   Count in how many of these outcomes the event in question occurs.

3   Divide the second number by the first.

For example, to find the probability that a card drawn from a conventional pack of playing cards will be less than 7, we note that there are 52 equally likely outcomes from the experiment, that is, 52 possible cards that can be drawn. The event in question (the card is less than 7) occurs in 24 of these. So the probability is 24 ÷ 52, or about 0.46.

WHAT ABOUT THE 'LAW OF AVERAGES'?

There is no such law in mathematics. A popular misconception about probability is that the more times an event does not occur, the greater the probability of it occurring next time. If the events are independent (see below) then this is *not* how probability works. The outcome of throwing a die has no effect on the outcome of throwing it again.

It is important to remember what I said at the beginning of this chapter about the meaning of probability. It is a measure of how strongly you believe an event will happen. So when I say the probability of a coin turning up heads is 0.5, I am making a statement about how strongly I believe that it will come up heads, based on the symmetry of the coin. To a logical person, this kind of theoretical probability, provided the argument based on symmetry is valid, does not change from one outcome to the next. So the result of one trial does not affect the probabilities of what will happen in the next. If I have just thrown a head, the probability of the next toss being a head is still 0.5. If I have just thrown 20 tails in succession (which is unlikely but not impossible), the probability of the next one being a head is still only 0.5. (Of course, there might be something peculiar about the coin, but I am assuming that it is not bent or weighted in any way that might distort the results.)

What the probability does tell me, however, is that *in the long run*, if you go on tossing the coin long enough, you will see the relative frequency of heads (and tails) gradually getting closer and closer to 50%. This does not mean that with a thousand tosses I would *expect* 500 of each; in fact, that would be very surprising. But I would expect the proportion of heads to be about 50% and getting closer to 50% the more experiments I perform. It is therefore important for children studying probability actually to do

LEARNING AND TEACHING POINT

Emphasize the idea that probability does not tell you anything about what will happen next, but predicts what will happen in the long run.

such experiments a large number of times, obtain the relative frequencies of various outcomes for which they have determined the theoretical probability, and observe and discuss the fact that the two are not usually exactly the same.

There is a wonderfully mystical idea here: that in an experiment with a number of equally likely possible outcomes we cannot know what will be the outcome of any given experiment, but we can predict with a high degree of confidence what will happen in the long run.

HOW DO YOU DEAL THEORETICALLY WITH TOSSING TWO COINS OR THROWING TWO DICE?

We do have to be careful when arguing theoretically about possible outcomes to ensure that they are really all equally likely. For example, one group of children decided there were three possible outcomes when you toss two coins – two heads, two tails, one of each – and determined the probabilities to be $^1/_3$ for each. Then, performing the experiment 1000 times between them (40 times each for 25 children), they found that two heads turned up 256 times, two tails turned up 234 times and one of each turned up 510 times. So the relative frequencies were 25.6%, 23.4% and 51%, obviously not getting close to the 'theoretical' 33.3%. The problem is that these three outcomes are *not equally likely*. Calling the two coins A and B, we can identify *four* possible outcomes: A and B both heads, A head and B tail, A tail and B head, A and B both tails. So the theoretical probabilities of two heads, two tails and one of each are 0.25, 0.25 and 0.50 respectively.

In this example, the outcome of tossing coin A and the outcome of tossing coin B are technically called **independent events**. This means simply that what happens to coin B is not affected in any way by what happens to coin A, and vice versa. With experiments involving two independent events, such as two coins being tossed or two dice being thrown, a useful device for listing all the possible outcomes is a **two-way table**. Figure 28.1 is such a two-way table, showing the four possible outcomes from tossing two coins. This is an application of the use of rectangular grids of data discussed in Chapter 20 (see Figure 20.1). For example, Figure 20.1(c) gives all 36 possible outcomes, shown as total scores in the table, when two dice are thrown. From that table, we can discover, for instance, that the probability of scoring seven (seven occurs 6 times out of 36; $^6/_{36}$ = 0.17 approximately) is much higher than, say, scoring eleven (11 occurs 2 times out of 36; $^2/_{36}$ = 0.06 approximately).

Second coin

		Head(H)	Tail(T)
First coin	Head(H)	H+H	H+T
	Tail(T)	T+H	T+T

Outcomes from tossing two coins

Figure 28.1   A two-way table for an experiment with two independent events

An important principle in probability theory is that the probability of both of two independent events occurring is obtained by multiplying the probabilities of each one occurring. For example, if I toss a coin the probability of obtaining a head is 0.5. If I throw a die, the probability of scoring an even number is 0.5. So, if I toss the coin and throw the die simultaneously, the probability of getting a head *and* an even number is $0.5 \times 0.5 = 0.25$. This principle can be expressed as a generalization as follows:

If the probabilities of two independent events A and B are p and q, then the probability of both A and B occurring is p × q.

WHAT ARE MUTUALLY EXCLUSIVE EVENTS?

Events that cannot possibly occur at the same time are said to be **mutually exclusive**. For example, if I throw two dice, getting a total score of 7 and getting a total score of 11 are two mutually exclusive outcomes – since you cannot score both 7 and 11 simultaneously. However, getting a total score of 7 and getting a total score that is odd are not mutually exclusive events, since clearly you can do both at the same time. The reader should note that mutually exclusive events are definitely not independent, because if one occurs then the other one cannot.

A second important principle of probability theory is that the probability of one or other of two mutually exclusive events occurring is the sum of their probabilities. So, for example, the probability of scoring 7 or 11 when I throw two dice is the sum of $^6/_{36}$ (the probability of scoring 7) and $^2/_{36}$ (the probability of scoring 11), that is, $^8/_{36}$, or 0.22, approximately. This principle can be expressed as a generalization as follows:

If the probabilities of two mutually exclusive events A and B are p and q, then the probability of either A or B occurring is p + q.

If you list a set of mutually exclusive events that might occur in a particular experiment that cover all possible outcomes, then the sum of all their probabilities must equal 1. For example, in throwing two coins we could identify these three mutually exclusive events: two heads (probability 0.25), two tails (probability 0.25), one head and one tail (probability 0.5). The sum of these probabilities is $0.25 + 0.25 + 0.5 = 1$.

> **LEARNING AND TEACHING POINT**
>
> One obvious application of probability is to betting and lotteries. Be aware that some parents will hold strong moral views about gambling, so handle discussion of probability in a way that is sensitive to different perspectives on this subject.

HOW DO YOU ASSESS RISK?

Taking a risk is when you invest some money or time or other resources into some action or option in the hope that the outcome will produce some reward. A simple mathematical model for assessing risk is as follows. For any individual event, the 'expected value' in taking a risk is obtained by multiplying the probability of the desired outcome occurring by the value of the reward associated with it. For example, imagine you bought a lottery ticket for £1 in the hope of winning a prize of £100, and there were 1000 lottery tickets sold. The probability of having the winning ticket is 0.001,

> **LEARNING AND TEACHING POINT**
>
> Children in primary school can discuss the risk associated with various actions and can begin to understand how an assessment of the probability of a particular outcome and the value of the reward associated with it might modify their behaviour and choices.

so the expected value of the ticket you purchased is £100 × 0.001 = £0.10. Given that you spent £1 on the ticket, this represents an expected loss of 90p. What this means is that, on average, over time, if you continue repeating this action, you will lose 90p in every £1 invested.

RESEARCH FOCUS

Schlottman (2001) investigated younger children's intuitive understanding of risk and whether they could simultaneously take into account both the likelihood of an outcome and the reward associated with it. Some six-year-olds were asked to judge how happy a puppet would be to play a game in which the puppet would win a large or a small prize (numbers of crayons) depending on where a marble finished up in a tube. She discovered that these young children seemed intuitively to have a sense of the probability of winning the prizes and how the probability of winning and the value of the prize were integrated multiplicatively into a sense of how good a game it was for the puppet to play. The evidence here is that young children demonstrate a functional understanding of probability and expected value.

LEARNING RESOURCES

Access activities for your **lesson plans** at: https://study.sagepub.com/haylock7e

Before trying the self-assessment questions below, you should complete the **self-assessment questions** for this chapter at: https://study.sagepub.com/haylock7e

28.1: The table below shows the frequency of various numbers of letters in a random sample of 100 words in this book (ignoring numerals):

No. of letters	1	2	3	4	5	6	7	8	9	10	11	12
Frequency		4	21	19	14	17	7	6	9	1	1	0	1

Using the data in this sample, estimate the probability that a word chosen at random in this book will have fewer than six letters in it.

28.2: What would be the most appropriate way to determine the probability that:

a. a drawing-pin will land point-up when tossed in the air?

b. a person aged 50–59 years in England will have two living parents?

c. the total score when two dice are thrown is an even number?

28.3: I throw two conventional dice. Write down an outcome that has a probability of 0 and another outcome that has a probability of 1.

28.4: If you draw a card at random from a pack of playing cards, the probability that the card will be an ace is $^1/_{13}$. The probability that it will be a black card is $^1/_2$. Are these two outcomes independent? Are they mutually exclusive? What is the probability of getting a black ace?

28.5: If a shoe is tossed in the air, the probability of it landing the right way up is found by experiment to be 0.35. The probability that it will land upside down is found to be 0.20. Are these two events independent? Are they mutually exclusive outcomes? What is the probability of the shoe landing either the right way up or upside down? The only other possible outcome is that it lands on one of its sides; what is the probability of this?

FURTHER PRACTICE

FROM THE STUDENT WORKBOOK

Questions 28.01–28.18: Checking understanding (probability)

Questions 28.19–28.30: Reasoning and problem solving (probability)

Questions 28.31–28.38: Learning and teaching (probability)

GLOSSARY OF KEY TERMS INTRODUCED IN CHAPTER 28

Probability   A mathematical measure of the strength of our belief that some event will occur, based on whatever evidence we can assemble; a measure of how likely an event is to happen.

Probability scale   A scale for measuring probability, ranging from 0 (impossible) to 1 (certain).

Evens   Where we judge an event to be as likely to happen as not to happen; probability = 0.5.

Subjective probability   An estimate of the probability of some event occurring based on subjective judgements of the available evidence.

Relative frequency   An estimate of the probability of an event occurring in the members of a population, obtained from the ratio of the number of times an event is recorded in a sample to the total number in the sample.

Experimental probability   An estimate of the probability of an event occurring, obtained from repeating an experiment a large number of times and finding the ratio of the number of times an event occurs to the total number of trials.

Theoretical probability   An estimate of probability based on theoretical arguments of symmetry and equally likely outcomes; if there are n equally likely outcomes from an experiment, the probability of each one occurring is $^{1}/_{n}$.

Independent events   Two (or more) events where whether or not one occurs is completely independent of the other; for example, throw 6 on the red die, throw 6 on the blue die. The probability of both of two independent events occurring is the product of their individual probabilities.

Two-way table   A systematic way of identifying in a rectangular array all the possible combinations of the values of two variables; used in probability to identify all the possible combinations of two independent events. See Figure 28.1 and Figure 20.1(c) in Chapter 20.

Mutually exclusive events   Two (or more) events such that if one occurs then the other cannot occur; for example, throw 6 on the blue die, throw 5 on the blue die. The probability that one or other of a number of mutually exclusive events will occur is the sum of their individual probabilities.

Expected value   A measure used in assessing risk. A simple model for expected value of an action is the product of the probability of success and the value of the reward associated with it.

SUGGESTIONS FOR FURTHER READING FOR SECTION H

1   Chapter 10 ('Understanding data handling') of Haylock and Cockburn (2017) uses data from a class of 7–8-year-olds to explain different kinds of data and various ways of representing it pictorially.

2   Read the entry on 'Cross-curricular mathematics' in Haylock with Thangata (2007). Handling data is one of the most obvious topics in the mathematics curriculum to be taught through a cross-curricular approach. The entry on this subject in this book clarifies the relationship between mathematics and other curriculum areas and provides some classroom examples.

3   Have a look at the interesting and useful analysis of children's errors and misconceptions in statistics in Chapter 9 of Hansen (2020).

4   Chapter 3 of Nickson (2004) is a good source for information on research into the learning of data handling and statistics by children in primary school.

5   Section 4 (Statistics) of Cooke (2007) will provide useful reinforcement of the mathematical ideas on data handling and comparing sets of data that have been outlined in Chapters 26 and 27. And, for an insightful chapter on probability, read Chapter 7 of this book.

6   For more on the ideas of theoretical probability, experimental probability, mutually exclusive events, independent and dependent events, see Hopkins, Pope and Pepperell (2005), section 3.3 on 'Probability'.

7   If you really want to get to grips with probability theory applied to everyday life problems, try working through the early chapters of Tijms (2012).

8   Finally, Nunes et al. (2015), in an extensive article on teaching mathematical reasoning in primary school, provide some insightful material on the teaching and learning of probability.

ANSWERS TO SELF-ASSESSMENT QUESTIONS

Particularly where the question asks for the invention of a sentence, a question, a method or a problem, the answers provided are only examples of possible valid responses.

CHAPTER 3: LEARNING HOW TO LEARN MATHEMATICS

3.1: (a) The formal mathematical language would be 'five add three equals eight'. (b) Putting up 5 fingers on one hand and 3 on the other, children might count all the fingers and say, 'five and three make eight altogether'. (c) Children might start at 5 on the number strip and count on 3 to get to 8.

3.2: Same: AB and DC are parallel to each other in both shapes; the areas are the same. Different: the diagonal line goes up from left to right in one and down from left to right in the other; AD is on the left of one shape and on the right of the other.

CHAPTER 4: KEY PROCESSES IN MATHEMATICAL REASONING

4.1: (a) A counter-example is 4, which is less than 10 but not greater than 5. (b) A counter-example is 9, which is not a multiple of 6 but is a multiple of 3.

4.2: (a) Incorrect because, for example, there are four multiples of 3 in the decade 21–30 (21, 24, 27 and 30). (b) True. In each third decade, the numbers ending in 1, 4, 7 and 0 are all multiples of 3 (for example, 51, 54, 57 and 60).

4.3: If there are, for example, 10 tiles along the edge, then you multiply this by 4 because there are 4 edges. But when you do this, the tiles in the corners get counted twice. So you have to subtract 4. It would be the same whatever the number of tiles along the edge.

4.4: The number of matches is double the number of triangles, plus 1. Explanation: put down 1 match, then 2 further matches are needed to make a triangle and each subsequent triangle. To make zero triangles does not require any matches, so this is a special case that does not fit the generalization.

4.5: This is because $7 \times 11 \times 13 = 1001$ and any six-digit number *abcabc* is the three-digit number *abc* multiplied by 1001. For example, the number 346,346 is equal to 346×1001.

4.6: I have tried to lead you astray here, expecting the answers: (a) 80°, (b) 100°, (c) 120°. The last of these is impossible because water boils at 100°. Also, the larger flask could be cooler than 50°. A little bit of flexibility in your thinking is required to obtain the correct answers: (a) 80° or 20°, (b) 100° or 0°, (c) −20° (the water is now ice, of course).

CHAPTER 5: MODELLING AND PROBLEM SOLVING

5.1: Mathematical model is $14.95 + 25.90 + 19.95$; mathematical solution, using a calculator, is 60.8; interpretation in the real world is £60.80; result is that the total cost of books is £60.80.

5.2: (a) A multi-buy of 3 cartons of juice costs £4.50. A single carton costs £1.80. How much do I save per carton by opting for the multi-buy? (Divide £4.50 by 3 and subtract the result from £1.80.) (b) Butter A costs £1.40 per pack and butter B costs £1.65 per pack. How much more does it cost me to buy 8 packs of butter B? (Subtract £1.40 from £1.65 and multiply the result by 8.)

5.3: Problem 1. The numbers are 7, 13 and 30. Hint: if you add the given numbers, 20, 43 and 37 (= 100), each of the boxes is counted twice, so the three numbers total 50.

Problem 2. See Figure A.

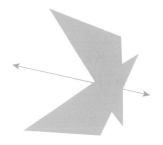

Figure A   Solution to SAQ 5.3, Problem 2

5.4:   Problem 8. Assuming you buy some of each, you could get: 5 snakes and 8 alligators; 10 snakes and 6 alligators; 15 snakes and 4 alligators; 20 snakes and 2 alligators.

Problem 9. Take 60 children, because this is the largest number (less than 80) that can be divided by 3, 4, 5 and 6.

CHAPTER 6: NUMBERS AND PLACE VALUE

6.1:   If you understand 'number' to mean integer (whole number) or natural number, the answer is 200. Otherwise there is no next number.

6.2:   (a) Impossible to say, or, if you like, an infinite number; (b) 19.

6.3:   CLVIII, CCLXVII, CCC, DCXIII, DCC (158, 267, 300, 613, 700).

6.4:   (a) 1954; (b) 1492.

6.5:   Four thousand one hundred (4099 + 1 = 4100).

6.6:   (a) $516 = (5 \times 10^2) + (1 \times 10^1) + 6$; (b) $3060 = (3 \times 10^3) + (6 \times 10^1)$; (c) $2,305,004 = (2 \times 10^6) + (3 \times 10^5) + (5 \times 10^3) + 4$.

6.7:   6 orange, 2 red, 4 white.

6.8:   (a) 101 > 98; (b) 998 < 1001; (c) 48 > 38 > 28.

6.9:   499 < 500 < 3500 < 3998 < 4002.

CHAPTER 7: ADDITION AND SUBTRACTION STRUCTURES

7.1:   I buy two books costing £5.95 and £6.95. What is the total cost?

7.2:   My monthly salary was £1750 and then I had a rise of £145. What is my new salary?

7.3:   The class's morning consists of 15 minutes registration, 25 minutes assembly, 55 minutes mathematics, 20 minutes break, 65 minutes English. What is the total time in minutes?

7.4:   The Australian Chardonnay is £6.95 and the Hungarian is £4.99. How much cheaper is the Hungarian?

7.5:   There are 250 pupils in a school; 159 have school lunches. How many do not?

7.6:   I want to buy a bicycle costing £989, but have only £650. How much more do I need?

CHAPTER 8: MENTAL STRATEGIES FOR ADDITION AND SUBTRACTION

8.1:   (a) $67 - (20 - 8) = (67 - 20) + 8$; general rule: $a - (b - c) = (a - b) + c$.
(b) $67 - (20 + 8) = (67 - 20) - 8$; general rule: $a - (b + c) = (a - b) - c$.

8.2:   The child may have (a) added, or done $2 - 0$ rather than $0 - 2$ in the tens column; (b) assumed that any subtraction involving zero gives the answer zero; (c) remembered decomposition recipe wrongly and written a little 9 instead of a little 1; (d) done $7 - 1$ instead of $1 - 7$ in units, and again mystified by the zero in the tens column; (e) consistently taken the smaller digit from the larger; (f) mis-remembered the decomposition recipe and failed to indicate that one of the two hundreds had been exchanged for 10 tens.

8.3: $500 + 200$ makes 700; $30 + 90$ makes 120, that's 820 in total so far; $8 + 4$ makes 12, add this to the 820, to get 832.

8.4: $423 + 98 = 423 + 100 - 2 = 523 - 2 = 521$.

8.5: $297 + 304 =$ double $300 - 3 + 4 = 601$.

8.6: $494 + 307 = 494 + 6 + 301 = 500 + 301 = 801$.

8.7: $26 + 77 = 25 + 75 + 1 + 2 = 100 + 3 = 103$.

8.8: $819 - 519 = 300$, so $819 - 523 = 300 - 4 = 296$.

8.9: $389 + 11 = 400$; add 300 to get to 700; then add 32 to get to 732. Then, $11 + 300 + 32 = 343$.

8.10: (a) 435; (b) 163; (c) 806; (d) 6994; (e) 2.

CHAPTER 9: WRITTEN METHODS FOR ADDITION AND SUBTRACTION

9.1:   Put out 2 orange and 8 white counters for 208; then 1 orange, 5 reds and 6 whites for 156; then 9 reds and 7 whites for 97. There are 21 whites; exchange 20 of these for 2 reds, leaving 1 white; there are now 16 reds; exchange 10 of these for 1 orange, leaving 6 reds. There are now 4 orange counters, 6 reds and 1 white, that is, 461.

9.2:   Put out 6 orange, 2 reds and 3 white counters. Take away 1 white, leaving 2 whites; not enough reds, so exchange 1 orange for 10 reds, giving 12 reds; take away 7 of these, leaving 5 reds; now take away 4 orange counters; the result is 1 orange, 5 reds and 2 whites, that is, 152.

9.3:   Put out 2 thousands and 6 units; not enough units to take away 8, no tens to exchange and no hundreds, so exchange 1 thousand for 10 hundreds, leaving 1 thousand; now exchange 1 hundred for 10 tens, leaving 9 hundreds; then exchange 1 ten for 10 units, leaving 9 tens and giving 16 units; can now take away 8 units, 3 tens and 4 hundreds; the result is 1 thousand, 5 hundreds, 6 tens and 8 units, that is, 1568.

9.4:   Add 2 to both numbers, to give 2008 – 440; add 60 to both, to give 2068 – 500; add 500 to both, to give 2568 – 1000; answer: 1568.

CHAPTER 10: MULTIPLICATION AND DIVISION STRUCTURES

10.1:   I bought 29 boxes of eggs with 12 eggs in each box … (29 lots of 12). There are 12 classes in the school with 29 children in each class … (12 lots of 29).

10.2:   I bought 12 kg of potatoes at 25p per kilogram. What was the total cost?

10.3:   The box can be seen as 4 rows of 6 yoghurts or 6 rows of 4 yoghurts.

10.4:   If the length of the wing in the model is 16 cm, how long is the length of the wing on the actual aeroplane? ($16 \times 25 = 400$ cm).

10.5:   Scale factor is 20 ($300 \div 15$); an example of the ratio structure.

10.6:   I need 25 months ($300 \div 12$); this is an example of the inverse-of-multiplication structure, using the idea of repeated addition.

10.7:   A packet of four chocolate bars costs 60p; how much per bar?

10.8:   How many toys costing £4 each can I afford if I have £60 to spend?

10.9:   A teacher earns £1950 a month, a bank manager earns £6240. How many times greater is the bank manager's monthly salary? ($6240 \div 1950 = 3.2$); the bank manager's monthly salary is 3.2 times that of the teacher.

CHAPTER 11: MENTAL STRATEGIES FOR MULTIPLICATION AND DIVISION

11.1:   16 lots of 25 is easier to calculate than 25 lots of 16; $4 \times 25 = 100$, so $16 \times 25 = 400$.

11.2: $25 \times 24 = 25 \times (4 \times 6) = (25 \times 4) \times 6 = 100 \times 6 = 600$.

11.3: $25 \times (20 + 4) = (25 \times 20) + (25 \times 4) = 500 + 100 = 600$.

11.4: $22 \times (40 – 2) = (22 \times 40) – (22 \times 2) = 880 – 44 = 836$.

11.5: $4 \times 90 = 360$; $40 \times 9 = 360$; $40 \times 90 = 3600$; $4 \times 900 = 3600$; $400 \times 9 = 3600$; $40 \times 900 = 36{,}000$; $400 \times 90 = 36{,}000$; $400 \times 900 = 360{,}000$.

11.6: $48 \times 25 = 12 \times 4 \times 25 = 12 \times 100 = 1200$.

11.7: $2 \times 103 = 206$; $4 \times 103 = 412$; $8 \times 103 = 824$; $16 \times 103 = 1648$. So, $26 \times 103 = 206 + 824 + 1648 = 2678$.

11.8: $10 \times 103 = 1030$ and $2 \times 103 = 206$. So, $26 \times 103 = 1030 + 1030 + 206 + 206 + 206 = 2678$.

11.9: (a) $154 \div 22$ is the same as $(88 + 66) \div 22$ which equals $(88 \div 22) + (66 \div 22)$; hence the answer is $4 + 3 = 7$.
(b) $154 \div 22$ is the same as $(220 - 66) \div 22$ which equals $(220 \div 22) - (66 \div 22)$; hence the answer is $10 - 3 = 7$.

11.10: $10 \times 21 = 210$; another 10×21 makes this up to 420; $2 \times 21 = 42$, which brings us to 462; 1 more 21 makes 483; answer is $10 + 10 + 2 + 1 = 23$.

11.11: $385 \div 55 = 770 \div 110$ (doubling both numbers) $= 7$.

CHAPTER 12: WRITTEN METHODS FOR MULTIPLICATION AND DIVISION

12.1: The four areas are 40×30, 40×7, 2×30 and 2×7, giving a total of $1200 + 280 + 60 + 14 = 1554$.

12.2: The six areas are 300×10, 300×7, 40×10, 40×7, 5×10 and 5×7, giving a total of $3000 + 2100 + 400 + 280 + 50 + 35 = 5865$.

12.3: From 126 take away 10 sevens (70), leaving 56, then 5 sevens (35), leaving 21, which is 3 sevens; answer is therefore $10 + 5 + 3$, that is, 18.

12.4: From 851 take away 20 lots of 23 (460), leaving 391, then 10 more (230), leaving 161, then 5 more (115), leaving 46, which is 2×23; answer is therefore $20 + 10 + 5 + 2$, that is, 37.

12.5: From 529 take away 50 lots of 8 (400), then 10 more (80), then 5 more (40), then 1 more (8), leaving a remainder of 1; answer is $50 + 10 + 5 + 1 = 66$, remainder 1.

12.6: 432×50 ($= 21{,}600$) and 432×7 ($= 3024$).

12.7: Answers only provided. (a) 448; (b) 18,392; (c) 23; (d) 72, remainder 9.

CHAPTER 13: NATURAL NUMBERS: SOME KEY CONCEPTS

13.1: 20 is a factor of 100; 21 is a triangle number; 22 is a multiple of 2 and 11; 23 is a prime number; 24 has eight factors; 25 is a square number; 26 is a multiple of 2 and 13; 27 is a cube number; 28 is a triangle number; 29 is a prime number.

13.2: $3 \times 37 = 111$, $6 \times 37 = 222$, $9 \times 37 = 333$, $12 \times 37 = 444$, $15 \times 37 = 555$, $18 \times 37 = 666$, $21 \times 37 = 777$, $24 \times 37 = 888$, $27 \times 37 = 999$; the pattern breaks down when 4-digit answers appear.

13.3: (a) 2652 is a multiple of 2 (ends in even digit), 3 (sum of digits is multiple of 3), 4 (last two digits multiple of 4), 6 (multiple of 2 and 3); (b) 6570 is a multiple of 2 (ends in even digit), 3 (sum of digits is multiple of 3), 5 (ends in 0), 6 (multiple of 2 and 3), 9 (digital root is 9); (c) 2401 is a multiple of none of these (it is 7 × 7 × 7 × 7).

13.4: A three-digit number is a multiple of 11 if the sum of the two outside digits is equal to the middle digit (for example, 561, 594 and 330) or 11 more than the middle digit (for example, 418 and 979).

13.5: 24 (the lowest common multiple of 8 and 12).

13.6: (a) Factors of 95 are 1, 5, 19, 95; (b) factors of 96 are 1, 2, 3, 4, 6, 8, 12, 16, 24, 32, 48 and 96; (c) factors of 97 are 1 and 97 (it is prime); clearly 96 is the most flexible.

13.7: Factors of 48 are 1, 2, 3, 4, 6, 8, 12, 16, 24, 48; factors of 80 are 1, 2, 4, 5, 8, 10, 16, 20, 40, 80; common factors are 1, 2, 4, 8, 16; the highest common factor is 16.

13.8: The sequence is 5, 7, 11, 13, 17, 19, 23, 25, 29, 31, 35, 37, 41, 43, 47, 49, 53, 55, 59, 61; they are all prime except 25, 35, 49 and 55.

13.9: The 1 by 1 square does fit this pattern: $1^2 = (2 \times 0) + 1$.

13.10: The differences between successive square numbers are 3, 5, 7, 9, 11 …, the odd numbers; these are the numbers of dots added to each square in Figure 13.6(a) to make the next one in the sequence.

13.11: The answers are the square numbers: 4, 9, 16, 25, 36 and so on; two successive triangles of dots in Figure 13.10 can be fitted together to make a square number.

CHAPTER 14: INTEGERS: POSITIVE AND NEGATIVE

14.1: The order is B (+3), A (–4), C (–5).

14.2: (a) The temperature one winter's day is 4 °C; that night it falls by 12 degrees; what is the night-time temperature? (Answer –8); (b) the temperature one winter's night is –6 °C; when it rises by 10 degrees what is the temperature? (Answer: 4.)

14.3: (a) If I am overdrawn by £5, how much must be paid into my account to make the balance £20? (Answer: 25); (b) if I am overdrawn by £15, how much must be paid into my account so that I am only overdrawn by £10? (Answer: 5); (c) if I am overdrawn by £10 and withdraw a further £20, what is my new balance? (Answer: –30.)

14.4: The mathematical model is for the difference between the two balances: 458.64 – (–187.85); the cheque paid in was £646.49.

CHAPTER 15: FRACTIONS AND RATIOS

15.1:   (a) A bar of chocolate is cut into 5 equal pieces and I have 4 of them; (b) $^4/_5$ of a class of 30 children is 24 children; (c) share 4 pizzas equally between 5 people; (d) if I earn £400 a week and you earn £500 a week, my earnings are $^4/_5$ of yours.

15.2:   (a) $^1/_4 = {}^2/_8 = {}^3/_{12}$; (b) $^1/_2 = {}^2/_4 = {}^3/_6 = {}^4/_8 = {}^6/_{12}$; (c) $^3/_4 = {}^6/_8 = {}^9/_{12}$; (d) $^4/_4 = {}^8/_8 = {}^{12}/_{12} = {}^6/_6 = {}^3/_3 = {}^2/_2 = 1$; (e) $^2/_{12} = {}^1/_6$; (f) $^4/_{12} = {}^2/_6 = {}^1/_3$; (g) $^8/_{12} = {}^4/_6 = {}^2/_3$; (h) $^{10}/_{12} = {}^5/_6$.

15.3: $^3/_5$ is $^{24}/_{40}$ and $^5/_8$ is $^{25}/_{40}$. The latter would be slightly more pizza.

15.4: $^1/_6$ ($^2/_{12}$), $^1/_3$ ($^4/_{12}$), $^5/_{12}$, $^2/_3$ ($^8/_{12}$), $^3/_4$ ($^9/_{12}$).

15.5:   Compare (by ratio) the prices of two pots, pot A costing £15, pot B costing £25.

15.6: $^9/_{24}$ or $^3/_8$.

15.7:   (a) $^1/_5$ of £100 is £20, so $^3/_5$ is £60; (b) £1562.50 (2500 ÷ 8 × 5).

15.8:   (a) $1^3/_8$; (b) $^7/_{12}$; (c) $^1/_4$; (d) $^6/_8$ (or $^3/_4$); (e) $^5/_{12}$.

CHAPTER 16: DECIMAL NUMBERS AND ROUNDING

16.1:   3.2 is 3 flats and 2 longs; 3.05 is 3 flats and 5 small cubes; 3.15 is 3 flats, 1 long and 5 small cubes; 3.10 is 3 flats and 1 long. In order: 3.05, 3.10, 3.15, 3.2.

16.2:   3.608 lies between 3 and 4; between 3.6 and 3.7; between 3.60 and 3.61; between 3.607 and 3.609.

16.3:   The mathematical model is: 124 × 5.95 (or 5.95 × 124); the mathematical solution is 737.8; interpretation: the total cost of the order will be £737.80; to the nearest ten pounds, the total cost of the order will be about £740; to three significant digits, the total cost of the order will be about £738.

16.4:   (a) 327 ÷ 40 = 8.175 (calculator), or 8 remainder 7. So 9 coaches are needed. We round *up* (otherwise we would have to leave 7 children behind); (b) 500 ÷ 65 = 7.6923076 (calculator), or 7 remainder 45. So we can buy 7 cakes. We round *down* (and have 45p change for something else).

16.5:   (a) How many buses holding 50 children do we need to transport 320 children? We need 7 buses. (b) How many tables costing £50 each can we afford with a budget of £320? We can afford 6 tables.

16.6:   Mathematical model is 27.90 ÷ 3; mathematical solution (calculator answer) is 9.3; this is an exact but slightly inappropriate answer, because of the convention of 2 digits after the point for money; interpretation is that each person pays £9.30.

16.7:   Mathematical model is 39.70 ÷ 3; mathematical solution (calculator answer) is 13.233333; this is an answer that has been truncated; interpretation is that each person owes £13.23 and a little bit; two people pay £13.23, but one has to pay £13.24.

16.8:   Calculator result is 131.88888; the average height is 132 cm to the nearest cm.

16.9:   (a) 3; (b) 3.2; (c) 3.16.

CHAPTER 17: CALCULATIONS WITH DECIMALS

17.1:   (a) 0.080 + 1.220 + 0.015 = 1.315; (b) 10.50 − 1.05 = 9.45.

17.2:   How much for 4 box files costing £3.99 each? Answer: £15.96. (Find £4 multiplied by 4, then compensate.)

17.3:   Divide a 4.40-m length of wood into 8 equal parts; each part is 0.55 m (55 cm) long. (Change the calculation to 440 cm divided by 8.)

17.4:   (a) 18.125; (b) 20.71.

17.5:   On a calculator, 3500 ÷ 17 = 205.882353. This rounds down to 205 books per shop; that's 205 × 17 = 3485 books altogether, so the remainder is 3500 − 3485 = 15 books.

17.6:   (a) Answer should be about 3 × 1 = 3, so 2.66; (b) answer should be about 10 × 0.1 = 1, so 0.964; (c) answer should be about 30 ÷ 1 = 30, so 31.

CHAPTER 18: PROPORTIONALITY AND PERCENTAGES

18.1:   Since one-third is about 33%, this is the greater reduction.

18.2:   (a) 13 out of 50 is the same proportion as 26 out of 100. So 26% do not achieve the standard and 74% do. (b) 57 out of 300 is the same proportion as 19 out of 100. So 19% do not achieve the standard and 81% do. (c) 24 out of 80 is the same proportion as 3 out of 10, or 30 out of 100. So 30% do not achieve the standard and 70% do. (d) 26 out of 130 is the same proportion as 2 out of 10, or 20 out of 100. So 20% do not achieve the standard and 80% do.

18.3:   English, about 42% vowels; Italian, about 49% vowels.

18.4: $^5/_8$ is 0.125 × 5 = 0.625, equivalent to 62.5%. $^7/_8$ is 0.125 × 7 = 0.875, equivalent to 87.5%.

18.5:   £271.04 (275 × 1.12 × 0.88).

CHAPTER 19: ALGEBRAIC REASONING

19.1: The relationship is $f = 3y$. Criticism: using f and y is misleading, since they look like abbreviations for a foot and a yard, instead of variables (the number of feet and the number of yards).

19.2: (a) The total number of pieces of fruit bought; (b) the cost of the apples in pence; (c) the cost of the bananas; (d) the total cost of the fruit. Criticism: using a and b is misleading, since they look like abbreviations for an apple and a banana; so $10a + 12b$ looks as though it means 10 apples and 12 bananas.

19.3: (a) Jenny can have 11 rides. Arithmetic steps: 12 divided by 2, add 5. (b) Total cost of n rides is $2(n - 5)$. (c) The number of rides Jenny can have is the solution to the equation $2(n - 5) = 12$.

19.4: (a) 60; (b) 10.

19.5: For Figure 19.4(c): (a) add 5; (b) 498; (c) multiply by 5, subtract 2; (d) $y = 5x - 2$.
For Figure 19.4(d): (a) subtract 1; (b) 0; (c) subtract from 100; (d) $y = 100 - x$.

19.6: (a) Add 2; (b) 204; (c) multiply by 2, add 4; (d) $y = 2x + 4$.

19.7: My number is 42; the equation is $x(2x + 3) = 3654$.

19.8: The first 10 triangle numbers are 1, 3, 6, 10, 15, 21, 28, 36, 45, 55.
Their doubles are $2 = 1 \times 2$, $6 = 2 \times 3$, $12 = 3 \times 4$, $20 = 4 \times 5$, $30 = 5 \times 6$, $42 = 6 \times 7$, $56 = 7 \times 8$, $72 = 8 \times 9$, $90 = 9 \times 10$, $110 = 10 \times 11$.
So the nth triangle number doubled is $n \times (n + 1)$.
Hence, the nth triangle number is $^1/_2 n \times (n + 1)$.
So the one-hundredth triangle number (which equals $1 + 2 + 3 + \ldots + 100$) is $50 \times 101 = 5050$.

19.9: The nth odd number is $2n - 1$.

CHAPTER 20: COORDINATES AND LINEAR RELATIONSHIPS

20.1: The points are (1, 2), (1, 4), (2, 5), (4, 5), (5, 4), (5, 2), (4, 1) and (2, 1).

20.2: They are all linear relationships, producing straight-line graphs.

20.3: (a) The total number of eggs; (b) the total number of beats (in waltz time, there are 3 beats in each bar); (c) the top number in the fractions in the set.

20.4: The point (2.5, 6) on the line shows that $2x + 1 = 6$ when $x = 2.5$.

20.5: Using the x-axis for weights in stones, the straight-line graph should pass through (0, 0) and (11, 70). Then, for example, the point (10, 64) on this line (approximately) converts 10 stone to about 64 kg, and the point (9.4, 60) gives 9.4 stone as the approximate equivalent of 60 kg.

20.6: (a) 6 km; (b) 110 minutes; (c) 6 km per hour and 3 km per hour respectively.

CHAPTER 21: CONCEPTS AND PRINCIPLES OF MEASUREMENT

21.1: It works as far as 55 miles, which is 89 km to the nearest km (88.5115). The next value, 89 miles, is 143 km to the nearest km (143.2277), rather than the Fibonacci number, 144.

21.2: (a) Yes; if A is earlier than B and B is earlier than C, then A must be earlier than C. (b) No; for example, 20 cm is half of 40 cm and 40 cm is half of 80 cm, but 20 cm is not half of 80 cm.

21.3: (a) 297 mm; (b) 29.7 cm; (c) 2.97 dm; (d) 0.297 m.

21.4: (a) Possible – 7 tonnes is about average for an adult male African elephant; (b) impossible – it would be much more than that, probably more than 400 litres; (c) impossible to make this claim – all measurements are approximate; (d) possible; (e) impossible – I might manage it in a month; (f) possible – you should get away with one first-class stamp.

CHAPTER 22: PERIMETER, AREA AND VOLUME

22.1: The rectangle with the largest perimeter will be 36 cm long and 1 cm wide. This has a perimeter of 74 cm.

22.2: (a) As a cube with side 3 units; total surface area = 54 square units. (b) As a cuboid, 4 units by 4 units by 3 units (the nearest you can get to a cube); total surface area = 80 square units (16 + 16 + 12 + 12 + 12 + 12).

22.3: The area is 25 mm², or 0.25 cm², or 0.000025 m².

22.4: Volume is 125 cm³ or 0.000125 m³.

22.5: 20 × 25 × 10 = 5000 cuboids needed.

22.6: About 78.5 cm (25 × 3.14); 80 cm to be on the safe side.

22.7: About 127 m (400 ÷ 3.14).

CHAPTER 23: ANGLE

23.1: (a) 135°; (b) 225°; (c) Most people can make an angle of about 35°.

23.2: $\frac{1}{8}$ of a turn (acute), 89° (acute), 90° (right), 95° (obtuse), 150° (obtuse), 2 right angles (straight), 200° (reflex), $\frac{3}{4}$ of a turn (reflex).

23.3: (a) 6 right angles; (b) 8 right angles; (c) 10 right angles. The sequential rule is 'add two right angles'. For a two-dimensional shape with N sides, the global rule for the sum of the angles is $2N - 4$. When $N = 100$, the sum of the angles is 196 right angles.

23.4: 360°, one complete rotation. The sum of the external angles of a quadrilateral is 360°, the same as that of a triangle.

CHAPTER 24: TRANSFORMATIONS AND SYMMETRY

24.1:   (a) Translation, moving –6 units in x-direction, 0 units in y-direction; (b) reflection in vertical line passing through (8, 0); (c) rotation through half-turn about (5, 2), clockwise or anticlockwise.

24.2:   (a) Shape R is enlarged to Shape P by a scale factor of 2, so the area (5 square units) will be increased by a factor of 4 (2^2); Shape P will therefore require 20 square units; Shape R is enlarged to Shape Q by a scale factor of 6, so the area will be increased by a factor of 36 (6^2); Shape Q will therefore require 180 square units. (b) Because R is transformed into Q by a scaling with factor 6, Q to R requires a scaling with factor $^1/_6$. These transformations are inverses of each other.

24.3:   A diagonal line passing through (11, 3) and (13, 1); a vertical line passing through (12, 2); a horizontal line passing through (12, 2). The order of rotational symmetry is four.

24.4:   E has two lines of symmetry and rotational symmetry of order two. F has one line of symmetry. G has five lines of symmetry and rotational symmetry of order five.

CHAPTER 25: CLASSIFYING SHAPES

25.1:   Because all three angles in an equilateral triangle must be 60°.

25.2:   (a) Square; (b) square; (c) equilateral triangle; (d) cuboid; (e) tetrahedron.

25.3:   The parallelogram tessellates, as do all quadrilaterals. The regular pentagon and regular octagon do not tessellate.

25.4:   Nine.

25.5:   See Figure B on page opposite.

CHAPTER 26: HANDLING DATA

26.1:   The four subsets are: 11-year-olds who walked; not 11-year-olds who walked; 11-year-olds who did not walk; and not 11-year-olds who did not walk. (a) Two overlapping circles, one representing 11-year-olds, the other those who walked. (b) A 2 by 2 grid, with columns labelled 11-year-olds and not 11-year-olds, and rows labelled walked, did not walk.

26.2:   (a) Which way of travelling to school is used by most children? How many fewer children walk than come by car? (b) How many children have fewer than five

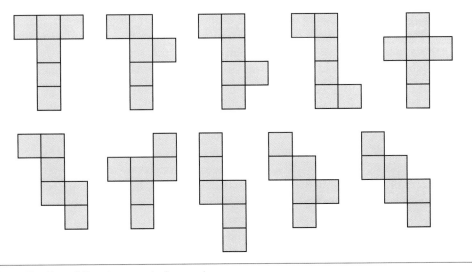

Figure B   Ten of the eleven nets for a cube

writing implements? Which group has no children in it? (c) How many have stride measurements in the range 60 to 64 cm, to the nearest centimetre? How many have stride measurements to the nearest centimetre that are greater than 89 cm?

26.3:   Fifty-pence intervals would produce 10 groups: £0.00–£0.49, £0.50–£0.99, £1.00–£1.49 and so on.

26.4:   (a) 140 degrees; (b) about 153 degrees (calculator answer: 360 ÷ 33 × 14 = 152.72727).

26.5:   Particularly for opinions about what time school should start, the first 50 pupils arriving in the morning are unlikely to be a representative sample. A systematic way of getting a representative sample (of 48) would be to select 12 children at random from each of the four year-groups.

CHAPTER 27: COMPARING SETS OF DATA

27.1:   The girls in the sample were generally better than the boys at recalling names of daily newspapers. A greater proportion of boys could name none. A greater proportion of girls could name three. More than half the girls could name two or three, compared with about a quarter of the boys.

27.2:   (a) Group A English, mean = 53.7, median = 53; group B English, mean = 67.5, median = 71. Both averages support the view that group B did better. (b) The mean for English for the two groups combined is 59.8 (1494 ÷ 25). This is less than the mean of the two separate means (the mean of 53.7 and 67.5 is 60.6). Because there are more pupils in group A, this mean has a greater weighting in the combined mean. (c) With a total of 25 in the set, the median is the 13th value when arranged in order; so, for English, median = 56; for mathematics, median = 53. The range for English is 45 (90 – 45) and the range for mathematics is 72 (95 – 23). (d) For mathematics, John's mark (49) is less than the mean (53.6) and less than the median (56.5), but well above the bottom of the range.

27.3:   P (about 68 cm per second) has a greater average speed than Q (65 cm per second).

27.4:   Since the figure of 69% achieving the specified level falls between the 40th percentile (72%) and the lower quartile (67%), it is fair to conclude that St Anne's is performing below average compared with other schools in this group. Because their percentage is less than the 40th percentile, it means that more than 60% of schools in the group did better than St Anne's.

CHAPTER 28: PROBABILITY

28.1:   75% in the sample have fewer than 6 letters; so estimate of probability is 0.75.

28.2:   (a) By experiment: finding the relative frequency of successful outcomes in a large number of trials. (b) By collecting data from a large and random sample of people aged 50–59 years in England. (c) Using an argument based on symmetry, considering all the possible, equally likely outcomes.

28.3:   Probability of 'scoring 1' is 0. Probability of 'scoring less than 13' is 1.

28.4:   They are independent but not mutually exclusive. Probability of a black ace is $2/52 = 1/26 \ (= 1/13 \times 1/2)$.

28.5:   The events are mutually exclusive but not independent. Probability of right way up or upside down is 0.35 + 0.20 = 0.55. Probability of landing on a side is 1 – 0.55 = 0.45.

REFERENCES

ACARA (Australian Curriculum, Assessment and Reporting Authority) (2022) *Foundation to Year 10 Australian Curriculum, version 9.* Available at: australian-curriculum.org/mathematics/curriculum/f-10.

Altiparmak, K. and Özdoğan, E. (2010) 'A study on the teaching of the concept of negative numbers', *International Journal of Mathematical Education in Science and Technology, 41*(1): 31–47. Available at: doi.org/10.1080/00207390903189179.

Anghileri, J. (2001) 'What are we trying to achieve in teaching standard calculating procedures?', *Proceedings of the 25th Conference of the International Group for the Psychology of Mathematics Education, 2*: 41–8.

Anghileri, J. (2008) *Developing Number Sense: Progression in the Middle Years.* London: Continuum.

Ashcraft, M. and Kirk, E. (2001) 'The relationships among working memory, math anxiety, and performance', *Journal of Experimental Psychology, 130*(2): 224–37.

Ashcraft, M. and Moore, A. (2009) 'Mathematics anxiety and the affective drop in performance', *Journal of Psychoeducational Assessment, 27*(3): 197–205.

Askew, M. (2015) *Transforming Primary Mathematics*, 2nd edn. Abingdon: Routledge.

Askew, M. et al. (2015) *Teaching for Mastery, Questions, Tasks and Activities to Support Assessment.* Oxford: Oxford University Press.

ATM/MA (2021) *Responding to the 2021 Mathematics Ofsted Research Review.* Available at: www.m-a.org.uk/resources.

Ball, D., Thames, M. and Phelps, G. (2008) 'Content knowledge for teaching: what makes it special?', *Journal of Teacher Education, 59*(5): 389–407.

Barmby, P., Harries, T., Higgins, S. and Suggate, J. (2009) 'The array representation and primary children's understanding and reasoning in multiplication', *Educational Studies in Mathematics, 70*: 217–41.

Barrett, J. and Clements, D. (1998) 'Analyzing children's length strategies with two-dimensional tasks: what counts for length?', *Proceedings of the 12th Annual Meeting of the North American Chapter of the International Group for the Psychology of Mathematics Education*: Columbus, Ohio.

Battista, M. (2007) 'The development of geometric and spatial thinking', in F. Lester Jr (ed.), *Second Handbook of Research on Mathematics Teaching and Learning.* Charlotte, NC: National Council of Teachers of Mathematics, Information Age Publishing.

Bellos, A. (2020) *Alex's Adventures in Numberland.* London: Bloomsbury.

Boaler, J. (2020) *The Elephant in the Classroom: Helping Children Learn and Love Mathematics*, revised and updated edn. London: Profile Books.

Boaler, J. (2022) *Mathematical Mindsets: Unleashing Students' Potential through Creative Math, Inspiring Messages and Innovative Teaching*, 2nd edn. Hoboken, New Jersey: Jossey-Bass a Wiley Brand.

Borthwick, A., Gifford, S. and Thouless, H. (2021) *The Power of Pattern – Patterning in the Early Years (ATM)*. Available at: www.atm.org.uk/shop/All-Books/The-Power-of-Pattern–Patterning-in-the-Early-Years/ACT133

Boulet, G. (1998) 'Didactical implications of children's difficulties in learning the fraction concept', *Focus on Learning Problems in Mathematics, 20*(4): 19–34.

Briggs, M. (1993) 'Bags and baggage revisited', *Mathematics Education Review, 2*: 16–20.

Brown, J. and Burton, R. (1978) 'Diagnostic models for procedural "bugs" in basic mathematical skills', *Cognitive Science, 2*: 155–92.

Brown, T. (2003) *Meeting the Standards in Primary Mathematics: A Guide to the ITT NC*. London: RoutledgeFalmer.

Burger, W. and Shaughnessy, J. (1986) 'Characterizing the van Hiele levels of development in geometry', *Journal for Research in Mathematics Education, 17*(1): 31–48.

Burgess, T. (2009) 'Statistical knowledge for teaching: exploring it in the classroom', *For the Learning of Mathematics, 29*(3): 18–21.

Burnett, S. and Wichman, A. (1997) *Mathematics and Literature: An Approach to Success*. Chicago, IL: Saint Xavier University and IRI/Skylight.

Burton, R. (1981) 'DEBUGGY: diagnosis of errors in basic mathematical skills', in D. Sleeman and J. Brown (eds), *Intelligent Tutoring Systems*. New York: Academic Press.

Carey, E., Hill, F., Devine, A. and Szucs, D. (2016) *The Chicken or the Egg? The Direction of the Relationship between Mathematics Anxiety and Mathematics Performance*. Available at: www.frontiersin.org/articles/10.3389/fpsyg.2015.01987.

Carraher, D. and Schliemann, D. (2007) 'Early algebra', in F. Lester Jr (ed.), *Second Handbook of Research on Mathematics Teaching and Learning*. Charlotte, NC: National Council of Teachers of Mathematics, Information Age Publishing.

Carraher, D., Schliemann, A., Briznela, B. and Earnest, D. (2006) 'Arithmetic and algebra in early mathematics education', *Journal for Research in Mathematics Education, 37*(2): 87–115. Available at: web.cortland.edu/andersmd/psy501/early.pdf.

Carraher, T., Carraher, D. and Schliemann, A. (1985) 'Mathematics in the streets and schools', *British Journal of Developmental Psychology, 3*: 21–9.

CCEA, Northern Ireland (2019a) *Key Stages 1 & 2 Curriculum, Using Mathematics*. Available at: ccea.org.uk/key-stages-1-2/curriculum/using-mathematics.

CCEA, Northern Ireland (2019b) *Statutory Requirements for Mathematics and Numeracy Key Stages 1 & 2*. Available at: ccea.org.uk/curriculum/key_stages_1_2/areas_learning/mathematics_and_numeracy.

Chestnut, E., Lei, R., Leslie, S. and Cimpian, A. (2018) *The Myth that Only Brilliant People Are Good at Math and Its Implications for Diversity*. Available at: doi.org/10.3390/educsci8020065.

Chval, K., Lannin, J. and Jones, D. (2016) *Putting Essential Understanding of Geometry and Measurement into Practice in Grades 3–5*. Reston, VA: NCTM.

Clarke, D. (2005) 'Written algorithms in the primary years: undoing the "good work"'. *Proceedings of the 20th Biennial Conference of the Australian Association of Mathematics Teachers*. Adelaide: AAMT.

Coben, D., with Colwell, D., Macrae, S., Boaler, J., Brown, M. and Rhodes, V. (2003) *Adult Numeracy: Review of Research and Related Literature*. London: National Research Centre for Adult Literacy and Numeracy, London Institute of Education.

Cockburn, A. and Littler, G. (eds) (2008) *Mathematical Misconceptions: Opening the Doors to Understanding.* London: Sage Publications.

Cooke, H. (2007) *Mathematics for Primary and Early Years: Developing Subject Knowledge,* 2nd edn. London: Sage Publications.

Cramer, K. and Post, T. (1993) 'Connecting research to proportional reasoning', *Mathematics Teacher,* 86(5): 404–7.

Daroczy, G., Wolska, M., Meurers, W. and Nuerk, H. (2015) 'Word problems: a review of linguistic and numerical factors contributing to their difficulty', *Frontiers in Psychology,* 6. Available at: www.ncbi.nlm.nih.gov/pmc/articles/PMC4381502.

Davey, G. and Pegg, J. (1991) 'Angles on angles: students' perceptions', *Proceedings of the 14th Annual Conference of The Mathematics Education Research Group of Australasia*: Perth.

Davis B. and the Spatial Reasoning Study Group (2015) *Spatial Reasoning in the Early Years.* Abingdon, Oxon: Routledge (Taylor & Francis).

De Corte, E., Verschaffel, L. and Van Coillie, V. (1988) 'Influence of number size, structure and response mode on children's solutions of multiplication word problems', *Journal of Mathematical Behavior,* 7: 197–216.

DFE (Department for Education) (2013) *Mathematics Programmes of Study, Key Stages 1 and 2: National Curriculum in England.* London: DfE.

DFE (Department for Education) (2021) *Statutory Guidance, National Curriculum in England: Mathematics Programmes of Study.* London: DfE.

Donaldson, M. (1978) *Children's Minds.* London: Fontana.

Doxiadis, A. (2001) *Uncle Petros and Goldbach's Conjecture.* London: Faber & Faber.

Drury, H. (2014) *Mastering Mathematics.* Oxford: Oxford University Press.

Du Sautoy, M. (2011) *The Number Mysteries.* London: Fourth Estate.

Education Scotland (2010a) *Curriculum for Excellence: Numeracy and Mathematics, Experiences and Outcomes.* Available at: www.education.gov.scot/Documents/numeracy-maths-eo.

Education Scotland (2010b) *Curriculum for Excellence: Numeracy and Mathematics, Principles and Practice.* Available at: www.education.gov.scot/Documents/mathematics-pp.

English, L. (2004) 'Mathematical modelling in the primary school', *Proceedings of the 27th Annual Conference of The Mathematics Education Research Group of Australasia.* Townsville, Queensland: James Cook University.

English, L. (2013) 'Modeling with complex data in the primary school', in R. Lesh, P. Galbraith, C. Haines and A. Hurford (eds), *Modeling Students' Mathematical Modeling Competencies: International Perspectives on the Teaching and Learning of Mathematical Modelling.* Dordrecht: Springer.

English, L. and Watters, J. (2005) 'Mathematical modelling in the early school years', *Mathematics Education Research Journal,* 16(3): 58–79.

English, R. (2013) *Teaching Arithmetic in Primary Schools.* London: Sage Publications.

Fenna, D. (2002) *A Dictionary of Weights, Measures and Units.* Oxford: Oxford University Press.

Fiori, C. and Zuccheri, L. (2005) 'An experimental research on error patterns in written subtraction', *Educational Studies in Mathematics,* 60: 323–31.

Fluellen, J. (2008) 'Algebra for babies: exploring natural numbers in simple arrays', *Proceedings of the 29th Ethnography and Education Research Forum*: University of Pennsylvania. Available at: www.eric.ed.gov.

Ford, S., Staples, P., Sheffield, D. and Vanono, L. (2005) 'Effects of maths anxiety on performance and serial recall', *Proceedings of the British Psychological Society Annual Conference, 2005:* Manchester.

Gelman, R. and Gallistel, C.R. (1986) *The Child's Understanding of Number*, 2nd edn. Cambridge, MA: Harvard University Press.

Germia, E. and Panorkou, N. (2020) 'Using Scratch programming to explore coordinates', *Mathematics Teacher: Learning & Teaching, pk–12*, 113(4): 293–300. Available at: par.nsf.gov/servlets/purl/10182025.

Goulding, M., Rowland, T. and Barber, P. (2002) 'Does it matter? Primary teacher trainees' subject knowledge in mathematics', *British Educational Research Journal, 28*(5): 689–704.

Greer, B. (1997) 'Modelling reality in mathematics classrooms: the case of word problems', *Learning and Instruction, 7*(4): 293–307.

Groves, S. (1993) 'The effect of calculator use on third graders' solutions of real world division and multiplication problems', *Proceedings of the 17th International Conference for the Psychology of Mathematics Education*, 2: 9–16.

Groves, S. (1994) 'The effect of calculator use on third and fourth graders' computation and choice of calculating device', *Proceedings of the 18th International Conference for the Psychology of Mathematics Education*, 3: 9–16.

Haighton, J., Holder, D. and Thomas, V. (2020) *Maths: The Basics, Functional Skills*, 3rd edn. Oxford: Oxford University Press.

Hansen, A. (ed.) (2020) *Children's Errors in Maths: Understanding Common Misconceptions*, 5th edn. London: Sage Publications.

Harcourt-Heath M. and Borthwick A. (2014) 'Calculating: how have Year 5 children's strategies changed over time?', in G. Adams (ed.), *Proceedings of The British Society for Research into Learning Mathematics 34*(3). Available at: https://bsrlm.org.uk/publications/proceedings-of-day-conference/ip34-3/

Hartnett, J. (2015) 'Teaching computation in primary school without traditional written algorithms', in *Proceedings of the 38th Annual Conference of the Mathematics Education Research Group of Australasia*. Sunshine Coast: MERGA.

Haylock, D. (1997) 'Recognising mathematical creativity in schoolchildren', *International Reviews on Mathematical Education*, 3: 68–74.

Haylock, D. and Cockburn, A. (2017) *Understanding Mathematics for Young Children*, 5th edn. London: Sage Publications.

Haylock, D. with Thangata, F. (2007) *Key Concepts in Teaching Primary Mathematics*. London: Sage Publications.

Haylock, D. and Warburton, P. (2013) *Mathematics Explained for Healthcare Practitioners*. London: Sage Publications.

Heirdsfield, A. and Cooper, T. (1997) 'The architecture of mental addition and subtraction', *Proceedings of the Annual Conference of the Australian Association of Research in Education:* Brisbane, January 1997.

Heirdsfield, A. and Cooper, T. (2004) 'Factors affecting the process of proficient mental addition and subtraction: case studies of flexible and inflexible computers', *Journal of Mathematical Behavior*, 23: 443–63.

Henderson, D. and Taimina, D. (2020) *Experiencing Geometry: Euclidean and Non-Euclidean with History*. Project Euclid: Books by Independent Authors. Available at: projecteuclid.

org/ebooks/books-by-independent-authors/Experiencing-Geometry/toc/10.3792/euclid/9781429799850.

Hok, T., Tarmizi, R., Yunus, A. and Ayub, A. (2015) 'Understanding the primary school students' van Hiele levels of geometry thinking in learning shapes and spaces: a Q-methodology', *Eurasia Journal of Mathematics, Science & Technology Education, 11*(4): 793–802. Available at: www.ejmste.com/download/understanding-the-primary-school-students-van-hiele-levels-of-geometry-thinking-in-learning-shapes-4409.pdf.

Hopkins, C., Pope, S. and Pepperell, S. (2005) *Understanding Primary Mathematics*. London: David Fulton.

Hourigan, M. and O'Donoghue, J. (2013) 'The challenges facing initial teacher education: Irish prospective elementary teachers' mathematics subject matter knowledge', *International Journal of Mathematical Education in Science and Technology, 44*(1): 36–58.

Hughes, M. and Donaldson, M. (1979) 'The use of hiding games for studying the coordination of viewpoints', *Educational Review, 31*(2): 133–40.

Irwin, K. (2001) 'Using everyday knowledge of decimals to enhance understanding', *Journal for Research in Mathematics Education, 32*(4): 399–420.

Jefferey, B. (2011) 'Aspects of numeracy', in V. Koshy and J. Murray (eds), *Unlocking Mathematics Teaching*, 2nd edn. Abingdon: Routledge.

Jones, G., Thornton, C., Langrall, C., Mooney, E., Perry, B. and Putt, I. (2000) 'A framework for characterizing students' statistical thinking', *Mathematical Thinking and Learning, 2*: 269–308.

Kaput, J., Carraher, D. and Blanton, M. (eds) (2008) *Algebra in the Early Grades*. Abingdon, Oxford: Lawrence Erlbaum Associates, Taylor & Francis Group.

Karika, T. (2020) 'Rational errors in learning fractions among 5th-Grade students', *Teaching Mathematics and Computer Science, 18*(4): 347–58. Available at: doi.org/10.5485/TMCS.2020.0479.

Kaufman, J. and Baer, J. (2004) 'Sure, I'm creative – but not in mathematics: self-reported creativity in diverse domains', *Empirical Studies of the Arts, 22*(2): 143–55.

Khoule, A., Bonsu, N. and El Houari, H. (2017) 'Impact of conceptual and procedural knowledge on students' mathematics anxiety', *International Journal of Educational Studies in Mathematics , 4*(1): 8–17.

Kilhamn, C. (2008) *'Making sense of negative numbers through metaphorical reasoning'*. PhD thesis, University of Gothenburg.

Koshy, V. and Murray, J. (eds) (2011) *Unlocking Mathematics Teaching*, 2nd edn. Abingdon: Routledge.

Lee, J.-E. (2007) 'Making sense of the traditional long division algorithm', *Journal of Mathematical Behaviour, 26*: 48–59.

Lim, C. (2002) 'Public images of mathematics', *Philosophy of Mathematics Education Journal, 15*. Available at: education.exeter.ac.uk/research/centres/stem/publications/pmej/pome15/contents.htm.

Mann, E.L. (2006) 'Creativity: the essence of mathematics', *Journal for the Education of the Gifted, 30*: 236–60.

Mason, J. (2008) 'Making use of children's powers to produce algebraic thinking', in J. Kaput, D. Carraher and M. Blanton (eds), *Algebra in the Early Grades*. Abingdon & Oxford: Lawrence Erlbaum Associates/Taylor & Francis Group.

Mitchelmore, M. (1998) 'Young students' concepts of turning and angle', *Cognition and Instruction*, *16*(3): 265–84.

Mitchelmore, M. and White, P. (2000) 'Development of angle concepts by progressive abstraction and generalisation', *Educational Studies in Mathematics*, *41*(3): 209–38.

Mulligan, J. and Mitchelmore, M. (2009) 'Awareness of pattern and structure in early mathematical development', *Mathematics Education Research Journal*, *21*(2): 33–49.

Murphy, C. (2011) 'Comparing the use of the empty number line in England and the Netherlands', *British Educational Research Journal*, *37*: 147–61.

NCETM (2014) *Mastery Approaches to Mathematics and the New National Curriculum*. Available at: www.ncetm.org.uk/public/files/19990433.

NCETM (2015) *Calculation Guidance for Primary Schools*. Available at: www.ncetm.org.uk/public/files/24756940.

NCETM (2017a) *Mastery Explained: What Mastery Means*. Available at: www.ncetm.org.uk/resources/49450.

NCETM (2017b) *Teaching for Mastery: Supporting Research, Evidence and Argument*. Available at: www.ncetm.org.uk/resources/50819.

NCETM (2017c) *Five Big Ideas in Teaching for Mastery*. Available at: www.ncetm.org.uk/teaching-for-mastery/mastery-explained/five-big-ideas-in-teaching-for-mastery.

NCETM (2019) *Early Years Materials: Measures*. Available at: www.ncetm.org.uk/classroom-resources/ey-measures.

NCETM (2022) *The Bar Model*. Available at: www.ncetm.org.uk/classroom-resources/ca-the-bar-model.

Newell, R. (2023) *Mastery Mathematics for Primary Teachers*, 2nd edn. London: Sage Publications.

Newstead, K. (1998) 'Aspects of children's mathematics anxiety', *Educational Studies in Mathematics*, *36*(1): 53–71.

Nickson, M. (2004) *Teaching and Learning Mathematics: A Guide to Recent Research and its Application*, 2nd edn. London: Continuum.

Nunes, T. and Bryant, P. (1996) *Children Doing Mathematics*. Oxford: Blackwell.

Nunes, T., Bryant, P., Evans, D., Gottardis, L. and Terlektsi, M. (2015) *Teaching Mathematical Reasoning: Probability and Problem Solving in Primary School*. Available at: www.nuffieldfoundation.org/sites/default/files/files/Nunes&Bryant2015_Teachingreasoning%20-%2028Jan15.pdf.

Nunes, T., Schliemann, A. and Carraher, D. (1993) *Street Mathematics and School Mathematics*. Cambridge: Cambridge University Press.

Ofsted (2021) *Research Review Series: Mathematics*. Available at: www.gov.uk/government/publications/research-review-series-mathematics.

Orton, A. (ed.) (2004) *Pattern in the Teaching and Learning of Mathematics*. London: Continuum.

Peters, G., de Smedt, B., Torbeyns, J., Ghesquière, P. and Verschaffel, L. (2012) 'Children's use of addition to solve two-digit subtraction problems', *British Journal of Psychology*, *104*(4): 495–511.

Piaget, J. (1952) *The Child's Conception of Number*. London: Routledge and Kegan Paul.

Piaget, J. (1953) *The Origin of Intelligence in the Child*. London: Routledge and Kegan Paul.

Piaget, J. and Inhelder, B. (1956) *The Child's Conception of Space*. London: Routledge and Kegan Paul.

Pound, L. (2006) *Supporting Mathematical Development in the Early Years*, 2nd edn. Maidenhead: Open University Press.

Price, P. (2001) 'The development of year 3 students' place-value understanding: representations and concepts'. PhD thesis, Faculty of Education, Queensland University of Technology.

Puteh, M. (1998) 'Factors associated with mathematics anxiety and its impact on primary teacher trainees in Malaysia'. PhD thesis, University of East Anglia, Norwich.

Rizvi, N. (2004) 'Prospective teachers' ability to pose word problems', *International Journal for Mathematics Teaching and Learning*, October: 166–88. Available at: www.cimt.org.uk/journal/rizvi.pdf.

Rowland, T. and Ruthven, K. (eds) (2011) *Mathematical Knowledge in Teaching*. London: Springer.

Rowland, T., Martyn, S., Barber, P. and Heal, C. (2000) 'Primary teacher trainees' mathematics subject knowledge and classroom performance', in T. Rowland and C. Morgan (eds), *Research in Mathematics Education*. Vol. *2*, Papers of the British Society for Research into Learning Mathematics. London: BSRLM.

Rowland, T., Turner, F., Thwaites, A. and Huckstep, P. (2009) *Developing Primary Mathematics Teaching*. London: Sage Publications.

Ruddock, G. and Sainsbury, M. (2008) *Comparison of the Core Primary Curriculum in England to Those of Other High Performing Countries, Research Report DCSF-RW048*. National Foundation for Educational Research. Available at www.nfer.ac.uk/media/1629/bpc01.pdf.

Sarama, J., Clements, D., Swaminathan, S., McMillen, S. and González-Gómez, M. (2003) 'Development of mathematical concepts of two-dimensional space in grid environments: an exploratory study', *Cognition and Instruction*, 21(3): 285–324.

Schlottman, A. (2001) 'Children's probability intuitions: understanding the expected value of complex gambles', *Child Development*, 72(1): 103–22.

Sharma, S. (2017) 'Definitions and models of statistical literacy: a literature review', *Open Review of Educational Research,* 4(1): 118–33. Available at: www.tandfonline.com/doi/full/10.1080/23265507.2017.1354313.

Siegler, R. and Pyke, A. (2013) 'Developmental and individual differences in understanding of fractions', *Developmental Psychology*, 49(10): 1994–2004.

Silver, A. and Burkett, M. (1994) *The posing of division problems by preservice elementary school teachers: conceptual knowledge and contextual connections*. Paper presented to the American Educational Research Association, New Orleans, LA, May 1994. Available at: www.eric.ed.gov/?id=ED381348.

Skinner, C. and Stevens, J. (2012) *Foundations of Mathematics: An Active Approach to Number, Shape and Measures in the Early Years*. London: Featherstone Education, Bloomsbury.

Smith, C.P., King, B. and Hoyte, J. (2014) 'Learning angles through movement: critical actions for developing understanding in an embodied activity', *The Journal of*

Mathematical Behavior, 36: 95–108. Available at: www.sciencedirect.com/science/article/pii/S0732312314000522.

Spencer, R. and Fielding, H. (2015) 'Using the Singapore Bar Model to support the interpretation and understanding of word problems in Key Stage 2', *Proceedings of BSRLM Conference, Reading University, 35*(3): 114–19.

Squire, S., Davies, C. and Bryant, P. (2004) 'Does the cue help? Children's understanding of multiplicative concepts in different problem contexts', *British Journal of Educational Psychology,* 74: 515–32.

Stripp, C. (2016) 'What is teaching for mastery in maths?', *Schools Week,* 20 April. Available at: schoolsweek.co.uk/what-is-teaching-for-mastery-in-maths.

Suggate, J., Davis, A. and Goulding, M. (2017) *Mathematical Knowledge for Primary Teachers,* 5th edn. London: David Fulton.

Sun, W. and Zhang, Y. (2001) 'Teaching addition and subtraction facts: a Chinese perspective', *Teaching Children Mathematics, 8*(1): 28–31.

Swain, J. (2004) 'Money isn't everything', *Education Guardian Weekly,* 8 June.

Szilágyi, J., Clements, D. and Sarama, J. (2013) 'Young children's understandings of length measurement: evaluating a learning trajectory', *Journal for Research in Mathematics Education, 44*(3): 581–620.

Tammet, D. (2013) *Thinking in Numbers: How Maths Illuminates Our Lives.* London: Hodder & Stoughton.

TES Resource Team (2016) *Measurement: Length, Weight, Area and Volume, Teaching for Mastery in Primary Maths.* London: TES (in partnership with Mathematics Mastery and The White Rose Hub). Available at: www.tes.com/teaching-resource/length-weight-area-and-volume-teaching-for-mastery-booklet-11460632.

Thompson, I. (2000) 'Teaching place value in the UK: time for a reappraisal?', *Educational Review, 52*(3): 291–8.

Thompson, I. and Bramald, R. (2002) *An Investigation of the Relationship between Young Children's Understanding of the Concept of Place Value and Their Competence at Mental Addition.* Newcastle: University of Newcastle, Department of Education.

Tijms, H. (2012) *Understanding Probability: Chance Rules in Everyday Life,* 3rd edn. Cambridge: Cambridge University Press.

Tucker, K. (2014) *Mathematics through Play in the Early Years,* 3rd edn. London: Sage Publications.

Van Hiele, P. (1999) 'Developing geometric thinking through activities that begin with play', *Teaching Children Mathematics, 5*(6): 310–16.

Verschaffel, L., Greer, B. and De Corte, E. (2007) 'Whole number concepts and operations', in F. Lester Jr (ed.), *Second Handbook of Research on Mathematics Teaching and Learning.* Charlotte, NC: National Council of Teachers of Mathematics, Information Age Publishing.

Welsh Government (2014) *National Literacy and Numeracy Framework (Statutory Guidance).* Available at: learning.gov.wales/resources/improvement areas/curriculum.

Welsh Government (2020) *Curriculum for Wales: Areas of Learning and Expertise, Mathematics and Numeracy, Introduction.* Available at: hwbgov.wales/curriculum-for-wales/mathematics-and-numeracy.

Williams, P. (2008) *Independent Review of Mathematics Teaching in Early Years Settings and Primary Schools, Final Report.* London: DCSF.

Witt, M. and Mansergh, J. (2008) 'Breaking the anxiety spiral: what can ITT providers do?' *Proceedings of the British Society for Research into Learning Mathematics, 28*(3). Available at: www.bsrlm.org.uk.

Zvonkin, A. (1992) 'Mathematics for little ones', *Journal of Mathematical Behaviour, 11*(2): 207–19.

INDEX

The page numbers in this index refer to the entries in the end-of-chapter glossaries. Further material related to these entries will be found in the preceding content of the chapter.

Acute angle 422
Ad hoc subtraction 215
Addend 123
Adhocorithm 74
Aesthetic aim in teaching mathematics 29
Aggregation (addition structure) 123
Aims of teaching mathematics 29
Algebra 351
Algebraic operating system 352
Algorithm 74
Angle 421
Ante meridiem and post meridiem 390
Application aim in teaching mathematics 29
Area 406
Areas method for multiplication 215
Associative law of addition 139
Associative law of multiplication 196
Augmentation (addition structure) 123
Average 490
Average speed 491
Axes (plural of axis) 365
Axiom 61

Bar chart 473
Bar-modelling 124
Base 103
Block graph 472

Cancelling 270
Capacity 390
Cardinal aspect of number 102
Carroll diagram 472
Carrying (one) 159
Celsius scale (°C) 391
Centi 391
Centimetre (cm) 391
Centre of rotational symmetry 436
Circle 407
Circumference 407
Classification 42

Column addition and column subtraction 159
Commutative law of addition 123
Commutative law of multiplication 178
Comparison (subtraction structure) 123
Compensation 140
Complement of a set 472
Composite (rectangular) number 240
Cone 450
Congruence 435
Congruent shapes 435
Conjecture 60
Connections model 40
Conservation (in measurement) 391
Conservation of number 41
Constant difference method (subtraction) 160
Constant ratio method for division 196
Continuous variable 473
Convergent thinking 61
Conversion graph 366
Coordinates 365
Counter-example 60
Creativity in mathematics 61
Cube (shape) 240
Cube number 240
Cube root 240
Cubed 240
Cubic centimetre (cm3) 407
Cubic metre (m3) 391
Cubic unit 240
Cuboid 451
Cylinder 450

Decagon 449
Deci 392
Decile 491
Decimal number 292
Decimal point 292
Decomposition (subtraction) 160
Deductive reasoning 61
Degree 421

Denominator 270
Dependent variable 352
Diameter 407
Dienes blocks 103
Digital root 240
Digital sum 240
Digits 103
Direct proportion 328
Discrete sets 123
Discrete variable 472
Distance–time graph 366
Distributive laws of division 196
Distributive laws of multiplication 196
Divergent thinking 61
Divisible 179
Divisible 239
Dodecahedron 450

Edge 450
Empty number line 140
Epistemological aim in teaching
    mathematics 29
Equal additions 160
Equal sharing between 179
Equation 351
Equilateral triangle 449
Equivalence 40
Equivalent fractions 270
Equivalent ratios 270
Evens 502
Exchange 103
Exchange 159
Expected value 503
Experimental probability 503

Face 450
Factor 196
Factor pair 240
Fahrenheit scale (°F) 391
Fibonacci sequence 392
First quadrant 365
Five-number summary 490
Formula 352
Fraction 270
Frequency 472
Frequency table 472
Friendly numbers 140
Front-end approach 140
Function 352

Generalization 60
Global generalization 352

Grid method for multiplication 215
Grouped discrete data 473

Heptagon 449
Hexagon 449
Hexahedron 450
Highest common factor 240
Hypothesis 60

Icosahedron 450
Imperial units 392
Improper fraction 270
Independent events 503
Independent variable 352
Inductive reasoning 61
Inequality 104
Integer 102
Intellectual development aim in teaching 29
Inter-quartile range 491
Intersection of two sets 472
Interval scale 391
Inverse of addition (subtraction structure) 123
Inverse of multiplication 179
Inverse processes 124
Investigation 74
Irrational number 103
Isosceles triangle 449

Kilo 391
Kilogram (kg) 391

Lies between 104
Line graph 473
Line of symmetry 436
Linear relationship 365
Litre 391
Long division 215
Long multiplication 215
Lower quartile (LQ) 490
Lowest common denominator 270
Lowest common multiple 240

Manipulatives 29
Mapping 352
Mass 390
Mastery in learning mathematics 29
Mathematical modeling 74
Mean (arithmetic mean) 490
Meaningful-learning mind set 40
Median 490
Metre (m) 391
Milli 391

Minuend 123
Minus 102
Mixed number 270
Mode 490
Multiples 61
Multiplicand 178
Multiplier 179
Mutually exclusive events 503

Natural numbers 102
Near-double 140
Negative integer 102
Net 451
Newton 390
Nonagon 449
Number line 103
Numeral 102
Numerator 270

Oblong rectangle 450
Obtuse angle 422
Octagon 449
Octahedron 450
Order of rotational symmetry 436
Ordinal aspect of number 102
Origin 365

Parallel lines 450
Parallelogram 450
Partitioning (subtraction structure) 123
Partitioning into 100s, 10s and ones 140
Pentagon 449
Per 178
Per cent (%) 328
Percentage increase or decrease 328
Percentile 491
Perimeter 406
Pi (π) 407
Pictogram 473
Pie chart 473
Place holder 104
Place value 103
Plane of symmetry 451
Plane surface 450
Plus 102
Polygon 449
Polyhedron (plural: polyhedra) 450
Population 471
Positive integer 102
Power 103
Precedence of operators 352
Prime factorization 240
Prime number 240

Prism 451
Pro rata increase 179
Probability 502
Probability scale 502
Problem 74
Product 178
Proof 60
Proof by exhaustion 61
Proper fraction 270
Proportion 270
Protractor 422
Pyramid 451

Quadrant 365
Quadrilateral 422
Quartiles 490
Quotient, dividend and divisor 196

Radius 407
Range 490
Ratio 179
Ratio scale 391
Rational number 102
Real number 103
Rectangle 450
Rectangular array 178
Recurring decimal 292
Reduction (subtraction structure) 123
Reference item 392
Reflection 436
Reflective symmetry 436
Reflex angle 422
Regular polygon 449
Regular polyhedron 450
Relative frequency 503
Remainder 179
Repeated addition (division) 179
Repeated aggregation 178
Repeated subtraction (division) 179
Rhombus 450
Right angle 421
Roman numerals 103
Rotation 435
Rotational symmetry 436
Rote-learning mind set 40
Rounding (to the nearest 10 or 100) 104
Rounding 292
Rules of divisibility 239

Sample 472
Scalene triangle 450
Scaling a shape 436
Scaling down 436

Scaling of quantity 178
Scaling up 436
Scatter diagram (scatter plot, scattergram, scattergraph) 365
Separator 292
Sequential generalization 352
Short division 215
Short multiplication 215
SI units 391
Significant digits 292
Similar shapes 436
Solution 352
Special case 60
Sphere 450
Square (shape) 240
Square centimetre (cm2) 406
Square metre (m2)
Square millimetre (mm2) 407
Square number 240
Square root 240
Square unit 240
Squared 240
Stem sentence 61
Stepping stone 140
Straight angle 421
Subitizing 102
Subjective probability 502
Subset 472
Subtrahend 123
Sum 123
Suppression of zero 473
Surface area 407

Tabulation 351
Tallying 472
Tens-frames 140
Tessellation 450
Tetrahedron 450
Theoretical probability 503
Transformation 40
Transitive property 239
Translation 435
Trapezium 407
Trial and improvement 352
Triangle 422
Triangle numbers 240
Truncation 292
Two-way table 503

Union of sets 123
Unit fraction 270
Universal set 471
Upper quartile (UQ) 490
Utilitarian aim in teaching mathematics 29

Variable (in statistics) 472
Variable 328
Venn diagram 472
Vertex (plural: vertices) 422
Volume 390

Weight 390

Zero hours 390